CAMBRIDGE   TRACTS   IN MATHEMATICS

General Editors

B. BOLLOBAS, P. SARNAK, C. T. C. WALL

111   Affine differential geometry

T0275846

Katsumi Nomizu

Brown University

Takeshi Sasaki

Kobe University

Affine differential geometry

Geometry of Affine Immersions

CAMBRIDGE
UNIVERSITY PRESS

CAMBRIDGE UNIVERSITY PRESS
Cambridge, New York, Melbourne, Madrid, Cape Town, Singapore, São Paulo

Cambridge University Press
The Edinburgh Building, Cambridge CB2 8RU, UK

Published in the United States of America by Cambridge University Press, New York

www.cambridge.org
Information on this title: www.cambridge.org/9780521441773

First published 1994
This digitally printed version 2008

A catalogue record for this publication is available from the British Library

Library of Congress Cataloguing in Publication data

Nomizu, Katsumi, 1924–
    Affine differential geometry: geometry of affine immersions /
Katsumi Nomizu and Takeshi Sasaki.
    p.   cm. – (Cambridge tracts in mathematics; 111)
Includes bibliographical references (p. 000–000) and indexes.
ISBN 0 521 44177 3
1. Geometry, Affine.   I. Sasaki, Takeshi, 1944– .   II. Title.
III. Series.
QA477.N66   1994
516.3´6–dc20 93-46712 CIP

ISBN 978-0-521-44177-3 hardback
ISBN 978-0-521-06439-2 paperback

CONTENTS

Preface vii

Introduction ix

Chapter I   Affine geometry and affine connections

1. Plane curves 1
2. Affine space 7
3. Affine connections 11
4. Nondegenerate metrics 18
5. Vector bundles 22

Chapter II   Geometry of affine immersions – the basic theory

1. Affine immersions 27
2. Fundamental equations. Examples 32
3. Blaschke immersions – the classical theory 40
4. Cubic forms 50
5. Conormal maps 57
6. Laplacian for the affine metric 64
7. Lelieuvre's formula 68
8. Fundamental theorem 73
9. Some more formulas 77
10. Laplacian of the Pick invariant 82
11. Behavior of the cubic form on surfaces 87

Chapter III   Models with remarkable properties

1. Ruled affine spheres 91
2. Some more homogeneous surfaces 95
3. Classification of equiaffinely homogeneous surfaces 102
4. $SL(n, \mathbf{R})$ and $SL(n, \mathbf{R})/SO(n)$ 106
5. Affine spheres with affine metric of constant curvature 113
6. Cayley surfaces 119

7. Convexity, ovaloids, ellipsoids 122
8. Other characterizations of ellipsoids 125
9. Minkowski integral formulas and applications 129
10. The Blaschke–Schneider theorem 138
11. Affine minimal hypersurfaces and paraboloids 141

Chapter IV   Affine-geometric structures

1. Hypersurfaces with parallel nullity 147
2. Affine immersions $\mathbf{R}^n \to \mathbf{R}^{n+1}$ 152
3. The Cartan–Norden theorem 158
4. Affine locally symmetric hypersurfaces 161
5. Rigidity theorem of Cohn-Vossen type 165
6. Extensions of the Pick–Berwald theorem 169
7. Projective structures and projective immersions 174
8. Hypersurfaces in $\mathbf{P}^{n+1}$ and their invariants 181
9. Complex affine geometry 187

Notes

1. Affine immersions of general codimension 196
2. Surfaces in $\mathbf{R}^4$ 198
3. Affine normal mappings 202
4. Affine Weierstrass formula 203
5. Affine Bäcklund transformations 209
6. Formula for a variation of ovaloid with fixed enclosed volume . . 213
7. Completeness and hyperbolic affine hyperspheres 215
8. Locally symmetric surfaces 217
9. Centro-affine immersions of codimension 2 221
10. Projective minimal surfaces in $\mathbf{P}^3$ 230
11. Projectively homogeneous surfaces in $\mathbf{P}^3$ 230

Appendices

1. Torsion, Ricci tensor, and projective invariants 235
2. Metric, volume, divergence, Laplacian 240
3. Change of immersions and transversal vector fields 242
4. Blaschke immersions into a general ambient manifold 243

Bibliography 246

List of symbols 256

Index 258

Preface

Affine differential geometry has undergone a period of revival and rapid progress in the last ten years. Our purpose in writing this book is to provide a systematic introduction to the subject from a contemporary viewpoint, to cover most of the important classical results, and to present as much of the recent development as possible within the limitation of space. Rather than an encyclopedic collection of facts, we wanted to write an account of what we consider significant features of the subject including, wherever possible, the relationship to other areas of differential geometry. The Introduction will give the reader a brief history of the subject and the range of topics taken up in this book.

We should like to acknowledge our indebtedness to various institutions that made possible our cooperation with each other and with many other mathematicians working in affine differential geometry. First of all, to the Max-Planck Institut für Mathematik, where both authors did earlier work in the subject and later returned to do joint work, to Brown University, where the second author was visiting professor in 1988, to the Technische Universität Berlin, where we both were visitors in 1990 and where the first author worked in 1991 with a research grant from the Alexander von Humboldt Foundation, to the Mathematisches Forschungsinstitut Oberwolfach for the two conferences in affine differential geometry in 1986 and 1991, to the Katholieke Universiteit Leuven, where the first author had the pleasure of visiting for lectures and research, to Sichuan University, where the second author visited in 1992, and finally to Kobe University, where the first author was warmly received during the time we worked on the last stage of preparation for the current book – to all these institutions we express our sincere gratitude. With equally warm feelings we salute many international colleagues with whom we have cooperated in these several years: U. Pinkall, U. Simon, An-Min Li, L. Verstraelen, F. Dillen, L. Vrancken, B. Opozda, F. Podestà, A. Martínez, A. Magid, and many others. T. Cecil and M. Yoshida read the draft and offered helpful suggestions. C. Lee produced graphics with Mathematica on which the illustrations in the book are based. To all these colleagues and friends, we express our thanks for sharing their ideas.

Finally, we should like to thank Mr David Tranah for inviting us to publish this monograph with the Cambridge University Press and also Mr Peter Jackson for providing much helpful advice while editing our manuscript.

Katsumi Nomizu
Takeshi Sasaki

Introduction

Affine differential geometry – its history and current status

I. Before 1950

In 1908 Tzitzéica [Tz] showed that for a surface in Euclidean 3-space the ratio of the Gaussian curvature to the fourth power of the support function from the origin o is invariant under an affine transformation fixing o. He defined an S-surface to be any surface for which this ratio is constant. These S-surfaces turn out to be what are now called proper affine spheres with center at o. (See Theorem 5.11 and Remark 5.1 of Chapter II.) Blaschke, in collaboration with Pick, Radon, Berwald, and Thomsen among others, developed a systematic study of affine differential geometry in the period from 1916 to 1923, his work culminating in the treatise [Bl]. We wish to call attention to the two C.R. notes by E. Cartan [Car1,2] on affine surfaces in 1924 and to the work by Ślebodziński [Sl1,2] on locally symmetric affine surfaces. (See Note 8.) The main stream opened up by the Blaschke School was, however, followed by Kubota, Süss, Su Buging, Nakajima and others, whose work was mostly published in *Tôhoku Math. Journal* and the *Japanese Journal of Math.* in the late twenties and the thirties.

The main object in classical affine differential geometry is the study of properties of surfaces in 3-dimensional affine space that are invariant under the group of unimodular affine (or equiaffine) transformations. The ambient space $\mathbf{R}^3$ has a flat affine connection D and the usual determinant function Det regarded as a parallel volume element. A generalization of the theory to hypersurfaces followed almost immediately, but relatively little has been achieved in the study of affine submanifolds with codimension greater than 1. Burstin and Mayer [BM] studied surfaces M^2 in $\mathbf{R}^4$. See also Weise [Wei], later followed by Klingenberg [Kl1, 2].

Before we indicate how to study a hypersurface M^n in $\mathbf{R}^{n+1}$ from the equiaffine point of view, we recall the Euclidean hypersurface theory. Using the Euclidean metric in $\mathbf{R}^{n+1}$ we choose a field of unit normal vectors N to M^n. If X and Y are vector fields on M^n, we take the covariant derivative

$D_X Y$ in $\mathbf{R}^{n+1}$ and decompose it in the form

$$D_X Y = \nabla_X Y + h(X, Y)N,$$

where $\nabla_X Y$ is the tangential component and h the second fundamental form. It turns out that $\nabla_X Y$ is nothing but the covariant derivative relative to the Levi-Civita connection of the Riemannian metric induced on M^n. Now in the affine case, we simply take an arbitrary transversal vector field ξ and define ∇ and h in the same way as above. We can easily check that h is determined up to a conformal factor, and hence the rank of h depends only on the surface and not on the choice of ξ. It makes sense to say that M^n is a nondegenerate surface if h has rank n. Now for a nondegenerate hypersurface M^n, we can prove that there is a unique choice of ξ (up to sign) such that the induced volume element

$$\theta(X_1, \ldots, X_n) = \det [X_1 \ldots X_n\, \xi]$$

is parallel relative to the induced connection ∇ and, moreover, is identical with the volume element of the nondegenerate metric h. We prove this fundamental fact in Section 3 of Chapter II, and provide the *procedure for finding the affine normal* for practical computation. Such ξ is called the affine normal field, and the corresponding h the affine metric. The affine shape operator S is defined by $D_X \xi = -SX$. We get an equiaffine structure (∇, θ) on M^n; together with h and S, it will be called the Blaschke structure.

Affine spheres are those surfaces for which S is a scalar multiple of the identity, but unlike the Euclidean case they are by no means simple or easy to determine. Another important object is the cubic form $C = \nabla h$; it satisfies the property that trace $_h(\nabla_X h) = 0$ for each tangent vector X, called apolarity. A classical theorem of Pick and Berwald states that a nondegenerate surface has vanishing cubic form if and only if it is a quadric. We give two proofs to this theorem in Section 4 of Chapter II and later provide generalizations in Sections 6 and 8 of Chapter IV. Blaschke also showed that a compact affine sphere is an ellipsoid (see Theorem 7.5 of Chapter III). An ellipsoid was also characterized as an ovaloid whose affine Gaussian curvature $K = \det S$ or affine mean curvature $H = \frac{1}{n}\text{trace}\, S$ is constant. (See Section 9 of Chapter III.) Radon [R] gave a version of the fundamental theorem for surfaces, that is, the realization theorem of a 2-manifold with given h and C as a nondegenerate surface. We give a modern formulation for an arbitrary dimension in Theorem 8.2 of Chapter II. Thomsen studied minimal surfaces in Euclidean 3-space which are also minimal in the sense that $H = 0$. Affine ruled surfaces and affine minimal surfaces were studied from the local point of view.

II. Before 1980

Throughout the 1960s and 1970s a few expository papers and books appeared on the subject. In particular, we mention Schirokow and Schirokow

[Schr], 1962, as a very useful source of extensive information. Guggenheimer [Gu], 1963, has a chapter on affine differential geometry, in which he provides a classification list of equiaffine homogeneous surfaces. (This list has been completed in [NS2] by adding one more model, see Section 3 of Chapter III.) Spivak's volume IV [Sp] in 1975 contains a good account of affine surface theory. Schneider [Schn1,2] obtained interesting global results, and Simon [Si1–3] emphasized the viewpoint of so-called relative geometry that originates with Süss. See also Chern [Ch3], Flanders [Fl], Hsiung and Shahin [HS].

It was Calabi [Cal1–3] who made several major contributions dealing, for the first time, with global problems concerning noncompact affine hyperspheres. When a locally strictly convex hypersurface (that is, the affine metric is definite at each point) is a proper affine hypersphere (that is, the affine normals meet at one point, called the center), it is said to be elliptic or hyperbolic depending on whether the center lies on the concave side or the convex side of the hypersurface. Calabi's major results are the following.

(i) An improper affine hypersphere whose affine metric is complete is an elliptic paraboloid. (See Theorem 11.5 of Chapter III.)

(ii) A complete elliptic affine hypersphere is compact and hence an ellipsoid. (See Theorem 7.6 of Chapter III. Refer also to [Pog].)

(iii) His conjecture on hyperbolic affine hyperspheres was studied by Cheng and Yau [CY1] and Gigena [Gi] as well as by Sasaki [S1]. The result can be stated as follows: Every closed hyperbolic affine hypersphere is asymptotic to the boundary of a convex cone. Conversely, every nondegenerate cone V determines a hyperbolic affine hypersphere. We are not, however, going to give a proof of this result in our book. (See Note 7.)

We also mention some results by Chern and Terng [ChT] on affine minimal surfaces and affine Bäcklund theory as well as on the following result: A surface in Euclidean space $\mathbf{R}^3$ that is isometric to part of the elliptic paraboloid has zero affine mean curvature. This peculiar mixture of Euclidean and affine properties seems so far to be an isolated result (except that [NS1] gave a similar result regarding a timelike surface in the Minkowski space with $ds^2 = dx^2 - dy^2 + dz^2$ isometric to $z = x^2 - y^2$). We should also mention the important progress being made by Calabi [Cal4] on the so-called affine Bernstein problem, formulated in [Ch3] as follows. If the graph $z = f(x, y)$ defined on the entire plane is a nondegenerate surface with zero affine mean curvature, is it affinely congruent to the graph of $z = x^2 + y^2$? See Section 11 of Chapter III. Most of the important aspects of the classical theory are treated in Chapters II and III, together with some of the results based on the new approach to the subject described in the next section.

III. After 1980

In the *Complete Works* of Blaschke, two survey articles on his work in affine differential geometry as well as later developments appeared; see Burau and Simon [BS] and Simon [Si4].

In the lecture entitled 'What is affine differential geometry?' at the Münster Conference in 1982, Nomizu formulated the starting point of affine differential geometry from the structural point of view. After discussing $SL(n, \mathbf{R})$ as an example of a nondegenerate affine hypersurface and observing that the induced connection on $SL(2, \mathbf{R})$ is locally symmetric, he posed a problem: Determine the nondegenerate hypersurfaces whose induced connections are locally symmetric. Verheyen and Verstraelen [VeVe] solved the problem by showing that such hypersurfaces M^n, $n \geq 3$, must have either $S = 0$ (hence the graph of a function) or $C = 0$ (hence a quadric). This result has now been extended, as stated in Section 4 of Chapter IV. Such a generalization was possible only after the notion of affine immersion, introduced in [NP2], gave a more general viewpoint to the subject. We treat this notion in Sections 1 and 2 in Chapter II. An affine immersion is totally geodesic if h is identically 0; it is a graph immersion if S is identically 0. An isometric immersion may be regarded as an affine immersion. It is possible to allow the affine fundamental form h to be degenerate; in fact, many classical results can be extended to the case where the rank of h is at least equal to 2. (For example, see the extensions of the Pick–Berwald theorem in Section 6 of Chapter IV.) We may study the extreme situation of an affine immersion $f : \mathbf{R}^n \to \mathbf{R}^{n+1}$ as an analogue to the problem of determining all isometric immersions between Euclidean spaces E^n and E^{n+1} or between Minkowski–Lorentzian spaces L^n and L^{n+1}. Actually, the affine method can unify the metric cases, as shown in Section 2 of Chapter IV. Another bridge from the affine theory to the metric geometry is the Cartan–Norden theorem in [NP2], which is contained in Section 3 of Chapter IV. Opozda [O7] has studied the structures of Blaschke surfaces in $\mathbf{R}^3$ with locally symmetric induced connections and further the problem of realizing locally symmetric (or more generally, projectively flat) connections on 2-manifolds as induced connections on nondegenerate surfaces; these results are sketched in Note 8.

From the structural point of view, we mention two problems. One is the formulation for a generalization of Radon's theorem. On a given M^n assume that there is a nondegenerate metric h and a cubic form C. Under what conditions can we realize h and C as the affine metric and the cubic form for a nondegenerate immersion of M^n into $\mathbf{R}^{n+1}$? Such a pair (h, C) corresponds to a pair (h, ∇) satisfying Codazzi's equation (that ∇h is symmetric in all three variables). In addition to apolarity, the condition is that the conjugate connection $\bar{\nabla}$ of ∇ relative to h is projectively flat. This was established in [DNV]. We also note that the notion of conjugate connection has proved to be independently important due to its applications to mathematical statistics (see Amari [Am1]). We cannot go into this application, but quote a few references by Amari [Am2] and Kurose [Ku2–5] motivated by such applications.

The other problem is the rigidity theorem of Cohn-Vossen type given in [NO1]. The classical Cohn-Vossen theorem says that two compact convex surfaces (ovaloids) in Euclidean space are congruent if they are isometric. Since the affine theory deals with the induced connection instead of the induced metric, it is natural to anticipate a result that says that two ovaloids are affinely congruent if their induced connections are isomorphic. We prove this theorem, under the assumption that the affine Gauss–Kronecker curvature K does not vanish, by a method similar to the Herglotz proof of Cohn-Vossen's theorem (see Section 5 of Chapter IV). We note that Simon [Si7] has a different approach that does not require the condition on K.

Going back to the classical Blaschke immersions, we mention a few new results. An-Min Li [L5, 8] has proved that a locally strictly convex hypersphere M^n with complete affine metric is closed in $\mathbf{R}^{n+1}$, thereby strengthening [CY1] in the proof of the Calabi conjecture. He later proved that the affine metric of a closed, locally strictly convex hypersurface is complete if the affine principal curvatures are bounded [L7]. He also has contributed to the affine Bernstein problem by characterizing an elliptic paraboloid as a locally strictly convex, affine-complete, affine-minimal surface whose affine normals omit five or more directions in general position [L4]. Martínez and Milán have characterized an elliptic paraboloid as a locally strictly convex, affine-complete, affine-minimal surface with K bounded from below by a constant [MMi2].

The affine spheres with affine metrics of constant curvature have been classified (among them, ruled affine spheres, proper and improper), see Section 5, Chaper III, and the references therein for these and related results. As an extension of Theorem 5.1, Magid and Ryan [MR2] proved: If an affine hypersphere M^3 in $\mathbf{R}^4$ has nonzero Pick invariant and its affine metric has constant sectional curvature α, then α must be 0 and M^3 is affinely congruent to an open subset of one of the following models:

(i) $x_1 x_2 x_3 x_4 = 1$; (ii) $(x_1^2 + x_2^2)(x_3^2 - x_4^2) = 1$; (iii) $(x_1^2 + x_2^2)(x_3^2 + x_4^2) = 1$.

We mention that Nomizu and Vrancken [NV] recently introduced the notion of nondegeneracy for surfaces M^2 in $\mathbf{R}^4$. Under this assumption, it is possible to induce an equiaffine connection on M^2 by choosing a suitable transversal bundle. There are good indications that this is the right choice of connection on nondegenerate surfaces with codimension 2 (see Note 2 and references therein). A good amount of parallel ideas appears between this affine theory and the projective theory in Sasaki [S7].

For other recent work, see many papers by Dillen, Verstraelen, Vrancken and their joint work, [Di1–4], [DVe], [DVr1–5], [DVV], [VeVr], and [Wan1, 2], [NR], [Ce].

Another application of the formalism of affine immersion is to the study of projective differential geometry. Three different approaches can be mentioned. The first is the study of projective immersion – an analogue of affine immersion. We study maps between two manifolds each with a projective structure; for example, automorphisms of (M^n, P) into itself or a hypersurface immersion of (M^n, P) into $(\tilde{M}^{n+1}, \tilde{P})$. They were studied in [NP3] and [NP4]. We deal with some of these results in Section 7 of Chapter III. The second is the study of nondegenerate immersions of an n-manifold M^n into the real projective space $\mathbf{P}^{n+1}$. By using an atlas of flat affine connections with parallel volume element (D, ω) on $\mathbf{P}^{n+1}$, we consider a neighborhood of each point of M^n as a Blaschke hypersurface. We look for "projective invariants" through this process, for example, the conformal class of h, the vanishing of the Pick invariant J, the projective metric when J is nonvanishing, etc. (See Section 8 of Chapter IV.) This makes it possible to define projective invariants that usually require the method of moving frames (see [S3, 4]). The third is the method of centro-affine immersions of codimension 2, which is applied to hypersurface theory in $\mathbf{P}^{n+1}$ through the canonical map $\mathbf{R}^{n+2} - \{0\} \to \mathbf{P}^{n+1}$ (see Note 9 and [NS5]). The theory of codimension-2 centro-affine immersions was developed also by Walter [Wal] without regard to projective hypersurface theory. The second and third methods are combined in order to classify projectively homogeneous surfaces in $\mathbf{P}^3$ (see Note 11 and [NS5]).

Finally, we mention that affine differential geometry has been extended to the complex case. Specifically, let $f : M^n \to \mathbf{C}^{n+1}$ be a holomorphic immersion of a complex manifold M^n of complex dimension n into complex affine space $\mathbf{C}^{n+1}$. K. Abe [Ab] and Dillen, Verstraelen and Vrancken [DVV] considered an analogue of the real case by choosing a holomorphic transversal vector field and getting an induced holomorphic connection on M^n. See also [Di1,2], [DVr1,2], [DVe]. On the other hand, Nomizu, Pinkall and Podestà [NPP] chose an anti-holomorphic transversal vector field and found on M^n an affine-Kähler connection, that is, a connection ∇ compatible with the complex structure J on M^n whose curvature tensor satisfies $R(JX, JY) = R(X, Y)$ for all tangent vectors X, Y. (We note that the Levi-Civita connection of a Kähler metric is affine-Kähler. If an affine-Kähler connection is also holomorphic, then it is flat.) For this topic, see also [NPo1,2], [Po1].

Opozda has dealt with a general theory that includes the holomorphic and anti-holomorphic choices of transversal vector fields. See [O5,6], and [O8]. We treat complex affine geometry in Section 9 of Chapter IV.

I

Affine geometry and affine connections

This chapter summarizes what will be needed in the main body of the book. Most of the material here can be recalled from the standard references on geometry and on differential geometry; basically, we shall suppose that the reader is familiar with the terminology in manifold theory. Some of the topics, however, are already part of an introduction to affine differential geometry: affine curves in Section 1, the notions of equiaffine connection and centro-affine surface in Section 3, and the notions of conjugate connection and cubic form in Section 4. In Section 5 we briefly discuss vector bundles and interpret the conjugate connection from the point of view of the cotangent bundle.

1. Plane curves

In this section, we provide a preview of affine differential geometry through elementary discussions of curves in the affine plane. The affine plane $\mathbf{R}^2$ is referred to in terms of a coordinate system $\{x, y\}$. By an allowable coordinate change, we mean $(x, y) \mapsto (\bar{x}, \bar{y})$ such that

(1.1) $$\bar{x} = ax + by + p, \qquad \bar{y} = cx + dy + q,$$

where

(1.2) $$\det \begin{bmatrix} a & b \\ c & d \end{bmatrix} = 1.$$

Equiaffine geometry is concerned with geometric properties expressed by conditions invariant under an allowable change of coordinates. Or we may consider (1.1) as an allowable transformation of the plane $\mathbf{R}^2$, called an *equiaffine transformation*; then equiaffine geometry is the study of properties invariant under equiaffine transformations. Geometrically, we may assume that the notion of parallelogram of unit area is given. A coordinate system is allowable if the parallelogram determined by $(1, 0)$ and $(0, 1)$ has unit area.

This being said, we consider a smooth curve

$$\mathbf{x}(t) = \begin{bmatrix} x(t) \\ y(t) \end{bmatrix},$$

where $x(t)$, $y(t)$ are smooth functions of t defined over a certain interval J. We denote by $\mathbf{a}_1(t)$ the tangent vector $\dot{\mathbf{x}} = d\mathbf{x}/dt$ and look for a second vector $\mathbf{a}_2(t)$ along the curve $\mathbf{x}(t)$ such that the vectors $\mathbf{a}_1(t)$ and $\mathbf{a}_2(t)$ span unit area, that is, $\det [\mathbf{a}_1(t)\ \mathbf{a}_2(t)] = |\mathbf{a}_1(t)\ \mathbf{a}_2(t)| = 1$.

We can actually try to do better. We look for a new parameter σ for the curve $\mathbf{x}(t)$ such that

$$[\mathbf{x}'\ \mathbf{x}''] \in SL(2, \mathbf{R}),$$

where $'$ denotes the derivative with respect to σ, and $SL(2, \mathbf{R})$ is the group of all 2×2 real matrices with determinant 1. We have

$$|\dot{\mathbf{x}}\ \ddot{\mathbf{x}}| = (\dot\sigma)^3 |\mathbf{x}'\ \mathbf{x}''| = \left(\frac{d\sigma}{dt} \right)^3.$$

Thus

(1.3)
$$\frac{d\sigma}{dt} = |\dot{\mathbf{x}}\ \ddot{\mathbf{x}}|^{\frac{1}{3}},$$

and hence

(1.4)
$$\sigma(t) = \int_{t_0}^{t} |\dot{\mathbf{x}}\ \ddot{\mathbf{x}}|^{\frac{1}{3}} dt.$$

In order for (1.4) to define a reparametrization $t \mapsto \sigma$, we shall assume that the given curve $\mathbf{x}(t)$ satisfies the condition

(1.5)
$$|\dot{\mathbf{x}}\ \ddot{\mathbf{x}}| \neq 0 \quad \text{for any } t.$$

The reader may easily verify that this condition is independent of parametrization. We say that the curve is *nondegenerate*. To see the geometric meaning of the condition we suppose $|\dot{\mathbf{x}}\ \ddot{\mathbf{x}}| > 0$ for all t. Then for small h we have

$$\mathbf{x}(t + h) - \mathbf{x}(t) = h\dot{\mathbf{x}}(t) + \frac{1}{2}h^2\ddot{\mathbf{x}}(\tau),$$

where τ is between t and $t + h$. Hence we get

$$|\dot{\mathbf{x}}(t)\ \ \mathbf{x}(t + h) - \mathbf{x}(t)| = \frac{1}{2}h^2|\dot{\mathbf{x}}(t)\ \ddot{\mathbf{x}}(\tau)| > 0.$$

This means that the secant from $\mathbf{x}(t)$ to $\mathbf{x}(t + h)$ lies on one side of the tangent, that is, all points of the curve near $\mathbf{x}(t)$ lie on one side of the tangent at $\mathbf{x}(t)$. Thus our condition means that the curve has no inflection points.

We have shown that a nondegenerate curve $\mathbf{x}(t)$ admits a parameter σ such that

$$A(\sigma) = [\mathbf{x}' \; \mathbf{x}''] \in SL(2, \mathbf{R}).$$

From (1.3) it is clear that such a parameter, called an *affine arclength parameter*, is unique up to a constant. We define the *affine curvature* of the curve by

(1.6) $$\kappa(\sigma) = |\mathbf{x}'' \; \mathbf{x}'''|.$$

Differentiating $|\mathbf{x}' \; \mathbf{x}''| = 1$, we obtain $|\mathbf{x}' \; \mathbf{x}'''| = 0$ and hence

(1.7) $$\mathbf{x}''' = -\kappa \mathbf{x}'.$$

Note that the affine arclength parameter and affine curvature are invariant under an equiaffine transformation of the plane: if $\mathbf{x}(\sigma)$ is a curve with affine arclength parameter σ, then for any equiaffine transformation f the curve $\mathbf{y}(\sigma) = f(\mathbf{x}(\sigma))$ is a nondegenerate curve with σ as its affine arclength parameter; moreover, the affine curvature of $\mathbf{y}(\sigma)$ is equal to the affine curvature of $\mathbf{x}(\sigma)$ for each value of σ.

To give a Lie group-theoretic interpretation, we define the matrix $C(\sigma) = A(\sigma)^{-1} dA/d\sigma$. By virtue of (1.7), we get

$$\frac{dA}{d\sigma} = A(\sigma) \begin{bmatrix} 0 & -\kappa \\ 1 & 0 \end{bmatrix},$$

that is,

(1.8) $$C(\sigma) = \begin{bmatrix} 0 & -\kappa \\ 1 & 0 \end{bmatrix}.$$

Theorem 1.1. *Given a smooth function $\kappa = \kappa(\sigma)$ on an interval J, there exists a plane curve with σ as affine arclength parameter and κ as affine curvature. Such a curve is unique up to equiaffine transformation of the plane.*

Proof. We consider the matrix-valued function $C(\sigma)$ as in (1.8) with the given $\kappa(\sigma)$. From the existence theorem for ordinary differential equations, we have a unique solution of the equation

(1.9) $$\frac{dA}{d\sigma} = A(\sigma)C(\sigma),$$

with the initial condition $A(0) = I_2$ (identity matrix of degree 2). Using the fact that $C(\sigma)$ has trace 0 we can verify that $A(\sigma)$ is a curve in the group $SL(2, \mathbf{R})$. By writing $A(\sigma) = [\mathbf{a}_1(\sigma) \; \mathbf{a}_2(\sigma)]$ and integrating $\mathbf{a}_1(\sigma)$ we obtain a desired curve $\mathbf{x}(\sigma)$. We leave the proof of uniqueness to the reader.

Remark 1.1. Theorem 1.1 is the affine analogue of the fundamental theorem for curves in the Euclidean plane, where the uniqueness is up to Euclidean isometry. We call $\kappa = \kappa(\sigma)$ a *natural equation* for the curve.

Let us consider the special case where κ is a constant function.

First, assume $\kappa = 0$. Thus $\mathbf{x}''' = 0$, from which we obtain

$$\mathbf{x} = \frac{1}{2}\sigma^2 \mathbf{b} + \sigma \mathbf{a} + \mathbf{c},$$

where $\mathbf{a}$, $\mathbf{b}$, and $\mathbf{c}$ are constant vectors such that $|\mathbf{a}\ \mathbf{b}| = 1$. By taking the coordinate system $\{x, y\}$ such that

$$\mathbf{a} = (1,0), \quad \mathbf{b} = (0,1), \quad \mathbf{c} = (0,0)$$

we see that the curve is a *parabola* $y = \frac{1}{2}x^2$.

Second, assume that $\kappa > 0$. We have then

$$(\mathbf{x}')'' + (\sqrt{\kappa})^2 \mathbf{x}' = 0.$$

We obtain

$$\mathbf{x}' = \cos(\sqrt{\kappa}\sigma)\mathbf{a} + \sin(\sqrt{\kappa}\sigma)\mathbf{b},$$

where $\mathbf{x}'(0) = \mathbf{a}$ and $\mathbf{x}''(0) = \sqrt{\kappa}\,\mathbf{b}$ so that $|\mathbf{a}\ \mathbf{b}| = 1/\sqrt{\kappa}$. Choosing a coordinate system such that $\mathbf{a} = (1,0)$ and $\sqrt{\kappa}\,\mathbf{b} = (0,1)$, we have

$$\mathbf{x} = \left(\frac{\sin(\sqrt{\kappa}\sigma)}{\sqrt{\kappa}}, -\frac{\cos(\sqrt{\kappa}\sigma)}{\kappa}\right).$$

The curve is an ellipse

$$(1.10) \qquad\qquad \kappa x^2 + \kappa^2 y^2 = 1, \quad \kappa > 0.$$

Finally, in the case where $\kappa < 0$, we obtain the curve in the form

$$\mathbf{x} = \left(\frac{\sinh(\sqrt{-\kappa}\sigma)}{\sqrt{-\kappa}}, \frac{\cosh(\sqrt{-\kappa}\sigma)}{-\kappa}\right),$$

that is, a hyperbola

$$(1.11) \qquad\qquad \kappa x^2 + \kappa^2 y^2 = 1, \quad \kappa < 0.$$

To sum up, we have

Theorem 1.2. *If a nondegenerate curve has constant curvature κ, it is a quadratic curve.*

We also state

Example 1.1. Relative to an allowable affine coordinate system, consider the curves

$$\mathbf{x} = (a \cos t, \; b \sin t)$$

and

$$\mathbf{x} = (a \cosh t, \; b \sinh t).$$

Their affine curvatures are $(ab)^{-\frac{2}{3}}$ and $-(ab)^{-\frac{2}{3}}$, respectively.

   We say that a plane curve is *homogeneous* if it is the orbit of a certain point under a 1-parameter group of equiaffine transformations. A homogeneous nondegenerate curve has constant affine curvature because from one point p to another point q of the curve there is an equiaffine transformation of the plane that takes p to q and leaves the curve invariant. Since affine curvature is an equiaffine property, it remains constant on the curve. Conversely, a nondegenerate curve of constant affine curvature is homogeneous. For example, an *ellipse*

$$\frac{x^2}{a^2} + \frac{y^2}{b^2} = 1$$

is the orbit of the point $(a, 0)$ under the 1-parameter group of equiaffine transformations

$$(1.12) \qquad A(t) = \begin{bmatrix} \cos t & -\lambda \sin t \\ \frac{1}{\lambda} \sin t & \cos t \end{bmatrix},$$

where $\lambda = a/b$. For a *hyperbola*

$$\frac{x^2}{a^2} - \frac{y^2}{b^2} = 1,$$

we take

$$(1.13) \qquad A(t) = \begin{bmatrix} \cosh t & \lambda \sinh t \\ \frac{1}{\lambda} \sinh t & \cosh t \end{bmatrix}.$$

For a parabola

$$y = \frac{1}{2} x^2,$$

we can take the 1-parameter group of equiaffine transformations

$$(1.14) \qquad \tilde{A}(t) = \begin{bmatrix} 1 & 0 & t \\ t & 1 & \frac{1}{2} t^2 \\ 0 & 0 & 1 \end{bmatrix} \quad \text{acting on} \quad \begin{bmatrix} x \\ y \\ 1 \end{bmatrix}$$

and consider the orbit of the origin $(0, 0)$.
   We can thus state

Theorem 1.3. *A nondegenerate curve has constant affine curvature if and only if it is the orbit of a point under a certain 1-parameter group of equiaffine transformations.*

We shall refer to [Bl] for the following result.

Theorem 1.4. *The 1-parameter groups of equiaffine transformations of the plane can be classified as follows, up to affine equivalence.*

(A)
$$\begin{bmatrix} \cos t & -\sin t & 0 \\ \sin t & \cos t & 0 \\ 0 & 0 & 1 \end{bmatrix} ; \quad \text{same as } (1.12) \text{ with } \lambda = 1.$$

(B)
$$\begin{bmatrix} \cosh t & \sinh t & 0 \\ \sinh t & \cosh t & 0 \\ 0 & 0 & 1 \end{bmatrix} ; \quad \text{same as } (1.13) \text{ with } \lambda = 1.$$

(C)
$$\begin{bmatrix} 1 & t & 0 \\ 0 & 1 & 0 \\ 0 & 0 & 1 \end{bmatrix} .$$

(D)
$$\begin{bmatrix} 1 & 0 & t \\ t & 1 & \frac{1}{2}t^2 \\ 0 & 0 & 1 \end{bmatrix} ; \quad \text{same as } (1.14).$$

(E)
$$\begin{bmatrix} 1 & 0 & t \\ 0 & 1 & 0 \\ 0 & 0 & 1 \end{bmatrix} .$$

We shall derive a few geometric results by using the equiaffine properties of these transformations.

Example 1.2. Let us consider two parabolas

$$y = \frac{1}{2}x^2 \quad \text{and} \quad y = \frac{1}{2}x^2 + c^2.$$

The transformation group (D) maps each parabola onto itself transitively. Take a point P on the second parabola and let the tangent at P meet the first parabola at M and N. Then the region G bounded by the first parabola and the line MN has constant area independent of the choice of P. To see this, let P' be any other point and construct the points M' and N'. The transformation (D) that takes P to P' will map M and N upon M' and N' and the region G onto the other region G'. Since (D) is equiaffine, we see that G and G' have the same area. See Figure 1(a).

Example 1.3. Consider two ellipses

$$\frac{x^2}{a^2} + \frac{y^2}{b^2} = 1 \quad \text{and} \quad \frac{x^2}{a^2} + \frac{y^2}{b^2} = k, \ k > 1.$$

For a point P on the first ellipse, let the tangent at P meet the second ellipse at M and N. Then the region G bounded by the second ellipse and the line

MN outside the first ellipse has constant area independent of P. We use the group (1.12). See Figure 1(b).

Example 1.4. A similar result is valid for two hyperbolas

$$\frac{x^2}{a^2} - \frac{y^2}{b^2} = 1 \quad \text{and} \quad \frac{x^2}{a^2} - \frac{y^2}{b^2} = k, \ k > 1.$$

Example 1.5. For the hyperbola $xy = 1$, $x > 0$, let the tangent at a point P meet the x-axis and the y-axis at M and N. The triangle $\triangle OMN$, O being the origin, has constant area independent of P. Here we use the group $\begin{bmatrix} a & 0 \\ 0 & 1/a \end{bmatrix}$, $a > 0$, that maps each of the hyperbola $xy = 1$, the half-line $y = 0$, $x > 0$, and the half-line $x = 0$, $y > 0$ onto itself transitively.

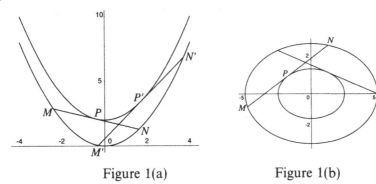

Figure 1(a)   Figure 1(b)

2. Affine space

In this section we first give a rigorous definition of affine space of an arbitrary dimension by using Weyl's axioms. We then proceed to define related concepts and results.

Definition 2.1. Let V be a real n-dimensional vector space. A non-empty set Ω is said to be an *affine space* associated to V if there is a mapping

$$\Omega \times \Omega \to V$$

denoted by

$$(p, q) \in \Omega \times \Omega \mapsto \overrightarrow{pq} \in V$$

satisfying the following axioms:

(2.1) for any $p, q, r \in \Omega$, we have $\overrightarrow{pr} = \overrightarrow{pq} + \overrightarrow{qr}$;

(2.2) for any $p \in \Omega$ and for any $x \in V$ there is one and only one $q \in \Omega$ such that $x = \overrightarrow{pq}$.

It is also convenient to write $q = p + x$ instead of $x = \overrightarrow{pq}$, particularly when we are given the point p and the vector x and wish to denote a point q as in (2.2). The *dimension* of Ω is defined as that of V. We also say that V is the *associated vector space* for the affine space Ω.

Example 2.1. Let V be a real vector space of dimension n. Consider V as a set and, for $(p,q) \in V \times V$, define $x = \overrightarrow{pq}$ to be the vector $q - p \in V$. In this way, V becomes an n-dimensional affine space.

Example 2.2. In particular, let V be the standard real n-dimensional vector space $\mathbf{R}^n$. Then regard it as an affine space in the manner of Example 2.1. We call it the *standard n-dimensional affine space*.

The simple axioms above, due to H. Weyl, are sufficient to derive all the properties of an affine space. We shall outline how they can be developed. First, observe that we have $\overrightarrow{pp} = 0$ and $\overrightarrow{qp} = -\overrightarrow{pq}$ for any $p, q \in \Omega$.

Definition 2.2. An affine coordinate system with origin $o \in \Omega$ can be defined as follows. Let $\{e_1, \ldots, e_n\}$ be a basis of V. For any point $p \in \Omega$, we write

$$(2.3) \qquad\qquad \overrightarrow{op} = \sum_{i=1}^{n} x^i(p)e_i,$$

where $(x^1(p), \ldots, x^n(p))$ is a uniquely determined n-tuple of real numbers, called the *coordinates* of p. The set of functions $\{x^1, \ldots, x^n\}$ is called an *affine coordinate system*.

If we have two affine coordinate systems $\{x^1, \ldots, x^n\}$ and $\{y^1, \ldots, y^n\}$, then they are related by

$$(2.4) \qquad\qquad y^i = \sum a^i_j x^j + c^i, \ 1 \le i \le n,$$

where $A = [a^i_j]$ is a nonsingular $n \times n$ matrix and $c = [c^i]$ is a vector. This relation may be expressed by the equation $y = Ax + c$, or in its expanded matrix form,

$$\begin{bmatrix} y \\ 1 \end{bmatrix} = \begin{bmatrix} A & c \\ 0 & 1 \end{bmatrix} \begin{bmatrix} x \\ 1 \end{bmatrix}.$$

We now define the notion of affine transformation. Let $f : \Omega \to \Omega$ be a one-to-one mapping of Ω onto itself. For each $p \in \Omega$ we define a mapping $F_p : V \to V$ as follows. For each $x \in V$, let $r \in \Omega$ be a uniquely determined point in Ω such that $\overrightarrow{pr} = x$. Then we set $F_p(x) = \overrightarrow{f(p)f(r)}$.

Definition 2.3. We say that f is an *affine transformation* if, for a certain $p \in \Omega$, the map F_p is a linear transformation of V onto itself (and it is nonsingular, since it is one-to-one together with f). In this case, it follows that for any point $q \in \Omega$ the map F_q coincides with F_p. We can therefore call this map the *associated linear transformation* and denote it simply by F.

Let $\{x^1, \ldots, x^n\}$ be an affine coordinate system with origin o and based on a basis $\{e_1, \ldots, e_n\}$. Let $\{y^1, \ldots, y^n\}$ be the affine coordinate system with origin $f(o)$ and based on the basis $\{F(e_1), \ldots, F(e_n)\}$. Then we have

$y_i(f(p)) = x_i(p)$ for $p \in \Omega$. We can write the relationship between the coordinate systems $\{x^1, \ldots, x^n\}$ and $\{y^1, \ldots, y^n\}$ in the form

(2.5) $$x^i = \sum b^i_j y^j + d^i, \ 1 \le i \le n.$$

Therefore, relative to the one coordinate system $\{x^1, \ldots, x^n\}$, the coordinates $\bar{x}^i = x^i(f(p))$ of the image $f(p)$ can be expressed in terms of the coordinates of p in the form

(2.6) $$\bar{x}^i = x^i(f(p)) = \sum b^i_j y^j(f(p)) + d^i = \sum b^i_j x^j(p) + d^i, \ 1 \le i \le n.$$

Again, this equation can be put in the form $\bar{x} = Bx + d$ or in its expanded matrix form,

$$\begin{bmatrix} \bar{x} \\ 1 \end{bmatrix} = \begin{bmatrix} B & d \\ 0 & 1 \end{bmatrix} \begin{bmatrix} x \\ 1 \end{bmatrix}.$$

We define the notion of affine subspace.

Definition 2.4. A nonempty subset Ω' of Ω is called an *affine subspace* if, for a certain point $p \in \Omega'$, the set of vectors $\{\overrightarrow{pq} : q \in \Omega'\}$ of Ω forms a vector subspace W of V. In this case, p in the condition can be replaced by any other point of Ω' with the same vector subspace W resulting. It follows that Ω' is an affine space associated to the vector space W. The dimension of an affine subspace is, by definition, the dimension of the associated vector space.

Example 2.3. Given two points $p, q \in \Omega$, the line pq is a 1-dimensional affine subspace $\{r \in \Omega : \overrightarrow{pr} = t\overrightarrow{pq} \text{ for } t \in \mathbf{R}\}$.

We now consider the standard real affine space $\Omega = \mathbf{R}^n$ as an n-dimensional differentiable manifold. For each point $p \in \mathbf{R}^n$ we may identify the tangent space $T_p(\Omega) = T_p(\mathbf{R}^n)$ with the vector space $V = \mathbf{R}^n$. This means that we consider each $x \in V$ as a geometric vector placed at p, that is, $\overrightarrow{pq}$ interpreted as the pair of initial point p and end point q. Furthermore, we may consider $x \in V$ as a vector field that assigns to each $p \in \Omega$ a tangent vector $\overrightarrow{pq}$ determined by x. Geometrically, all these vectors determined by x are parallel. From the construction of an affine coordinate system $\{x^1, \ldots, x^n\}$ it follows that $\partial/\partial x^i$ as a vector field corresponds to e_i, where $\{e_1, \ldots, e_n\}$ is a basis of V on which the affine coordinate system is based.

Let us now consider an arbitrary vector field Y on $\Omega = \mathbf{R}^n$ and let $x_t, a < t < b$, be an arbitrary smooth curve in Ω. We define the *covariant derivative* $D_t Y$ of Y along the curve x_t by

$$D_t Y = \lim_{h \to 0} \frac{1}{h} (\Pi Y_{x_{t+h}} - Y_{x_t}),$$

where Π denotes the identification map $T_{x_{t+h}}(\mathbf{R}^n) \to T_{x_t}(\mathbf{R}^n)$ through V in the manner explained above. If we take any affine coordinate system $\{x^1, \ldots, x^n\}$ and write

$$Y = \sum_{i=1}^{n} Y^i \frac{\partial}{\partial x^i} \quad \text{and} \quad x_t = (x^1(t), \ldots, x^n(t)),$$

then

$$D_t Y = \sum_{i=1}^{n} \frac{dY^i(x_t)}{dt} \frac{\partial}{\partial x^i} = \sum_{i,j=1}^{n} \frac{\partial Y^i}{\partial x^j} \frac{dx^j}{dt} \frac{\partial}{\partial x^i}.$$

Thus $D_t Y$ is a generalization of the directional derivative of functions to vector fields. If X is a tangent vector at a point x_0, then $D_X Y$ is defined by $D_X Y = (D_t Y)_t$, where x_t is a curve with initial point x_0 and initial tangent vector X. From this definition, it is clear that covariant differentiation D has the following properties:

(1) $D_{X_1 + X_2} Y = D_{X_1} Y + D_{X_2} Y$;
(2) $D_{\phi X} Y = \phi D_X Y$;
(3) $D_X (Y_1 + Y_2) = D_X Y_1 + D_X Y_2$;
(4) $D_X (\phi Y) = (X\phi) Y + \phi D_X Y$;

where ϕ is a smooth function and X, Y, X_1, X_2, Y_1, and Y_2 are vector fields on Ω.

We now consider the notion of parallel volume element in $\Omega = \mathbf{R}^n$. First we fix a *volume element* ω in the vector space $V = \mathbf{R}^n$. This is nothing but a nonzero alternating n-form; once an orientation of V is fixed, it is determined up to a positive constant factor, that is, for any oriented basis $\{e_1, \ldots, e_n\}$, the value $\omega(e_1, \ldots, e_n)$ can be assigned to be an arbitrary positive number c, which determines ω uniquely.

A volume element ω in V determines a volume element on the manifold Ω, that is, a nonvanishing differential n-form, denoted by the same letter ω, such that

$$\omega \left(\frac{\partial}{\partial x^1}, \ldots, \frac{\partial}{\partial x^n} \right) = c,$$

where the vector fields $\partial/\partial x^1, \ldots, \partial/\partial x^n$ correspond to $e_1, \ldots, e_n$ as explained above. It is obvious that $\omega(X_1, \ldots, X_n) = \omega(Y_1, \ldots, Y_n)$ if each $Y_i \in T_y(\mathbf{R}^{n+1})$ is parallel to $X_i \in T_x(\mathbf{R}^{n+1})$. Hence ω is said to be *parallel*.

Once a parallel volume element is fixed in the affine space Ω, an affine transformation f is said to be *equiaffine* (or *unimodular*) if it preserves the volume element, that is, the associated linear transformation F preserves the corresponding volume element in V. It is easy to verify that this is the case if and only if f is expressed by (2.6), where the matrix $[b_j^i]$ has determinant 1. The set of all equiaffine transformations forms a subgroup of the group of all affine transformations.

The geometry of submanifolds of an affine space is called *affine differential geometry*. We study the properties that are invariant under the group of affine transformations, just as Euclidean differential geometry is the geometry of submanifolds of a Euclidean space in which we study the properties invariant under Euclidean isometries. In affine differential geometry, particularly important is the study of properties invariant under equiaffine transformations. We have already seen an elementary theory of curves from this point of view; for example, we have characterized the quadrics in $\mathbf{R}^2$ by constancy of their affine curvature. We shall study the differential-geometric properties of quadrics in $\mathbf{R}^3$ and, more generally, in $\mathbf{R}^{n+1}$ in Chapter II. In Euclidean differential geometry there are many characterizations of a sphere but no characterization of an elliptic paraboloid $z = x^2 + y^2$, for example. In affine differential geometry we can characterize paraboloids and other quadratic surfaces very nicely.

3. Affine connections

In this section, we summarize the basic notions concerning affine connections. Let M be a differentiable manifold of class C^∞. When we want to emphasize its dimension, we write M^n. The most convenient way to define the notion of an affine (or linear) connection is through covariant differentiation of a vector field with respect to another. We shall denote by $\mathfrak{F}(M)$ the set of all differentiable functions and by $\mathfrak{X}(M)$ the set of all smooth vector fields on M.

Definition 3.1. By a rule of *covariant differentiation* on M we mean a mapping

$$(X, Y) \in \mathfrak{X}(M) \times \mathfrak{X}(M) \mapsto \nabla_X Y \in \mathfrak{X}(M)$$

satisfying the following conditions:

$$(3.1) \qquad \nabla_{X_1 + X_2} Y = \nabla_{X_1} Y + \nabla_{X_2} Y;$$
$$(3.2) \qquad \nabla_{\phi X} Y = \phi \nabla_X Y;$$
$$(3.3) \qquad \nabla_X (Y_1 + Y_2) = \nabla_X Y_1 + \nabla_X Y_2;$$
$$(3.4) \qquad \nabla_X (\phi Y) = (X\phi)Y + \phi \nabla_X Y;$$

here $X, Y, X_1, X_2, Y_1, Y_2 \in \mathfrak{X}(M)$ and $\phi \in \mathfrak{F}(M)$. An *affine connection* on M is nothing but a rule of covariant differentiation on M and we denote it by ∇.

An affine connection on M induces an affine connection on any open submanifold U of M in the natural way. In particular, if U is a coordinate neighborhood with local coordinates $\{x^1, \ldots, x^n\}$, then we may write

$$(3.5) \qquad \nabla_{\frac{\partial}{\partial x^i}} \frac{\partial}{\partial x^j} = \sum_{k=1}^{n} \Gamma_{ij}^k \frac{\partial}{\partial x^k},$$

where the system of functions Γ_{ij}^k $(i, j, k = 1, \ldots, n)$ are called the *Christoffel symbols* for the affine connection relative to the local coordinate system at hand.

If Y is a vector field defined along a curve x_t, then the covariant derivative $\nabla_t Y$ can be defined. In terms of local coordinates, we have

$$(3.6) \qquad \nabla_t Y = \sum_{k=1}^{n} \left[\frac{dY^k}{dt} + \sum_{i,j=1}^{n} \Gamma_{ij}^k \frac{dx^i}{dt} Y^j \right] \frac{\partial}{\partial x^k},$$

where $Y = \sum Y^k (\partial/\partial x^k)$ and $x(t) = (x^k(t))$. We say that a curve $x(t)$ is a *pregeodesic* if

$$\nabla_t \dot{x}_t = \phi(t)\dot{x}_t,$$

where $\dot{x}_t$ is the tangent vector field of $x(t)$ and $\phi(t)$ is a certain function. If we take another parameter $s = s(t)$, it is easily verified that

$$\nabla_s \overrightarrow{x_s} = \frac{1}{\left(\frac{ds}{dt}\right)^2} \left[\frac{-\frac{d^2 s}{dt^2}}{\frac{ds}{dt}} \dot{x}_t + \phi(t)\dot{x}_t \right].$$

Thus choosing s such that

$$\frac{ds}{dt} = \exp\left(\int \phi(t)dt \right)$$

we obtain $\nabla_s \overrightarrow{x_s} = 0$. Such a parameter s is called an *affine parameter* (uniquely determined up to affine change $s \mapsto as + b, a \neq 0$). We speak of a geodesic when a curve is parametrized by an affine parameter.

Let us recall that an affine connection ∇ leads to covariant differentiation of tensor fields relative to a vector field. Specifically, let K be a tensor field of type (r, s), that is, with *contravariant* degree r and *covariant* degree s, and let X be a vector field. Then $\nabla_X K$ is a tensor field of the same type (r, s). We may also regard ∇K as a tensor field of type $(r, s + 1)$, i.e., a linear mapping

$$X \in \mathfrak{X}(M) \mapsto \nabla_X K.$$

For a general discussion of this matter, especially for the interpretation of ∇_X as a derivation of the algebra of tensor fields that extends covariant differentiation of vector fields and also commutes with contraction, we refer the reader to [KN1, Chapter III]. Here we mention the following two special cases:

(1) If K is a covariant tensor field of degree s, that is, an s-linear map $\mathfrak{X}(M) \times \mathfrak{X}(M) \times \cdots \times \mathfrak{X}(M)$ (taken s times) $\to \mathfrak{F}(M)$, then for any $X \in \mathfrak{X}(M)$ we have

$$(\nabla_X K)(X_1, \ldots, X_s) = X(K(X_1, \ldots, X_n)) - \sum_{i=1}^{s} K(X_1, \ldots, \nabla_X X_i, \ldots, X_s).$$

(2) If K is a tensor field of type $(1, s)$, that is, an s-linear map $\mathfrak{X}(M) \times \mathfrak{X}(M) \times \cdots \times \mathfrak{X}(M)$ (taken s times) $\to \mathfrak{X}(M)$, then

$$(\nabla_X K)(X_1, \ldots, X_s)$$
$$= \nabla_X(K(X_1, \ldots, X_n)) - \sum_{i=1}^{s} K(X_1, \ldots, \nabla_X X_i, \ldots, X_s).$$

We define a few tensor fields associated to a given affine connection ∇. The *torsion tensor field* T is defined by

$$(3.7) \qquad T(X, Y) = \nabla_X Y - \nabla_Y X - [X, Y];$$

it is a tensor of type $(1, 2)$ that associates to a pair of vector fields (X, Y) the vector field defined by (3.7). Indeed, it induces for each point $x \in M$ a skew-symmetric bilinear mapping $T_x(M) \times T_x(M) \to T_x(M)$. The components of the torsion tensor T in local coordinates are

$$(3.7a) \qquad T_{ij}^k = \Gamma_{ij}^k - \Gamma_{ji}^k.$$

The *curvature tensor field* R, which is of type $(1, 3)$, is defined by

$$(3.8) \qquad R(X, Y)Z = \nabla_X \nabla_Y Z - \nabla_Y \nabla_X Z - \nabla_{[X,Y]} Z.$$

Given $X, Y \in T_x(M)$, $x \in M$, $R(X, Y)$ is a linear transformation of $T_x(M)$. The components in local coordinates,

$$R\left(\frac{\partial}{\partial x^k}, \frac{\partial}{\partial x^l}\right) \frac{\partial}{\partial x^j} = \sum_i R_{jkl}^i \frac{\partial}{\partial x^i},$$

are given by

$$(3.9) \qquad R_{jkl}^i = \left(\frac{\partial \Gamma_{lj}^i}{\partial x^k} - \frac{\partial \Gamma_{kj}^i}{\partial x^l}\right) + \sum_m (\Gamma_{lj}^m \Gamma_{km}^i - \Gamma_{kj}^m \Gamma_{lm}^i).$$

From now on, we shall assume that the torsion tensor of a given connection ∇ is 0. We also say that ∇ has *zero torsion* or that ∇ is *torsion-free*.

In this case, we have the *first* and *second Bianchi identities*:

$$(3.10\ \mathrm{I}) \qquad R(X, Y)Z + R(Y, Z)X + R(Z, X)Y = 0;$$

$$(3.10\ \mathrm{II}) \qquad (\nabla_X R)(Y, Z) + (\nabla_Y R)(Z, X) + (\nabla_Z R)(X, Y) = 0.$$

Note that the left-hand side of each of these equations is the cyclic sum over X, Y, and Z of its first term.

If R is identically 0 on M, we say that ∇ is a *flat affine connection*. Thus an affine connection is flat if $T = 0$ and $R = 0$. It is known that ∇ is flat if and only if around each point there exists a local coordinate system such that $\Gamma_{ij}^k = 0$ for all i, j and k. If $T = 0$ and $\nabla R = 0$, then we say that ∇ is a *locally symmetric affine connection*. Geometrically, ∇ is locally symmetric if and only if at each point the geodesic symmetry is a connection-preserving transformation. See [KN1, p. 303].

We define the *Ricci tensor* Ric, of type $(0,2)$, by

(3.11) $\text{Ric}(Y,Z) = \text{trace}\,\{X \mapsto R(X,Y)Z\}.$

The components in local coordinates are given by

(3.11a) $R_{jk} = \text{Ric}\left(\dfrac{\partial}{\partial x^j}, \dfrac{\partial}{\partial x^k}\right) = \sum_i R^i_{kij}.$

The reader acquainted with Riemannian geometry knows that the Levi-Civita connection of a Riemannian metric has symmetric Ricci tensor, that is, $\text{Ric}(Z,Y) = \text{Ric}(Y,Z)$. But this property is not necessarily valid for an arbitrary affine connection with zero torsion. In fact, the property is closely related to the concept of parallel volume element, as we now explain.

We shall say that an affine connection ∇ is *locally equiaffine* if around each point x of M there is a parallel volume element, that is, a nonvanishing n-form ω such that $\nabla\omega = 0$.

Proposition 3.1. *An affine connection ∇ with zero torsion has symmetric Ricci tensor if and only if it is locally equiaffine.*

Proof. From (3.10) and $R(Z,X) = -R(X,Z)$ we obtain

$$\text{Ric}\,(Y,Z) - \text{Ric}\,(Z,Y) = -\text{tr}\;R(Y,Z).$$

Hence Ric is symmetric if and only if tr $R(Y,Z) = 0$ for all Y and Z. Now suppose we pick any local volume element ω. Then we have a 1-form τ such that $\nabla_X\omega = \tau(X)\omega$. Since

$$\nabla_X(\nabla_Y\omega) = \nabla_X(\tau(Y)\omega) = X(\tau(Y))\omega + \tau(\nabla_X Y)\omega + \tau(Y)\tau(X)\omega,$$

we find

$$R(X,Y) \cdot \omega = 2(d\tau)(X,Y)\,\omega$$

On the other hand, we have $R(X,Y) \cdot \omega = -(\text{tr}\,R(X,Y))\omega$.

We are now in a position to prove that Ric is symmetric if and only if ∇ is locally equiaffine. If $\nabla\omega = 0$, then $\tau = 0$ and hence tr $R(X,Y) = 0$ for all $X,Y \in T_x(M)$. Conversely, suppose tr $R(X,Y) = 0$ for all $X,Y \in T_x(M)$. Then, using any local volume element ω, we find $R(X,Y) \cdot \omega = 0$ and therefore $d\tau = 0$. Now we can find a local function ϕ such that $d\log\phi = -\tau$. We may easily verify that the volume element $\phi\omega$ is ∇-parallel. This proves Proposition 3.1.

We give the following definition.

Definition 3.2. By an *equiaffine connection* ∇ on M we mean a torsion-free affine connection that admits a parallel volume element ω on M. If ω is a volume element on M such that $\nabla\omega = 0$, then we say that (∇, ω) is an *equiaffine structure* on M.

We note that if M is simply connected, then for any locally equiaffine connection ∇ on M there exists a volume element ω such that (∇, ω) is an equiaffine structure on M. A manifold M with an equiaffine structure is a generalization of the affine space with a fixed determinant function as volume element.

In the remainder of this section, we shall discuss the notion of projective change of a connection. We begin with a geometric discussion of a centro-affine (hyper)surface.

Let $\mathbf{R}^{n+1}$ be the affine space with a fixed point o. A hypersurface M imbedded in $\mathbf{R}^{n+1}$ is said to be a *centro-affine hypersurface* if the position vector x (from o) for each point $x \in M$ is transversal to the tangent plane of M at x. We let $-x$ play a role similar to a normal vector in ordinary surface theory in Euclidean space, that is, for any tangent vector fields X and Y to M, we consider the covariant derivative $D_X Y$ taken in $\mathbf{R}^{n+1}$ and decompose it in the form

$$(3.12) \qquad (D_X Y)_x = (\nabla_X Y)_x + h(X, Y)(-x)$$

at each point $x \in M$, where $(\nabla_X Y)_x$ is tangent to M.

We can verify that the mapping $(X, Y) \mapsto \nabla_X Y$ satisfies the conditions (3.1)–(3.4) in Definition 3.1. For example, for any differentiable function ϕ we have (3.4) and the identity $h(X, \phi Y) = \phi h(X, Y)$, because the equations

$$D_X(\phi Y) = (X\phi)Y + \phi(D_X Y) = (X\phi)Y + \phi(\nabla_X Y) + \phi h(X, Y)(-x)$$

and

$$D_X(\phi Y) = \nabla_X(\phi Y) + h(X, \phi Y)(-x)$$

imply

$$\nabla_X(\phi Y) = (X\phi) + \phi \nabla_X Y$$

as well as

$$h(X, \phi Y) = \phi h(X, Y).$$

Similarly, from

$$D_X Y - D_Y X - [X, Y] = 0$$

we get

$$\nabla_X Y - \nabla_Y X - [X, Y] = 0 \quad \text{and} \quad h(Y, X) = h(X, Y).$$

It follows that ∇ is a torsion-free affine connection and that h is a symmetric covariant tensor of degree 2. We call ∇ the *induced connection* and h the *centro-affine fundamental form* for M. The reader may prove

Proposition 3.2. *The curvature tensor R of the induced connection on a centro-affine hypersurface can be expressed by*

$$R(X, Y)Z = h(Y, Z)X - h(X, Z)Y$$

and the Ricci tensor by

$$\operatorname{Ric}(Y,Z) = (n-1)h(Y,Z)$$

which is symmetric.

Now we take a positive differentiable function, say λ, on M and define a new hypersurface $\bar{M}$ by setting

$$\bar{M} = \{\bar{x} : \bar{x} = \lambda x, x \in M\}.$$

In order to describe the geometry of $\bar{M}$, we use the mapping

(3.13) $$f : x \in M \mapsto f(x) = \bar{x} = \lambda x \in \mathbf{R}^{n+1}.$$

For $X, Y \in \mathfrak{X}(M)$ we get

$$f_*(Y) = D_Y(\bar{x}) = (Y\lambda)x + \lambda Y,$$

which shows that f is an immersion. Furthermore, we have

$$\begin{aligned}
D_X f_*(Y) &= D_X((Y\lambda)x + \lambda Y) \\
&= XY(\lambda)x + (Y\lambda)X + (X\lambda)Y + \lambda(D_X Y) \\
&= XY(\lambda)x + (Y\lambda)X + (X\lambda)Y + \lambda(\nabla_X Y) + \lambda h(X,Y)(-x),
\end{aligned}$$

which is equal mod $\bar{x}$ to

$$(Y\lambda)X + (X\lambda)Y + \lambda(\nabla_X Y).$$

On the other hand, we get

$$\begin{aligned}
D_X f_*(Y) &= f_*(\bar{\nabla}_X Y) + \bar{h}(X,Y)(-\bar{x}) \\
&= (\bar{\nabla}_X Y)\lambda + \lambda \bar{\nabla}_X Y + \bar{h}(X,Y)(-\bar{x}),
\end{aligned}$$

which is equal mod $\bar{x}$ to $\lambda(\bar{\nabla}_X Y)$, where $\bar{\nabla}$ is the connection induced by f on M (or geometrically speaking, the induced connection on $\bar{M}$). Therefore the two connections ∇ and $\bar{\nabla}$ on M are related by

$$\bar{\nabla}_X Y = \nabla_X Y + \frac{1}{\lambda}(X\lambda) + \frac{1}{\lambda}(Y\lambda),$$

that is,

(3.14) $$\bar{\nabla}_X Y = \nabla_X Y + \rho(X)Y + \rho(Y)X,$$

where ρ is the 1-form $d(\log \lambda)$.

If x_t is a ∇-pregeodesic, that is, $\nabla_t \dot{x}_t = \phi(t)\dot{x}_t$, then $\bar{\nabla}_t \dot{x}_t = (\phi(t)+2\rho(\dot{x}_t))\dot{x}_t$, hence x_t is a $\bar{\nabla}$-pregeodesic. The converse also holds. For the two centro-affine hypersurfaces M and $\bar{M}$ we have a direct geometric argument as follows. A curve x_t on M is a pregeodesic if and only if $d^2 x_t / dt^2$ is 0 mod x_t and dx_t / dt. Using this fact, we easily see that the curve $\bar{x}_t = f(x_t)$ is a pregeodesic, and $f : M \to \bar{M}$ is what is called a projective transformation from M onto $\bar{M}$ in the sense that it maps every pregeodesic on M to a pregeodesic on $\bar{M}$. In particular, if we take $\bar{M}$ to be a hyperplane by choosing a suitable function λ on M, the pregeodesics $\bar{x}_t$ on $\bar{M}$ through a point x_0 are nothing but the intersections of $\bar{M}$ with a plane containing the line from 0 to x_0. It also follows that every centro-affine hypersurface admits locally a projective transformation to a hyperplane (with flat affine connection). See Figure 2.

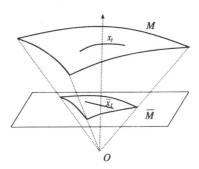

Figure 2

These discussions motivate the following definition.

Definition 3.3. Two torsion-free locally equiaffine connections ∇ and $\bar{\nabla}$ on a differentiable manifold M are said to be *projectively equivalent* if there is a 1-form ρ such that (3.14) holds. In this case, ρ is necessarily closed, as the reader may wish to verify. We also say that $\nabla \to \bar{\nabla}$ is a *projective change*.

Finally, we say that a locally equiaffine connection ∇ is *projectively flat* if around each point there is a projective change of ∇ to a flat affine connection $\bar{\nabla}$. For the given connection ∇ the *Weyl projective curvature tensor* is defined by

$$(3.15) \quad W(X,Y)Z = R(X,Y)Z - \frac{1}{n-1}[\mathrm{Ric}\,(Y,Z)X - \mathrm{Ric}\,(X,Z)Y].$$

It is convenient to introduce $\gamma = \frac{1}{n-1}\mathrm{Ric}$ and call it the *normalized Ricci tensor*. Then

$$(3.16) \quad W(X,Y)Z = R(X,Y)Z - [\gamma(Y,Z)X - \gamma(X,Z)Y].$$

The following is a classical result (for the proof, see [Ei, p. 88, p. 96]).

Theorem 3.3. *A torsion-free affine connection ∇ with symmetric Ricci tensor on a differentiable manifold M is projectively flat if and only if one of the following conditions holds:*

(1) *$\dim M \geq 3$ and W is identically 0;*

(2) *$\dim M = 2$ and $(\nabla_X \gamma)(Y,Z) = (\nabla_Y \gamma)(X,Z)$.*

Remark 3.1. If $\dim M = 2$, W is identically 0. If $\dim M \geq 3$ and $W = 0$, then γ satisfies $(\nabla_X \gamma)(Y,Z) = (\nabla_Y \gamma)(X,Z)$.

Remark 3.2. For more general discussions on projective change and projective flatness, see Appendix 1.

4. Nondegenerate metrics

We recall the basic knowledge about inner products on real vector spaces. Let V be an n-dimensional real vector space. For a given bilinear function $f : (x,y) \mapsto f(x,y) \in \mathbf{R}$, there exists a basis in V

$$\{e_1,\ldots,e_p,e_{p+1},\ldots,e_{p+q},e_{p+q+1},\ldots,e_n\}$$

such that

$$f(e_i,e_j) = 0 \quad \text{for all} \quad i \neq j;$$

$$f(e_i,e_i) = 1 \quad \text{for all} \quad i,\ 1 \leq i \leq p;$$

$$f(e_j,e_j) = -1 \quad \text{for all} \quad j,\ p+1 \leq j \leq p+q;$$

$$f(e_k,e_k) = 0 \quad \text{for all} \quad k,\ p+q+1 \leq k \leq n.$$

Moreover, the integers p and q are uniquely determined (although a basis of the kind above is not unique). This fact is often called *Sylvester's law of inertia*.

For a given f, the subspace $V_0 = \{x \in V : f(x,y) = 0 \text{ for all } y \in V\}$ is called the *null space* (or, *kernel*). Its dimension is equal to $n - (p+q)$. We say that f is *nondegenerate* if the null space is $\{0\}$; thus f is nondegenerate if and only if $p + q = n$. A nondegenerate, symmetric bilinear function is called an *inner product* on V. The pair (p,q) is called its *signature*. If $p = n$, then $f(x,x) \geq 0$ for all x, and the equality holds if and only if $x = 0$. In this case, f is said to be *positive-definite*. If $q = n$, then f is *negative-definite*. If f is positive- or negative-definite, we say f is *definite*. Otherwise, we say f is *indefinite*.

Example 4.1. In the vector space $\mathbf{R}^n$, for $0 < p < n$,

$$\langle x,y \rangle = \sum_{i=1}^{p} x^i y^i - \sum_{j=p+1}^{n} x^j y^j$$

is an indefinite inner product of signature $(p, n-p)$. In particular, when $p = n - 1$, it is called a *Lorentzian inner product*. For an indefinite inner product, usually denoted by $\langle \, , \, \rangle$, a vector x is said to be *spacelike, timelike,* or *null (lightlike)* according as $\langle x, x \rangle$ is positive, negative, or 0 (and $x \neq 0$).

Example 4.2. In the Minkowski space $\mathbf{R}^4$ with Lorentzian inner product, the vector $(1, 0, 0, 0)$ is spacelike, $(0, 0, 0, 1)$ timelike, and $(1, 0, 0, 1)$ null.

Now let M be an n-dimensional differentiable manifold. By a *nondegenerate metric*, say g, on M we mean a symmetric tensor field of type $(0, 2)$ that is nondegenerate, that is, if $g(X, Y) = 0$ for all $Y \in T_x(M)$, then $X = 0$. We also call it a *pseudo-Riemannian* (or *semi-Riemannian*) *metric*. If g is positive-definite, we say that g is a Riemannian metric. If $f : M \to E^m$ is an immersion of M into a Euclidean space E^m, then f induces a Riemannian metric g:

$$g(X, Y) = \langle f_*(X), f_*(Y) \rangle \quad \text{for all} \quad X, Y \in T_x(M),$$

where $\langle \, , \, \rangle$ denotes the Euclidean inner product.

Suppose M is given a nondegenerate metric g. We say that an affine connection ∇ on M is *metric* if $\nabla g = 0$. Geometrically, this property implies that parallel displacement preserves the inner product given by g. Indeed, let Y_t and Z_t be vector fields parallel along a curve x_t. If ∇ is metric, we get

$$0 = (\nabla_X g)(Y, Z) = Xg(Y, Z) - g(\nabla_X Y, Z) - g(Y, \nabla_X Z),$$

thus

$$(4.1) \qquad Xg(Y, Z) = g(\nabla_X Y, Z) + g(Y, \nabla_X Z)$$

for all $X, Y, Z \in \mathfrak{X}(M)$. This implies in particular

$$\frac{d}{dt} g(Y_t, Z_t) = g(\nabla_t Y_t, Z_t) + g(Y_t, \nabla_t Z_t) = 0,$$

which shows that $g(Y_t, Z_t)$ is constant.

We have

Theorem 4.1. *Given a nondegenerate metric g on M, there is an affine connection with zero torsion that is metric: $\nabla g = 0$. Such a connection is unique. (It is called the Levi-Civita connection for g.)*

Proof. If there is such a connection ∇, then equation (4.1) and the condition of zero torsion $\nabla_X Y - \nabla_Y X - [X, Y] = 0$ lead to

$$(4.2) \qquad 2g(\nabla_X Y, Z) = Xg(Y, Z) + Yg(X, Z) - Zg(X, Y)$$
$$+ g([X, Y], Z) + g([Z, X], Y) - g([Y, Z], X).$$

Conversely, given $X, Y \in \mathfrak{X}(M)$, we define $\nabla_X Y$ as a vector field such that (4.2) holds for every $Z \in \mathfrak{X}(M)$; it is uniquely determined because g is nondegenerate. Then we can verify that ∇ is an affine connection with zero torsion and that $\nabla g = 0$.

In terms of local coordinates $\{x^1, \ldots, x^n\}$, the components of g are

$$g_{ij} = g\left(\frac{\partial}{\partial x^i}, \frac{\partial}{\partial x^j}\right).$$

The Christoffel symbols of the Levi-Civita connection ∇ are given by

(4.3)
$$\Gamma_{ij}^k = \frac{1}{2}\sum_{\ell=1}^{n} g^{k\ell}\left(\frac{\partial g_{i\ell}}{\partial x^j} + \frac{\partial g_{j\ell}}{\partial x^i} - \frac{\partial g_{ij}}{\partial x^\ell}\right).$$

Now we consider a more general situation. Suppose an n-manifold M is provided with a nondegenerate metric, say h, and an affine connection ∇. We define a new affine connection $\bar{\nabla}$ by requiring that the following holds for all $X, Y, Z \in \mathfrak{X}(M)$:

(4.4)
$$Xh(Y,Z) = h(\nabla_X Y, Z) + h(Y, \bar{\nabla}_X Z).$$

This equation determines a unique affine connection $\bar{\nabla}$, which we call the *conjugate connection* of ∇ relative to h. Should the given ∇ be metric relative to h, then $\bar{\nabla}$ coincides with ∇. We are thus interested in the case where ∇ is not necessarily metric relative to h. If we consider the conjugate connection of $\bar{\nabla}$ relative to h, then, obviously, we get back ∇.

We have

Proposition 4.2. *The torsion tensors T and $\bar{T}$ of ∇ and $\bar{\nabla}$, respectively, satisfy*

(4.5) $(\nabla_X h)(Y,Z) + h(Y, T(X,Z)) = (\nabla_Y h)(X,Z) + h(Y, \bar{T}(X,Z))$

for all $X, Y, Z \in \mathfrak{X}(M)$.

Proof. From the definition of $(\nabla_X h)(Y,Z)$ and by (4.4) we get

$$(\nabla_X h)(Y,Z) = h(Y, \bar{\nabla}_X Z) - h(Y, \nabla_X Z)$$

and similarly
$$(\nabla_Z h)(Y,X) = h(Y, \bar{\nabla}_Z X) - h(Y, \nabla_Z X).$$

These two equations lead to

$$\begin{aligned}
(\nabla_X h)(Y,Z) &- (\nabla_Z h)(Y,X) \\
&= h(Y, \bar{\nabla}_X Z - \bar{\nabla}_Z X) - h(Y, \nabla_X Z - \nabla_Z X) \\
&= h(Y, [X,Z] + \bar{T}(X,Z)) - h(Y, [X,Z] + T(X,Z)) \\
&= h(Y, \bar{T}(X,Z) - T(X,Z))
\end{aligned}$$

by virtue of $T(X,Z) = \nabla_X Z - \nabla_Z X - [X,Z]$ and $\bar{T}(X,Z) = \nabla_X Z - \nabla_Z X - [X,Z]$.

Corollary 4.3. *Assume that ∇ is torsion-free. Then $\bar{\nabla}$ is torsion-free if and only if (∇, h) satisfies Codazzi's equation*

$$(4.6) \qquad (\nabla_X h)(Y,Z) = (\nabla_Z h)(Y,X) \quad \text{for all} \quad X,Y,Z \in \mathfrak{X}(M).$$

If (∇, h) satisfies Codazzi's equation, then $(X,Y,Z) \mapsto C(X,Y,Z) = (\nabla_X h)(Y,Z)$ is a symmetric 3-linear function. We call it the *cubic form* for (∇, h). In this case, we shall also say that (∇, h) is *compatible*.

Corollary 4.4. *If ∇ is torsion-free and (∇, h) is compatible, then*

    (1) *$(\bar{\nabla}, h)$ satisfies Codazzi's equation;*

    (2) *$\hat{\nabla} = \frac{1}{2}(\nabla + \bar{\nabla})$ is the Levi-Civita connection for h.*

Proof.

    (1) We know that $\bar{\nabla}$ has torsion 0 by Corollary 4.3. Since the conjugate connection ∇ of $\bar{\nabla}$ is torsion-free, we see that $(\bar{\nabla}, h)$ satisfies Codazzi's equation by the same corollary.

    (2) The connection $\hat{\nabla}$ has torsion 0, together with ∇ and $\bar{\nabla}$. By adding (4.4) and the equation with ∇ and $\bar{\nabla}$ interchanged, we obtain

$$Xh(Y,Z) = h(\hat{\nabla}_X Y, Z) + h(Y, \hat{\nabla}_X Z),$$

which shows that $\hat{\nabla} h = 0$, that is, $\hat{\nabla}$ is metric relative to h. Hence $\hat{\nabla}$ is the Levi-Civita connection for h.

We shall give a geometric interpretation of the equation (4.4).

Proposition 4.5. *Let x_t, $0 \leq t \leq 1$, be a (piecewise differentiable) curve in M, and let $X,Y \in T_{x_0}$. We denote by X_t and $\bar{Y}_t$ the vector fields along the curve obtained by parallel displacement from X and Y relative to the connections ∇ and $\bar{\nabla}$, respectively. Then $h(X_t, \bar{Y}_t)$ is constant along the curve.*

Proof. We have by (4.4)

$$\frac{d}{dt}[h(X_t, \bar{Y}_t)] = h(\nabla_t X_t, \bar{Y}_t) + h(X_t, \bar{\nabla}_t \bar{Y}_t) = 0,$$

which proves Proposition 4.5.

We continue to assume that (∇, h) is compatible and consider the conjugate connection $\bar{\nabla}$.

Proposition 4.6. *The curvature tensors R and $\bar{R}$ of ∇ and $\bar{\nabla}$ are related by*

$$(4.7) \qquad h(R(X,Y)Z, U) = -h(Z, \bar{R}(X,Y)U).$$

Proof. From $Y h(Z, U) = h(\nabla_Y Z, U) + h(Z, \bar{\nabla}_Y U)$ we obtain

$$X Y h(Z, U) = h(\nabla_X \nabla_Y Z, U)$$
$$+ h(\nabla_Y Z, \bar{\nabla}_X U) + h(\nabla_X Z, \bar{\nabla}_Y U) + h(Z, \bar{\nabla}_X \bar{\nabla}_Y U).$$

Subtracting from this equation the one obtained by interchanging X and Y as well as the equation

$$[X, Y] h(Z, U) = h(\nabla_{[X,Y]} Z, U) + h(Z, \bar{\nabla}_{[X,Y]} U),$$

we obtain the formula.

Corollary 4.7. $R = 0$ *if and only if* $\bar{R} = 0$.

More relationships between $\nabla, \bar{\nabla}$, and $\hat{\nabla}$ will be given in Section 9 of Chapter II.

The notion of conjugate connection (sometimes called the *dual connection*, whose name will be justified in the next section) has an origin in affine differential geometry. If (∇, h) is the pair of induced affine connection and the affine metric on a nondegenerate hypersurface M in the affine space $\mathbf{R}^{n+1}$, then the conjugate connection $\bar{\nabla}$ is the affine connection induced on M by an immersion of M into the dual affine space, called the conormal immersion (see Section 5 of Chapter II). The interplay of these two connections that arose in affine differential geometry can be formulated in an abstract way as we have presented in this section. When can one realize a given compatible pair (∇, h) by an immersion of M into $\mathbf{R}^{n+1}$? This was essentially the problem studied by Radon for the case $n = 2$ in 1918 (see Section 8 of Chapter II). It is worth noting that this notion does play an important role also in mathematical statistics (see [Am1]). It was introduced into the subject, apparently without any regard to what had been known in classical affine differential geometry, a fact that makes the relationship more intriguing.

5. Vector bundles

In the preceding sections we have been using the term affine connection. It is also called a linear connection, and the notion of linear connection can be generalized. Here we shall give a sketch of the basic concepts and examples related to vector bundles.

Definition 5.1. A *vector bundle* (E, M, π, F) consists of a total space E, a base manifold M, a projection $\pi : E \to M$, and a standard fiber F. We assume that E and M are differentiable manifolds of dimension $n + r$ and n, respectively, that π is differentiable, and that F is a real r-dimensional vector space. (Later, in Chapter IV, we may consider a complex vector bundle for which F is a complex vector space of dimension $\frac{1}{2} r$, with even r.) We further assume that each point $x \in M$ has a neighborhood U such that $\pi^{-1}(U)$ is diffeomorphic to $U \times F$; more precisely, there is a diffeomorphism

$$(5.1) \qquad \hat{\phi}_U : u \in \pi^{-1}(U) \mapsto \hat{\phi}_U(u) = (\pi(u), \phi_U(u)) \in U \times F.$$

In the intersection of two such neighborhoods U and V, there is a differentiable map ψ_{VU} of $U \cap V$ into the group $GL(F)$ of all linear transformations such that

$$(5.2) \qquad \phi_V(u) = \psi_{VU} \cdot \phi_U(u) \quad \text{for all} \quad u \in \pi^{-1}(U \cap V).$$

From these assumptions it follows that, for each $x \in M$, the fiber $F_x = \pi^{-1}(x)$ over x is diffeomorphic to the standard fiber F. Moreover, the diffeomorphism $u \in \pi^{-1}(x) \mapsto \phi_U(u)$ induces a vector space structure on the fiber F_x that does not depend on the choice of neighborhood U; this fact follows from (5.2) and from the fact that $\psi_{VU} \in GL(F)$.

A differentiable map $s : U \to E$, where U is an open subset of M, is called a *section* over U if $\pi \cdot s$ is the identity map on U. If $U = M$, it is called simply a *(global) section*. On the other hand, a *local section* is a section defined on a neighborhood of a point. From the definition of a vector bundle, every point $x \in M$ has a local section defined in its neighborhood. If U is an open subset of M, then the set $\mathfrak{S}(U)$ of all sections over U is a real vector space, and, in fact, a module over the algebra of all differentiable functions on U. We list a few basic examples.

Example 5.1. The *tangent bundle* $T(M)$ of a differentiable manifold M. The total space E is the set of all tangent vectors at all points x of M, the projection π takes a tangent vector $X \in T_x(M)$ to the point x. The standard fiber F is an n-dimensional standard real vector space $\mathbf{R}^n$. For each $x \in M$, the fiber over x is nothing but the tangent space $T_x(M)$ at x. For each point $x \in M$, let $\{x^1, \ldots, x^n\}$ be a system of local coordinates in a neighborhood U. Then we define $\hat{\phi}_U : \pi^{-1}(U) \to U \times \mathbf{R}^n$ as follows. For each $X \in \pi^{-1}(U)$, write X as a linear combination

$$X = \sum_{i=1}^{n} a^i \frac{\partial}{\partial x^i}$$

and set

$$\phi(X) = (a^1, \ldots, a^n) \in F.$$

We note that, for two coordinate neighborhoods U and V of x, ψ_{VU} is the Jacobian matrix between the two coordinate systems. We may regard $\partial/\partial x^i$ as a local section.

Example 5.2. The *cotangent bundle* $T^*(M)$ of a differentiable manifold M. This vector bundle is obtained in Example 5.1 by replacing tangent vectors with tangent covectors, the fibers $T_x(M)$ by $T_x^*(M)$, and the standard fiber $\mathbf{R}^n$ by its dual space $\mathbf{R}_n$. If $\{x^1, \ldots, x^n\}$ is a coordinate system in U, then dx^i, $1 \leq i \leq n$, are sections over U.

Example 5.3. The *tensor bundle* $T_s^r(M)$ over a differentiable manifold M. This generalizes Examples 5.1 and 5.2. For each $x \in M$ we denote by $T_s^r(x)$ the tensor space of type (r, s) over $T_x(M)$. Just as in Examples 5.1 and 5.2, we can make the total space $\bigcup_{x \in M} T_s^r(x)$ into a vector bundle with standard fiber T^r_s of all tensors of contravariant degree r and covariant degree s. Here again, a tensor field of type (r, s) is a section of the bundle $T^r_s(M)$.

We now discuss the notion of linear connection for a vector bundle (E, M, π, F).

Definition 5.2. By a *linear connection* on E we mean a rule of covariant differentiation, namely, a mapping

$$(X, s) \in \mathfrak{X}(M) \times \mathfrak{S}(M) \mapsto \nabla_X s \in \mathfrak{S}(M)$$

satisfying the following conditions:

(1) $\nabla_{X+Y} s = \nabla_X s + \nabla_Y s$;

(2) $\nabla_{\phi X} s = \phi \nabla_X s$;

(3) $\nabla_X (s + t) = \nabla_X s + \nabla_X t$;

(4) $\nabla_X (\phi s) = (X\phi)s + \phi \nabla_X s$.

Here $X, Y \in \mathfrak{X}(M), s, t \in \mathfrak{S}(M)$, and $\phi \in \mathfrak{F}(M)$.

When we take the tangent bundle $T(M)$ as E, we recover Definition 3.1. If s is a section defined only along a curve x_t, we can define the covariant derivative $\nabla_t s$ along the curve x_t. We say that s is parallel along x_t if $\nabla_t s = 0$ for every t.

We extend the notion of curvature tensor in the following way. Let $X, Y \in \mathfrak{X}(M)$ and let s be a section. We set

$$R(X, Y)s = \nabla_X \nabla_Y s - \nabla_Y \nabla_X s - \nabla_{[X,Y]} s,$$

which is a section of E. Now we can easily verify that $R(X, Y)s$ is linear over $\mathfrak{F}(M)$ in each of the variables X, Y and s. It follows that we have a 3-linear mapping

$$(X, Y, s) \in T_x(M) \times T_x(M) \times F_x \mapsto R(X, Y)s \in F_x.$$

If we fix $X, Y \in T_x(M)$, then $R(X, Y)$ can be regarded as a linear map of F_x into itself. This is the curvature operator $R(X, Y)$ at $x \in M$.

When an affine connection, that is, a linear connection in $T(M)$, is given, we may obtain a linear connection in any tensor bundle $T^r_s(M)$ by naturally extending covariant differentiation ∇_X to tensor fields of type (r, s); we have illustrated this in the cases $r = 0$ and $r = 1$ in Section 3. As a very special case, we have

Example 5.4. The dual connection. Let ∇ be an affine connection. Then we have a linear connection in the cotangent bundle $T^*(M)$, called the *dual connection* and denoted by ∇^*. That is, if α is a 1-form i.e., a section of $T^*(M)$, then for any $X \in \mathfrak{X}(M)$, $\nabla^*_X \alpha$ is defined as the 1-form given by

(5.3) $$(\nabla^*_X \alpha)(Y) = X(\alpha(Y)) - \alpha(\nabla_X Y).$$

Let M be a manifold with a nondegenerate metric h. We may identify the tangent bundle $T(M)$ and the cotangent bundle $T^*(M)$ by means of h in the following fashion. For each point $x \in M$, we have a linear isomorphism λ_h of the tangent space $T_x(M)$ onto the cotangent space $T_x^*(M)$:

(5.4) $\qquad \lambda_h(X)(Y) = h(X, Y) \quad \text{for all} \quad X, Y \in T_x(M).$

Thus we get

$$\lambda_h : X \in T(M) \mapsto \lambda_h(X) \in T^*(M),$$

which is a vector bundle isomorphism, that is, it commutes with projections $T(M) \to M$ and $T^*(M) \to M$ and gives an isomorphism of each fiber onto the corresponding fiber.

We now show that the conjugate connection introduced in Section 4 is essentially the dual connection when $T(M)$ is identified with $T^*(M)$ by λ_h.

Proposition 5.1. *Let (∇, h) be a compatible structure on M. The conjugate connection $\bar{\nabla}$ of ∇ relative to h and the dual connection ∇^* in $T^*(M)$ correspond to each other by the isomorphism $\lambda_h : T(M) \to T^*(M)$.*

Proof. For simplicity, we write λ for λ_h. Then we have

$$
\begin{aligned}
\lambda(\bar{\nabla}_X Y)(Z) &= h(Z, \bar{\nabla}_X Y) \\
&= X h(Z, Y) - h(\nabla_X Z, Y) \\
&= X(\lambda(Y)(Z)) - \lambda(Y)(\nabla_X Z) \\
&= (\nabla_X^* \lambda(Y))(Z),
\end{aligned}
$$

that is,

$$\lambda(\bar{\nabla}_X Y) = \nabla_X^*(\lambda(Y)).$$

We take this opportunity to explain the classical index notation for tensors g_{ij} and g^{ij}. For a nondegenerate metric g on a manifold M, g_{ij} are the components of g relative to a local coordinate system $\{x^1, \ldots, x^n\}$, and g^{ij} are usually introduced as the components of the inverse matrix $[g^{ij}]$ of the matrix $[g_{ij}]$. From the point of view of the tangent and cotangent bundles, g gives an inner product on each fiber of $T(M)$; likewise, we can get an inner product, say g^*, in each fiber of $T^*(M)$ such that

(5.5) $\qquad\qquad g(X, Y) = g^*(\lambda_g(X), \lambda_g(Y)),$

where λ_g is the linear isomorphism $T(M) \to T^*(M)$ by means of the metric g. We shall call g^* the *dual metric*. More generally, we speak of a *fiber metric* in a vector bundle if there is a differentiable family of inner products in each fiber: $x \in M \to \langle \, , \, \rangle_x$. It is then easy to verify that if a linear connection ∇ is metric relative to g, then the dual connection is metric relative to the dual metric and vice versa.

The linear isomorphism λ_g is specifically given by

(5.6)
$$\lambda_g \left(\frac{\partial}{\partial x^i} \right) = \sum_{j=1}^{n} g_{ij} dx^j.$$

From this we obtain

(5.7) $g^{ij} = g^*(dx^i, dx^j), \quad \text{where} \quad [g^{ij}] = [g_{ij}]^{-1},$

as can be verified.

Theorem 5.2. *Let (∇, h) be a compatible structure on a manifold M. If $\nabla g = 0$ for some nondegenerate metric g, then there is a nondegenerate metric $\bar{g}$ such that $\bar{\nabla}\bar{g} = 0$ for the conjugate connection $\bar{\nabla}$.*

Proof. Observe that the dual connection ∇^* is metric relative to the dual metric g^*. Now by means of λ_h we transfer g^* to a certain metric, say $\bar{g}$, in $T(M)$. Then it clearly follows that the conjugate connection $\bar{\nabla}$ is metric relative to $\bar{g}$.

Remark 5.1. Theorem 5.2 is useful in the proof of the Cartan–Norden theorem treated in Section 3 of Chapter IV. The proof above is more conceptual than the original technical proof.

II

Geometry of affine immersions
– the basic theory

We introduce the notion of affine immersion, special cases and examples. The fundamental concepts and equations are established. Emphasis is on the presentation of the basic theory of classical affine differential geometry from a new point of view and in a generalized form. In particular, Section 3 sets forth the classical theory of Blaschke with the general background prepared in Sections 1 and 2. In Section 4, we discuss cubic forms and apolarity and give two proofs to the classical theorem of Pick and Berwald that characterizes quadratic hypersurfaces by the vanishing of the cubic form. We study conormal maps in Section 5 and the Laplacians of various mappings in Section 6. Section 7 treats generalizations of the classical Lelieuvre formula and Section 8 a new formulation of Radon's theorem, which we may consider the fundamental theorem for affine hypersurfaces. In Section 9, we obtain more formulas relating the induced connection ∇ and the Levi-Civita connection $\hat{\nabla}$ for the affine metric. In Section 10, we obtain the formulas for the Laplacian $\triangle J$ of the Pick invariant and for $\triangle \log |J|$, which will be useful in proving a classical theorem of Blaschke in Chapter III. Finally, in Section 11, we study the behavior of cubic forms in the case of affine surfaces and its relation to ruled surfaces.

1. Affine immersions

We introduce the general notion of affine immersion. We shall always assume that the given affine connections have zero torsion.

Consider two differentiable manifolds with affine connections $(\tilde{M}, \tilde{\nabla})$ and (M, ∇) of dimensions m and n, respectively. Let $k = m - n$.

Definition 1.1. A differentiable immersion $f : M \to \tilde{M}$ is said to be an *affine immersion* if the following condition is satisfied. There is a k-dimensional differentiable distribution N along $f : x \in M \mapsto N_x$, a subspace of $T_{f(x)}(\tilde{M})$, such that

(1.1) $$T_{f(x)}(\tilde{M}) = f_*(T_x(M)) + N_x \quad \text{(direct sum)},$$

and such that for all $X, Y \in \mathfrak{X}(M)$ we have

(1.2) $(\tilde{\nabla}_X f_*(Y))_x = (f_*(\nabla_X Y))_x + (\alpha(X, Y))_x,$ where $\alpha(X, Y)_x \in N_x,$

at each point $x \in M$. Thus (1.2) gives the decomposition of $(\tilde{\nabla}_X Y)_x$ into the tangential component and the component in N_x according to (1.1).

A few explanations are in order. Since the given distribution $x \in M \mapsto N_x$ is differentiable, each point x has a local basis, namely, a system of k differentiable vector fields $\xi_1, \ldots, \xi_k$ on a neighborhood U of x that span N_y at each point $y \in U$. This distribution may be regarded as a bundle of *transversal k-subspaces*. In the case where $f : M \to \tilde{M}$ is an immersion of a manifold M into a Riemannian manifold $\tilde{M}$ with positive-definite Riemannian metric g, we can certainly choose the normal space at each point, namely,

$$N_x = \{\xi \in T_{f(x)}(\tilde{M}) : g(\xi, X) = 0 \quad \text{for all} \quad X \in T_x(M)\}.$$

In our original setup, there is no such natural choice of transversal subspaces. In any case, we assume that there is a certain distribution of transversal k-subspaces and that the connection ∇ on M is related to the connection $\tilde{\nabla}$ on $\tilde{M}$ by (1.2).

In this situation, it is easy to show that the map $(X, Y) \in \mathfrak{X}(M) \mapsto \alpha(X, Y)$ actually defines for each point $x \in M$ a symmetric bilinear map

$$T_x(M) \times T_x(M) \to N_x.$$

In the case of an immersion into a Riemannian manifold $\tilde{M}$, this map α is called the *second fundamental form*, whereas the *first fundamental form* refers to the induced Riemannian metric on M. In the geometry of affine immersions, we shall call α simply the *affine fundamental form*.

Suppose a differentiable manifold M, not provided with any particular affine connection, is immersed into a manifold with an affine connection $(\tilde{M}, \tilde{\nabla})$. If we take a distribution of transversal subspaces (1.1), then we can get a torsion-free affine connection ∇ with (1.2) as defining equation. To prove this, we have to verify that the map $\nabla : \mathfrak{X}(M) \times \mathfrak{X}(M) \to \mathfrak{X}(M)$ satisfies the conditions for covariant differentiation in Definition 3.1 of Chapter I. The verification, however, is straightforward and hence omitted. We might say that ∇ is defined as the tangential part of the connection $\tilde{\nabla}$; it is called the *induced connection*. As a result, $(M, \nabla) \to (\tilde{M}, \tilde{\nabla})$ is an affine immersion.

For the remainder of this section, we shall concentrate on the case of codimension $k = 1$ and also assume that the manifold $(\tilde{M}, \tilde{\nabla})$ is an affine space $\mathbf{R}^{n+1}$ with its usual flat affine connection D. We shall use a fixed parallel volume element $\tilde{\omega}$ on $\mathbf{R}^{n+1}$ whenever necessary, but with a word of caution. Most of what we do in the rest of this section can be stated in the case of an arbitrary manifold with an affine connection $(\tilde{M}^{n+1}, \tilde{\nabla})$, further provided with a parallel volume element $\tilde{\omega}$ if we want to talk about the induced volume element in M.

Thus we consider an n-dimensional manifold M together with an immersion $f : M \to \mathbf{R}^{n+1}$. We call M a *hypersurface* and f a *hypersurface immersion*. We may also call M an immersed hypersurface. For each point $x_0 \in M$ we choose a local field of transversal vectors $\xi : x \in U \mapsto \xi_x$, where U is a neighborhood of x_0. *Transversality* means in this case that

$$(1.3) \qquad T_{f(x)}(\mathbf{R}^{n+1}) = f_*(T_x(M)) + \text{Span}\{\xi_x\},$$

where Span $\{\xi_x\}$ means the 1-dimensional subspace spanned by ξ_x. The existence of such ξ can be directly verified by using a local representation of the immersion by means of coordinate systems. That is, if $\{x^1, \ldots, x^n\}$ is a local coordinate system around x_0, we write f as

$$y^i = f^i(x^1, \ldots, x^n), \quad 1 \le i \le n + 1,$$

where $\{y^1, \ldots, y^{n+1}\}$ is an affine coordinate system for $\mathbf{R}^{n+1}$. Since f is an immersion, the $(n + 1) \times n$ Jacobian matrix $[\partial y^i / \partial x^j]$ has rank n at x_0 (and in its neighborhood). This means that the n column vectors $f_*(\partial/\partial x^1), \ldots, f_*(\partial/\partial x^n)$ are linearly independent. We may now choose, for example, a column vector with constant components $a^1, \ldots, a^{n+1}$ that is linearly independent of those n vectors at x_0 and hence in its neighborhood. Now if we set $\xi_x = \sum_{i=1}^{n+1} a^i (\partial/\partial y^i)|_{f(x)}$, we get a local field ξ of transversal vectors. (The reader should be able to modify this argument in the case where $\mathbf{R}^{n+1}$ is replaced by an arbitrary manifold $\tilde{M}^{n+1}$.)

We repeat for the present situation what we said earlier.

Proposition 1.1. *For a hypersurface immersion $f : M \to \mathbf{R}^{n+1}$, suppose we have a transversal vector field ξ on M. Then we have a torsion-free induced connection ∇ satisfying*

$$(1.4) \qquad D_X f_*(Y) = f_*(\nabla_X Y) + h(X, Y)\xi \quad \text{(formula of Gauss)},$$

where h is a symmetric bilinear function on the tangent space $T_x(M)$.

The reader may recall this formula from the case of a centro-affine hypersurface in Section 3 of Chapter I.

Definition 1.2. The symmetric bilinear function h is called the *affine fundamental form* (relative to the transversal vector field ξ).

Definition 1.3. Let (M, ∇) be an n-dimensional manifold with an affine connection ∇. An immersion $f : M \to \mathbf{R}^{n+1}$ is called an *affine immersion* if there is a transversal vector field ξ on M such that equation (1.4) holds.

Thus for an immersed hypersurface $f : M \to \mathbf{R}^{n+1}$ a choice of transversal vector field ξ provides an induced connection ∇ in such a way that f becomes an affine immersion. We also state

Proposition 1.2. *For all $X \in \mathfrak{X}(M)$ we have*

(1.5) $D_X \xi = -f_*(SX) + \tau(X)\xi$ *(formula of Weingarten),*

where S is a tensor of type $(1,1)$, called the (affine) shape operator, and τ is a 1-form, called the transversal connection form.

Proof. We decompose $D_X \xi$ into the tangential component, which is linear in X, and the transversal component, which is also linear in X. We can express the former in the form $-f_*(SX)$, defining S in this manner, and the latter as a multiple of ξ by $\tau(X)$, thereby defining a differential 1-form τ.

We have

Proposition 1.3. *Let (M, ∇) be an n-dimensional manifold equipped with an affine connection ∇ and let $f : (M, \nabla) \to \mathbf{R}^{n+1}$ and $\bar{f} : (M, \nabla) \to \mathbf{R}^{n+1}$ be affine immersions relative to transversal vector fields ξ and $\bar{\xi}$, respectively. The objects h, S, and τ for $\bar{\xi}$ are denoted by $\bar{h}$, $\bar{S}$, and $\bar{\tau}$. Assume that*

$$h = \bar{h}, \quad S = \bar{S}, \quad \tau = \bar{\tau}.$$

Then there is an affine transformation A such that $\bar{f} = Af$.

Proof. For each $x \in M$, define a linear mapping L_x from $T_{f(x)}\mathbf{R}^{n+1}$ to $T_{\bar{f}(x)}\mathbf{R}^{n+1}$ by

$$L(f_*X) = \bar{f}_*X \quad \text{for} \quad X \in T_x M,$$
$$L(\xi_x) = \bar{\xi}_x.$$

By the usual identification of each tangent space $T_y(\mathbf{R}^{n+1})$, $y \in \mathbf{R}^{n+1}$, to $\mathbf{R}^{n+1}$ with the vector space $\mathbf{R}^{n+1}$, we can regard the $(1,1)$-tensor field L over f as a mapping of M into the space of endomorphisms of $\mathbf{R}^{n+1}$. Let us see that L is constant on M. Indeed, for $X \in T_x M$ and $Y \in \mathfrak{X}(M)$, we have

$$
\begin{aligned}
(D_X L)(f_*Y) &= D_X(L(f_*Y)) - L(D_X(f_*Y)) \\
&= D_X(\bar{f}_*Y) - L(f_*\nabla_X Y + h(X,Y)\xi) \\
&= \bar{f}_*(\nabla_X Y) + \bar{h}(X,Y)\bar{\xi} - (\bar{f}_*\nabla_X Y + h(X,Y)\bar{\xi}) \\
&= (\bar{h}(X,Y) - h(X,Y))\bar{\xi},
\end{aligned}
$$

which vanishes by the assumption. Similarly, for $X \in T_x(M)$, we have

$$
\begin{aligned}
(D_X L)(\xi) &= D_X(\bar{\xi}) - L(D_X \xi) \\
&= -\bar{f}_*\bar{S}X + \bar{\tau}(X)\bar{\xi} - (-\bar{f}_*SX + \tau(X)\bar{\xi}) \\
&= -\bar{f}_*(\bar{S}X - SX) + (\bar{\tau}(X) - \tau(X))\bar{\xi},
\end{aligned}
$$

which also vanishes. Now, consider the mapping $t : M \to \mathbf{R}^{n+1}$ defined by

$$t(x) = \bar{f}(x) - L(f(x)).$$

Then $D_X t = 0$ by the computation above, that is, t is a constant vector. By defining an affine transformation A with L as the linear part and with t as the translation part, we have $\bar{f} = Af$.

In these situations we are also interested in equiaffine connections. So we take a fixed parallel volume element $\tilde{\omega}$ in $\mathbf{R}^{n+1}$. For a hypersurface immersion $f : M \to \mathbf{R}^{n+1}$, let ξ be a transversal vector field. In addition to the induced connection ∇ and the affine fundamental form h, we consider the following volume element θ on M:

$$(1.6) \qquad \theta(X_1,\ldots,X_n) = \tilde{\omega}(f_*(X_1),\ldots,f_*(X_n),\xi).$$

Clearly, θ is a volume element on M, called the *induced volume element*. We are interested in the question whether (∇,θ) defines an equiaffine structure (Definition 3.2 of Chapter I), that is, whether $\nabla\theta = 0$ holds. This question is answered as follows.

Proposition 1.4. *We have*

$$(1.7) \qquad \nabla_X \theta = \tau(X)\theta \quad \text{for all} \quad X \in T_x(M).$$

Consequently, the following two conditions are equivalent:

(1) $\nabla\theta = 0$;

(2) $\tau = 0$, that is, $D_X\xi$ is tangential for every $X \in \mathfrak{X}(M)$.

Proof. The volume element $\tilde{\omega}$ is parallel: $D_X\tilde{\omega} = 0$. For any basis $\{X_1,\ldots,X_n\}$ in $T_x(M)$, we have

$$(D_X\tilde{\omega})(X_1,\ldots,X_n,\xi) = X\tilde{\omega}(X_1,\ldots,X_n,\xi)$$
$$- \sum_{i=1}^{n} \tilde{\omega}(X_1,\ldots,D_X X_i,\ldots,X_n,\xi)$$
$$- \tilde{\omega}(X_1,\ldots,X_n,D_X\xi),$$

where we omit f_* to simplify notation. Now

$$X\tilde{\omega}(X_1,\ldots,X_n,\xi) = X\theta(X_1,\ldots,X_n)$$

and

$$\tilde{\omega}(X_1,\ldots,D_X X_i,\ldots,X_n,\xi) = \theta(X_1,\ldots,\nabla_X X_i,\ldots,X_n),$$

because $D_X X_i = \nabla_X X_i + h(X,X_i)\xi$. We have also

$$\tilde{\omega}(X_1,\ldots,X_n,D_X\xi) = \tau(X)\theta(X_1,\ldots,X_n),$$

because $D_X\xi = -S(X) + \tau(X)\xi$ and $\tilde{\omega}(X_1,\ldots,X_n,SX) = 0$. From these equations we obtain

$$(\nabla_X\theta)(X_1,\ldots,X_n) = X\theta(X_1,\ldots,X_n) - \sum_{i=1}^{n} \theta(X_1,\ldots,\nabla_X X_i,\ldots,X_n)$$
$$= \tau(X)\theta(X_1,\ldots,X_n),$$

as we wanted to show.

In view of Proposition 1.4, it is convenient to make the following definition.

Definition 1.4. For a hypersurface immersion $f : M \to \mathbf{R}^{n+1}$, a transversal vector field ξ is said to be *equiaffine* if $D_X\xi$ is tangent to M for each $X \in T_x(M), x \in M$.

With an equiaffine transversal vector field ξ, we have an equiaffine structure (∇, θ) on M. Thus we may call $f : (M, \nabla, \theta) \to \mathbf{R}^{n+1}$ an *equiaffine immersion*. The equation (1.5) is simplified: $D_X\xi = -f_*(SX)$. We shall see more consequences of this condition $\tau = 0$ in the fundamental equations for affine immersions. An equiaffine transversal vector field is sometimes called a *relative normalization* (see [SiSV]).

Remark 1.1. For any hypersurface $f : M \to \mathbf{R}^{n+1}$, we can locally find an equiaffine transversal vector field. In fact, if we introduce any Euclidean metric in $\mathbf{R}^{n+1}$ and take a unit normal vector field N, then N is an equiaffine transversal vector field.

We shall see how the theory of plane curves can be viewed as a special case of hypersurface theory. Let $\mathbf{x}(\sigma)$ be a nondegenerate curve with affine arclength parameter σ in Section 1 of Chapter I. Since $|\mathbf{x}'\ \mathbf{x}''| = 1$, we may take $\mathbf{x}''$ as a transversal vector field. Since $\mathbf{x}'$ is the tangent vector field, the formula (1.4) applied to the curve simply reduces to

$$\nabla_\sigma \frac{d}{d\sigma} = 0 \quad \text{and} \quad h\left(\frac{d}{d\sigma}, \frac{d}{d\sigma}\right) = 1.$$

The first says that σ is a flat coordinate system for the induced connection ∇. The second means that h is a metric for which $d/d\sigma$ has unit length (i.e., affine arclength). On the other hand, we see in the formula (1.5) that $d\mathbf{x}''/d\sigma$ is tangential, in fact, equal to $-\kappa\mathbf{x}'$, where κ is the affine curvature. Thus $\xi = \mathbf{x}''$ is equiaffine and the affine shape operator S in this case is simply $\kappa \cdot I$. We also get $\theta(d/d\sigma) = |\mathbf{x}'\ \mathbf{x}''| = 1$, that is, the induced volume (length) element coincides with that of affine arclength.

2. Fundamental equations. Examples

We begin by deriving more fundamental equations for a hypersurface immersion $f : M \to \mathbf{R}^{n+1}$. First, we consider the case where the given transversal vector field ξ is arbitrary. We have

Theorem 2.1. *For an arbitrary transversal vector field ξ the induced connection ∇, the affine fundamental form h, the shape operator S, and the transversal connection form τ satisfy the following equations:*

(2.1) Gauss: $R(X, Y)Z = h(Y, Z)SX - h(X, Z)SY$;

(2.2) Codazzi for h:

$$(\nabla_X h)(Y, Z) + \tau(X)h(Y, Z) = (\nabla_Y h)(X, Z) + \tau(Y)h(X, Z);$$

(2.3) Codazzi for S: $\quad (\nabla_X S)(Y) - \tau(X)SY = (\nabla_Y S)(X) - \tau(Y)SX$;

(2.4) Ricci: $\quad\quad\quad h(X,SY) - h(SX,Y) = d\tau(X,Y)$.

Proof. Since D is a flat connection, we have

(1) $(D_X D_Y - D_Y D_X - D_{[X,Y]})Z = 0$,

(2) $(D_X D_Y - D_Y D_X - D_{[X,Y]})\xi = 0$,

for all $X, Y, Z \in \mathfrak{X}(M)$. We now compute (1) and take the tangential and transversal components. To simplify notation, we omit f_* in front of vector fields. Then

$$D_Y Z = \nabla_Y Z + h(Y,Z)\xi$$

and

$$\begin{aligned}
D_X(D_Y Z) &= D_X(\nabla_Y Z) + D_X(h(Y,Z)\xi) \\
&= \nabla_X(\nabla_Y Z) + h(X, \nabla_Y Z)\xi \\
&\quad + (\nabla_X h)(Y,Z)\xi + h(\nabla_X Y,Z)\xi + h(Y,\nabla_X Z)\xi \\
&\quad + h(Y,Z)(-SX + \tau(X)\xi) \\
&= \nabla_X \nabla_Y Z - h(Y,Z)SX \\
&\quad + [h(X, \nabla_Y Z) + (\nabla_X h)(Y,Z) + h(\nabla_X Y, Z) \\
&\quad + h(Y, \nabla_X Z) + \tau(X)h(Y,Z)]\xi.
\end{aligned}$$

Interchanging X and Y we get another equation. We have also one more equation

$$D_{[X,Y]}Z = \nabla_{[X,Y]}Z + h([X,Y],Z)\xi.$$

Subtracting the second and third equations from the first and separately setting the tangential and transversal components equal to 0, we obtain

$$R(X,Y)Z - h(Y,Z)SX + h(X,Z)SY = 0$$

and

$$(\nabla_X h)(Y,Z) - (\nabla_Y h)(X,Z) + \tau(X)h(Y,Z) - \tau(Y)h(X,Z) = 0,$$

which prove (2.1) and (2.2).

In order to prove (2.3) and (2.4) we compute (2) in the beginning of the proof. We have

$$\begin{aligned}
D_X D_Y \xi &= D_X(-SY + \tau(Y)\xi) \\
&= -\nabla_X(SY) - h(X,SY)\xi + X(\tau(Y))\xi + \tau(Y)[-SX + \tau(X)\xi] \\
&= -(\nabla_X S)(Y) - S(\nabla_X Y) - \tau(Y)SX \\
&\quad + [-h(X,SY) + X(\tau(Y)) + \tau(X)\tau(Y)]\xi.
\end{aligned}$$

Again interchanging X and Y we get another equation. We have also one more equation

$$D_{[X,Y]}\xi = -S([X,Y]) + \tau([X,Y])\xi.$$

Subtracting the second and third equations from the first and separating the tangential and transversal components, we obtain

$$(\nabla_Y S)X - (\nabla_X S)Y - \tau(Y)SX + \tau(X)SY = 0$$

and

$$-h(X,SY) + h(SX,Y) + X\tau(Y) - Y\tau(X) - \tau([X,Y]) = 0,$$

which prove (2.3) and (2.4) in view of

$$d\tau(X,Y) = X\tau(Y) - Y\tau(X) - \tau([X,Y]).$$

Corollary 2.2. *The Ricci tensor of the induced connection is given by*

$$(2.5) \qquad\qquad \mathrm{Ric}\,(Y,Z) = h(Y,Z)\,\mathrm{tr}\,S - h(SY,Z).$$

Proof. This follows from the definition of Ric, $\mathrm{Ric}(Y,Z) = \mathrm{trace}\{X \mapsto R(X,Y)Z\}$, and the equation of Gauss (2.1).

Remark 2.1. From (2.5) and (2.4) it follows that Ric is symmetric if and only if $d\tau = 0$.

In the equation of Codazzi (2.2) we see that the left-hand side is symmetric in X and Y as well as in Y and Z. Therefore if we set

$$(2.6) \qquad\qquad C(X,Y,Z) = (\nabla_X h)(Y,Z) + \tau(X)h(Y,Z),$$

we see that it is symmetric in all three variables. We call C the *cubic form* for the affine immersion.

The equations in Theorem 2.1 are not all independent. We have

Proposition 2.3. *Let M be a differentiable manifold with a torsion-free affine connection ∇, a symmetric covariant tensor field h of degree 2, a $(1,1)$-tensor field S, and a 1-form τ that together satisfy the equation of Gauss (2.1) and the equation of Codazzi for h (2.2). If rank $h \geq 3$, then the equation of Codazzi for S (2.3) is satisfied.*

Proof. Covariantly differentiating (2.1) relative to a vector field W we obtain

$$(\nabla_W R)(X,Y)Z = (\nabla_W h)(Y,Z)SX - (\nabla_W h)(X,Z)SY$$
$$+ h(Y,Z)(\nabla_W S)X - h(X,Z)(\nabla_W S)Y.$$

Denoting by $\mathscr{S}$ the cyclic sum over the variables X, Y, and W we have

$$
\begin{aligned}
\mathscr{S}(\nabla_W R)(X,Y)Z &= \mathscr{S}[(\nabla_W h)(Y,Z)SX - (\nabla_W h)(X,Z)SY] \\
&\quad + \mathscr{S}[h(Y,Z)(\nabla_W S)X - h(X,Z)(\nabla_W S)Y] \\
&= \mathscr{S}[(\nabla_W h)(Y,Z) - (\nabla_Y h)(W,Z)]SX \\
&\quad + \mathscr{S}h(Y,Z)[(\nabla_W S)X - (\nabla_X S)W] \\
&= \mathscr{S}[\tau(Y)h(W,Z) - \tau(W)h(Y,Z)]SX \\
&\quad + \mathscr{S}h(Y,Z)[(\nabla_W S)X - (\nabla_X S)W]
\end{aligned}
$$

by virtue of the Codazzi equation for h in the form

$$
(\nabla_W h)(Y,Z) - (\nabla_Y h)(W,Z) = \tau(Y)h(W,Z) - \tau(W)h(Y,Z)
$$

and two similar equations. Now by the second identity of Bianchi (3.10 II) of Chapter I, we have

$$
\mathscr{S}[\tau(Y)h(W,Z) - \tau(W)h(Y,Z)]SX + \mathscr{S}h(Y,Z)[(\nabla_W S)X - (\nabla_X S)W] = 0.
$$

Now, given any two vectors X and Y, we can find Z such that $h(X,Z) = h(Y,Z) = 0$ and $h(Z,Z) \neq 0$, because h has rank ≥ 3 by assumption. Setting $W = Z$ in the equation above, we obtain

$$
(\nabla_X S)Y - \tau(X)SY = (\nabla_Y S)(X) - \tau(Y)SX,
$$

namely, the Codazzi equation for S.

Proposition 2.3 (the case where $\tau = 0$ and h has rank $n \geq 3$) will play an important role in Section 8.

We shall now assume that a transversal vector field is equiaffine, that is, $\tau = 0$, and see how the fundamental equations (2.2)–(2.4) are simplified.

Theorem 2.4. *If $f : M \to \mathbf{R}^{n+1}$ is an affine immersion with an equiaffine transversal vector field, then the equations of Codazzi and Ricci are as follows:*

(2.2*) $\qquad\qquad (\nabla_X h)(Y,Z) = (\nabla_Y h)(X,Z);$

(2.3*) $\qquad\qquad (\nabla_X S)(Y) = (\nabla_Y S)(X);$

(2.4*) $\qquad\qquad h(X, SY) = h(SX, Y).$

Remark 2.2. When ξ is equiaffine, we have $C(X,Y,Z) = (\nabla_X h)(Y,Z)$. This is what we saw in a more abstract setting in Section 4 of Chapter I.

We shall now consider the change of a transversal vector field for a given immersion f.

Proposition 2.5. *Suppose we change a transversal vector field ξ to*

(2.7) $\qquad\qquad \bar{\xi} = \phi\xi + f_*(Z),$

where Z is a tangent vector field on M and ϕ is a nonvanishing function. Then the affine fundamental form, the induced connection, the transversal connection form, and the affine shape operator change as follows:

(2.8)
$$\bar{h} = \frac{1}{\phi}h;$$

(2.9)
$$\bar{\nabla}_X Y = \nabla_X Y - \frac{1}{\phi}h(X,Y)Z;$$

(2.10)
$$\bar{\tau} = \tau + \frac{1}{\phi}h(Z,\cdot) + d\log|\phi|;$$

(2.11)
$$\bar{S} = \phi S - \nabla.Z + \bar{\tau}(\cdot)Z;$$

$h(Z,\cdot), \nabla.$ *and* $\bar{\tau}(\cdot)Z$ *are the 1-forms whose values on* X *are* $h(Z,X), \nabla_X Z$, *and* $\bar{\tau}(X)Z$, *respectively.*

Proof. Omitting f_* in (1.4) we have

$$\begin{aligned}
D_X Y &= \nabla_X Y + h(X,Y)\xi \\
&= \bar{\nabla}_X Y + \bar{h}(X,Y)\bar{\xi} = \bar{\nabla}_X Y + \bar{h}(X,Y)Z + \bar{h}(X,Y)\phi\xi,
\end{aligned}$$

from which we obtain (2.8) and (2.9).

Next we compute

$$\begin{aligned}
D_X \bar{\xi} &= D_X(Z + \phi\xi) \\
&= \nabla_X Z + h(X,Z)\xi + (X\phi)\xi + \phi(-SX + \tau(X)\xi) \\
&= \nabla_X Z - \phi SX + [h(X,Z) + X\phi + \phi\tau(X)]\xi.
\end{aligned}$$

We have also

$$D_X \bar{\xi} = -\bar{S}X + \bar{\tau}(X)\bar{\xi} = -\bar{S}(X) + \bar{\tau}(X)Z + \bar{\tau}(X)\phi\xi.$$

Comparing these two equations, we obtain (2.10) and (2.11).

Definition 2.1. (2.8) in Proposition 2.5 shows that the rank of the affine fundamental form is independent of the choice of transversal vector field. We define it as the *rank* of the hypersurface or the hypersurface immersion. In particular, if the rank is n, that is, if h is nondegenerate, then we say that the hypersurface or the hypersurface immersion is *nondegenerate*.

Theorem 2.6. *Let* (M,∇) *be an n-dimensional manifold equipped with an affine connection* ∇ *and let* $f : (M,\nabla) \to \mathbf{R}^{n+1}$ *and* $\bar{f} : (M,\nabla) \to \mathbf{R}^{n+1}$ *be affine immersions relative to transversal vector fields* ξ *and* $\bar{\xi}$, *respectively. Assume that the rank of* f *is* ≥ 2 *everywhere and* $h = \phi\bar{h}$ *for a nowhere vanishing function* ϕ. *Then there exists an affine transformation* A *such that* $\bar{f} = Af$ *and* $\bar{\xi} = \phi L\xi$ *where* L *is the linear part of* A.

Proof. Since the replacement of ξ with $\phi\xi$ changes h to h/ϕ, we can assume $\phi = 1$. Since $\nabla h = \bar{\nabla} h$, the Codazzi equation shows

$$\tau(X)h(Y,Z) - \tau(Y)h(X,Z) = \bar{\tau}(X)h(Y,Z) - \bar{\tau}(Y)h(X,Z).$$

Let $\{X_1, \ldots, X_n\}$ be a basis of $T_x M$ such that $h(X_i, X_i) = \pm 1$ or 0 and $h(X_i, X_j) = 0$ for $i \neq j$. Since $\operatorname{rank} h \geq 2$, we can let $h(X_1, X_1) \neq 0$ and $h(X_2, X_2) \neq 0$. Then we have

$$\tau(X_i)h(X_1, X_1) = \bar{\tau}(X_i)h(X_1, X_1) \quad \text{for} \quad i > 1$$

and

$$\tau(X_1)h(X_2, X_2) = \bar{\tau}(X_1)h(X_2, X_2).$$

Therefore, $\tau = \bar{\tau}$. Combining the Gauss equations for h and $\bar{h}$, we have

$$h(Y,Z)SX - h(X,Z)SY = h(Y,Z)\bar{S}X - h(X,Z)\bar{S}Y,$$

which, when applied to the basis above, gives

$$h(X_1, X_1)SX_i = h(X_1, X_1)\bar{S}X_i \quad \text{for} \quad i > 1,$$
$$h(X_2, X_2)SX_1 = h(X_2, X_2)\bar{S}X_1.$$

Hence, $S = \bar{S}$. Now Proposition 1.3 proves the assertion.

Remark 2.3. Theorem 2.6 appears in [O3]. Our proof for Proposition 1.3 is direct, although there are various proofs, for example, [Di3], [DNV], which deal with equivalence theorems in affine differential geometry.

In the rest of the section we shall discuss a few examples of affine immersions.

Example 2.1. *Isometric immersion.* Let M be a Riemannian manifold of dimension n with positive-definite metric g and Levi-Civita connection ∇. Let $\tilde{M}$ be a Riemannian manifold of dimension $n + 1$ with positive-definite metric $\tilde{g}$ and Levi-Civita connection $\tilde{\nabla}$. If $f : (M, g) \to (\tilde{M}, \tilde{g})$ is an isometric immersion, then $(M, \nabla) \to (\tilde{M}, \tilde{\nabla})$ is an affine immersion with a unit normal vector field as an equiaffine transversal field. In order to prove this, we show that the induced connection by means of the unit normal field has torsion 0 and is metric relative to g, as is done in hypersurface theory in Riemannian geometry. (See [KN2, p. 11].)

Example 2.2. *Centro-affine hypersurface.* Let o be a point of the affine space $\mathbf{R}^{n+1}$ chosen as origin. We consider an immersion f of an n-manifold M into $\mathbf{R}^{n+1} - \{o\}$ such that the position vector $\overrightarrow{of(x)}$ for $x \in M^n$ is always transversal to $f(M)$. Take $\xi = -\overrightarrow{of(x)}$. We already considered the formula of Gauss (3.12) in Section 3 of Chapter I. We have $D_X\xi = -X$ so that $\tau = 0$ and $S = I$. The equation of Gauss reads

$$R(X, Y)Z = h(Y, Z)X - h(Z, Y)X.$$

The normalized Ricci tensor is given by $\gamma(Y,Z) = h(Y,Z)$. We showed that the induced connection on a centro-affine hypersurface is projectively flat. Conversely, we have

Proposition 2.7. *Let (M,∇) be an n-manifold with a projectively flat affine connection ∇ with symmetric Ricci tensor. Then (M,∇) can be locally immersed as a centro-affine hypersurface.*

Proof. Let U be an open neighborhood in M on which there is a 1-form ρ such that

$$\bar{\nabla}_X Y = \nabla_X Y + \rho(X)Y + \rho(Y)X$$

is flat. We may then assume that $(U,\bar{\nabla})$ is an open ball with center at the origin of the hyperplane, say, $x^{n+1} = 1$ of the affine space $\mathbf{R}^{n+1}$, and $\bar{\nabla}$ is the usual flat connection. As remarked in Definition 3.3 of Chaper I, the form ρ is closed and hence we may assume that $\rho = d\log\lambda$ for a certain positive function λ on U. (A direct proof of this fact will be indicated below.) We now define a centro-affine hypersurface $f : U \to \mathbf{R}^{n+1} - \{o\}$, where o is the origin of $\mathbf{R}^{n+1}$, by $f(x) = \frac{1}{\lambda} x$. Then the affine connection, say $\tilde{\nabla}$, induced on U by f is given by

$$\tilde{\nabla}_X Y = \bar{\nabla}_X Y + \mu(X)Y + \mu(Y)X,$$

where $\mu = d\log\frac{1}{\lambda}$ as in the formula (3.14) of Chapter I. But then $\mu = -\rho$ and hence $\tilde{\nabla} = \nabla$, proving that U with ∇ is realized as a centro-affine hypersurface.

As promised above, we now prove that if two torsion-free connections with symmetric Ricci tensors are projectively related, that is,

$$\bar{\nabla}_X Y = \nabla_X Y + \rho(X)Y + \rho(Y)X,$$

then ρ is locally of the form $d\log\lambda$ for a certain positive function λ. Since the Ricci tensors of ∇ and $\bar{\nabla}$ are symmetric, we have local volume elements ω and $\bar{\omega}$ such that $\nabla\omega = \bar{\nabla}\bar{\omega} = 0$. There is a positive function ϕ such that $\bar{\omega} = \phi\omega$. We have

$$0 = \bar{\nabla}_X\bar{\omega} = (X\phi)\omega + \phi(\bar{\nabla}_X\omega).$$

We have also

$$(\bar{\nabla}_X\omega)(Y_1,\ldots,Y_n) = (\nabla_X\omega)(Y_1,\ldots,Y_n) - \sum_{i=1}^n \omega(Y_1,\ldots,\rho(X)Y_i,\ldots,Y_n)$$

$$- \sum_{i=1}^n \omega(Y_1,\ldots,\rho(Y_i)X,\ldots,Y_n)$$

$$= -n\rho(X)\omega(Y_1,\ldots,Y_n) - \rho(X)\omega(Y_1,\ldots,Y_n)$$

$$= -(n+1)\rho(X)\omega(Y_1,\ldots,Y_n),$$

where $\{Y_1, \ldots, Y_n\}$ is a basis in the tangent space. From these we get

$$(X\phi)\omega = (n+1)\phi\rho(X)\omega, \quad \text{that is,} \quad \rho = \frac{1}{n+1} d\log\phi,$$

hence

$$\rho = d\log\lambda \quad \text{with} \quad \lambda = \phi^{\frac{1}{n+1}}.$$

Example 2.3. *Equiaffine immersions.* Let $f : (M, \nabla) \to \mathbf{R}^{n+1}$ be an affine immersion with a transversal vector field ξ. Suppose there exist a parallel volume element $\tilde{\omega}$ in $\mathbf{R}^{n+1}$ and a ∇-parallel volume element ω on M. Then we can find a function ϕ such that $\bar{\xi} = \phi\xi$ is equiaffine and $f : (M, \nabla, \omega) \to \mathbf{R}^{n+1}$ is an equiaffine immersion. In fact, let θ be the volume element induced by the given ξ. We have a nonzero function ϕ such that $\omega = \phi\theta$. Then it easily follows that the volume element induced by $\bar{\xi} = \phi\xi$ is equal to ω, whereas the connection induced by $\bar{\xi}$ remains unchanged as seen from (2.9) in Proposition 2.5.

Example 2.4. *Graph immersions.* Let Ω be a domain in $\mathbf{R}^n$ and F a real-valued differentiable function on Ω. By the *graph immersion* by F we mean the immersion f defined by

$$(2.12) \qquad x = (x^1, \ldots, x^n) \mapsto f(x) = (x^1, \ldots, x^n, F(x)) \in \mathbf{R}^{n+1}$$

together with a transversal vector field

$$\xi = (0, \ldots, 0, 1).$$

Since ξ is a constant vector field (i.e., a parallel vector field in $\mathbf{R}^{n+1}$), we have $D_X\xi = 0$ and hence $S = 0$ and $\tau = 0$.

We may compute the affine fundamental form h. To simplify notation, we write ∂_i for $\partial/\partial x^i$ and F_i for $\partial F/\partial x^i$ as well as F_{ij} for $\partial^2 F/\partial x^i \partial x^j$. Then we have

$$f_*(\partial_1) = (1, 0, \ldots, 0, F_1),$$
$$f_*(\partial_2) = (0, 1, \ldots, 0, F_2),$$

$$\cdots \quad \cdots$$

$$f_*(\partial_n) = (0, 0, \ldots, 1, F_n).$$

Therefore we get

$$(2.13) \qquad D_{\partial_i} f_*(\partial_j) = (0, 0, \ldots, 0, F_{ij}),$$

which implies that

$$h(\partial_i, \partial_j) = F_{ij}$$

and

$$\nabla_{\partial_i} \partial_j = 0.$$

This means that $\{x^1,\ldots,x^n\}$ is a flat coordinate system for the flat induced connection ∇.

Conversely, we have

Proposition 2.8. *Let* $f : M^n \to \mathbf{R}^{n+1}$ *be an affine immersion such that* ∇ *is flat. Then it is locally affinely equivalent to the graph immersion for a certain function* $F : M \to \mathbf{R}$.

Proof. Since ∇ is flat, there is a local parallel volume element θ and we may choose an equiaffine transversal vector field ξ. From the Gauss equation we have $S = 0$, which implies that $D_X\xi = 0$, that is, ξ is a constant vector field. Let $H = \mathbf{R}^n$ be a hyperplane in $\mathbf{R}^{n+1}$ that is transversal to ξ. Let $\pi : \mathbf{R}^{n+1} \to H$ be the projection along the direction of ξ so that $\pi \circ f : M \to \mathbf{R}^{n+1}$ is an affine immersion with image W, an open subset of H. We can find a differentiable function $F : M \to \mathbf{R}$ such that $f(x) = (\pi \circ f)(x) + F(x)\xi$. Thus f is a graph immersion.

3. Blaschke immersions – the classical theory

Let $f : M \to \mathbf{R}^{n+1}$ be a nondegenerate hypersurface immersion. Recalling Definition 2.1, we know that no matter which transversal field ξ we may choose, the affine fundamental form h has rank n, and can be treated as a nondegenerate metric on M. This is the basic assumption on which Blaschke developed affine differential geometry of hypersurfaces. In this section, we shall give a rigorous foundation from a structural point of view. We pick a fixed volume element on $\mathbf{R}^{n+1}$ (given by the determinant function, say).

If we choose an arbitrary transversal vector field ξ, then we obtain on M the affine fundamental form h, the induced connection ∇, and the induced volume element θ. We want to achieve, by an appropriate choice of ξ, the following two goals:

(I) (∇, θ) is an equiaffine structure, that is, $\nabla\theta = 0$;

(II) θ coincides with the volume element ω_h of the nondegenerate metric h.

Theorem 3.1. *Let* $f : M \to \mathbf{R}^{n+1}$ *be a nondegenerate hypersurface immersion. For each point* $x_0 \in M$, *there is a transversal vector field defined in a neighborhood of* x_0 *satisfying the conditions* (I) *and* (II) *above. Such a transversal vector field is unique up to sign.*

Proof. We find it convenient to formulate the following concepts. Suppose θ is an arbitrary volume element and h a nondegenerate metric, both defined in a neighborhood of a point. We define the *determinant* $\det_\theta h$ of a symmetric covariant tensor h of degree 2 *relative to* θ as follows. Let $\{X_1,\ldots,X_n\}$ be a *unimodular basis* for θ, that is, a basis in $T_x(M)$ such that $\theta(X_1,\ldots,X_n) = 1$. If we set $h_{ij} = h(X_i,X_j)$, then the determinant of the matrix $[h_{ij}]$ is independent of the choice of unimodular basis $\{X_1,\ldots,X_n\}$. We denote this number by $\det_\theta h$.

The proof of the following lemma is easy and hence omitted.

Lemma 3.2.

(1) *If $\bar{\theta} = \phi\theta$ for a nonvanishing function ϕ, then*

$$\det {}_{\bar{\theta}} h = \phi^{-2} \det {}_{\theta} h$$

for any symmetric covariant tensor of degree 2.

(2) *If $\bar{h} = h/\phi$ for a nonvanishing function ϕ, then*

$$\det {}_{\theta} \bar{h} = \phi^{-n} \det {}_{\theta} h.$$

(3) *If $\bar{\theta} = \phi\theta$ and $\bar{h} = h/\phi$ for a nonvanishing function ϕ, then*

$$\det {}_{\bar{\theta}} \bar{h} = \phi^{-(n+2)} \det {}_{\theta} h.$$

Now we are in a position to prove Theorem 3.1. Let ξ be an arbitrarily chosen transversal vector field with the corresponding ∇, θ, and h. Recall that the volume element ω_h is defined by

$$\omega_h(X_1, \ldots, X_n) = |\det [h_{ij}]|^{\frac{1}{2}}, \quad \text{where} \quad h_{ij} = h(X_i, X_j).$$

It follows therefore that $\omega_h = \theta$ if and only if $|\det {}_{\theta} h| = 1$.

Now starting with the given ξ, we consider a change

$$\bar{\xi} = \phi\xi + Z,$$

where the function ϕ and the vector field Z are to be chosen appropriately. By Proposition 2.5 we see that $\bar{h} = h/\phi$. We also have $\bar{\theta} = \phi\theta$. By Lemma 3.2, we get

$$\det {}_{\bar{\theta}} \bar{h} = \phi^{-(n+2)} \det {}_{\theta} h.$$

Therefore, in order to achieve $\omega_{\bar{h}} = \bar{\theta}$ it is necessary and sufficient to choose $\phi = |\det {}_{\theta} h|^{\frac{1}{n+2}}$.

So far the vector field Z is not subject to any condition. With $\bar{\xi}$ satisfying condition (II), we now want to choose Z in such a way that condition (I) is also satisfied. Again from Proposition 2.5, we have (2.10), where the function ϕ has been chosen. Since h is nondegenerate, it is clear that we can find Z in such a way that the right-hand side of (2.10) is equal to 0. In this way, $\bar{\xi}$ satisfies both conditions (I) and (II).

It remains to show that such ξ is unique up to sign. Suppose ξ and $\bar{\xi}$ satisfies (I) and (II). From the argument above, we have $|\det {}_{\theta} h| = 1 = |\det {}_{\bar{\theta}} \bar{h}|$. Then by (3) in Lemma 3.2 we see that $|\phi| = 1$, that is, $\phi = \pm 1$. From condition (II) for ξ and $\bar{\xi}$ we see that $\tau = \bar{\tau} = 0$. Since $\phi = \pm 1$, (2.10) shows that Z must be 0. Thus $\bar{\xi} = \pm \xi$, as we wanted to show. We have completed the proof of Theorem 3.1.

Definition 3.1. A transversal vector field satisfying (I) and (II) is called the *affine normal field* or *Blaschke normal field*. Locally, it is uniquely determined up to sign. For each point $x \in M$ we take the line through x in the direction of the affine normal vector ξ_x. This line, which is independent of the choice of sign for ξ, is called the *affine normal* through x.

Definition 3.2. By fixing an affine normal field ξ we have the induced connection ∇, the affine fundamental form h, which is traditionally called the *affine metric*, and the affine shape operator S determined by the formulas of Gauss and Weingarten (1.1) and (1.2). We shall call (∇, h, S) the *Blaschke structure* on the hypersurface M. The affine immersion $f : (M, \nabla) \to (\mathbf{R}^{n+1}, D)$ with affine normal field ξ is called a *Blaschke immersion*. We shall also speak of M as a *Blaschke hypersurface*. The induced connection ∇ is independent of the choice of sign for ξ and is called the *Blaschke connection*.

Remark 3.1. From the uniqueness in Theorem 3.1 it follows that the Blaschke structure is invariant under every equiaffine transformation of the ambient space $\mathbf{R}^{n+1}$. If a nondegenerate hypersurface M is orientable, then there is a globally defined affine normal field, which is unique up to sign. In fact, we may orient ξ using an orientation of M.

As a special case of Theorem 2.1, we have

Theorem 3.3. *For a Blaschke hypersurface M, we have the following fundamental equations:*

Gauss equation:	$R(X, Y)Z = h(Y, Z)SX - h(X, Z)SY$;
Codazzi equation for h:	$(\nabla_X h)(Y, Z) = (\nabla_Y h)(X, Z)$;
Codazzi equation for S:	$(\nabla_X S)Y = (\nabla_Y S)X$;
Ricci equation:	$h(SX, Y) = h(X, SY)$;
Equiaffine condition:	$\nabla \theta = 0$;
Volume condition:	$\theta = \omega_h$;
Apolarity condition:	$\nabla \omega_h = 0$.

Proposition 3.4. *For a nondegenerate hypersurface $(n \geq 2)$ with Blaschke structure, we have:*

(1) $S = 0$ *if and only if* $R = 0$.

(2) *The Ricci tensor is given by* $\mathrm{Ric}(Y, Z) = \mathrm{tr}\, S\, h(Y, Z) - h(SY, Z)$, *as in Corollary 2.2, and* Ric *is 0 if and only if* $S = 0$.

(3) *If $S = \lambda I$ where λ is a scalar function, then λ is a constant.*

Proof. (1) $S = 0$ implies $R = 0$ by the equation of Gauss. To prove the converse, assume $R = 0$ at x. Let $\{X_1, \ldots, X_n\}$ be an orthonormal basis relative to h, that is,

$$h(X_i, X_j) = \epsilon_i \delta_{ij}, \quad \epsilon_i = \pm 1.$$

For any $i, 1 \leq i \leq n$, take $k \neq i$. We get

$$0 = R(X_i, X_k)X_k = h(X_k, X_k)SX_i = \epsilon_k SX_i,$$

which implies $SX_i = 0$. Hence $S = 0$.

(2) If $S = 0$, we have Ric $= 0$. Conversely, assume that Ric $= 0$. Then since $h(\operatorname{tr} S\, Y - SY, Z) = 0$ for every Z and since h is nondegenerate, it follows that $SY = \lambda Y$ for every Y, where $\lambda = \operatorname{tr} S$. Then $\operatorname{tr} S = n\lambda = \lambda$, and thus $\lambda = 0$ $(n \geq 2)$. Hence $S = 0$.

(3) We have

$$(\nabla_X S)(Y) = \nabla_X(SY) - S(\nabla_X Y) = (X\lambda)Y + \lambda(\nabla_X Y) - \lambda(\nabla_X Y) = (X\lambda)Y.$$

Since this is symmetric in X and Y by the Codazzi equation for S, we obtain $(X\lambda)Y = (Y\lambda)X$, which implies $X\lambda = 0$ for every X, and thus λ is a constant.

Definition 3.3. A Blaschke hypersurface M is called an *improper affine hypersphere* if S is identically 0. If $S = \lambda I$, where λ is a nonzero constant, then M is called a *proper affine hypersphere*.

Proposition 3.5. *If M is an improper affine hypersphere, then the affine normals are parallel in $\mathbf{R}^{n+1}$. If M is a proper affine hypersphere, then the affine normals meet at one point in $\mathbf{R}^{n+1}$ (called the center). The converse of each of these statements is also valid.*

Proof. The first assertion follows from $D_X \xi = -f_*(SX) = 0$. Conversely, if the affine normals are parallel, we can find a nonvanishing differentiable function λ on M such that $\lambda\xi$ is a parallel vector field. Hence

$$0 = D_X(\lambda\xi) = (X\lambda)\xi - \lambda f_*(SX),$$

which implies that $SX = 0$ and hence $S = 0$.

To prove the second, assume that $SX = \lambda X$. Consider the map $x \in M \mapsto y = f(x) + \frac{1}{\lambda}\xi$. Then for each $X \in T_x(M)$ we get

$$D_X y = f_*(X) + \frac{1}{\lambda}D_X\xi = f_*(X) - \frac{1}{\lambda}f_*(SX) = 0,$$

which shows that y is a fixed point, say y_0. Thus the affine normals meet at y_0. Conversely, if the affine normals meet, we can find a function λ such that $y = f(x) + \lambda\xi$ is independent of $x \in M$. Thus for each $X \in T_x(M)$ we have

$$0 = f_*(X) + (X\lambda)\xi - \lambda f_*(SX) = 0,$$

which implies that $SX = \frac{1}{\lambda}X$, that is, M is a proper affine hypersphere.

An affine hypersphere has the following characteristic property.

Proposition 3.6. *Let $f : M^n \to \mathbf{R}^{n+1}$ be a nondegenerate immersion with Blaschke structure. Then f is an affine hypersphere if and only if every ∇-geodesic on M^n lies on a certain 2-plane in $\mathbf{R}^{n+1}$.*

Proof. Assume that f is an affine hypersphere, $S = \lambda I$, and let x_t, $0 \leq t \leq c$ $(c > 0)$, be a ∇-geodesic. For each t, let Σ_t be the 2-plane through the point $f(x_t)$ and spanned by the vectors $f_*(\dot{x}_t)$ and $\xi_t = \xi_{x_t}$. We have

$$D_t f_*(\dot{x}_t) = h(\dot{x}_t, \dot{x}_t)\xi_t \quad \text{and} \quad D_t \xi_t = -\lambda(\dot{x}_t),$$

both of which belong to Σ_t. Thus the plane field Σ_t is parallel in $\mathbf{R}^{n+1}$. We easily see that the geodesic x_t lies entirely in the 2-plane Σ_0 through x_0.

To prove the converse, let $X_0 \in T_{x_0}(M^n)$ such that $h(X_0, X_0) \neq 0$. Take the geodesic x_t with the initial condition (x_0, X_0). By assumption, there is a plane, say Σ_0, that contains x_t. We have for ξ_t as before

$$D_t f_*(\dot{x}_t) = h(\dot{x}_t, \dot{x}_t)\xi_t.$$

For small values of t such that $h(\dot{x}_t, \dot{x}_t) \neq 0$ we see that ξ_t is parallel to Σ_0 because $D_t \dot{x}_t$ is. In particular, $\xi_{x_0} \in \Sigma_0$ and Σ_0 is spanned by ξ_{x_0} and $f_*(X_0)$. Since $D_t \xi_t|_{t=0} = -f_*(SX_0)$ lies in Σ_0, we have

$$f_*(SX_0) = \lambda f_*(X_0) + \mu \xi_0$$

for some scalars λ and μ and hence $SX_0 = \lambda X_0$ and $\mu = 0$. It now remains to show that λ is independent of X_0 and, in fact, $S = \lambda I$. We get this from the following.

Lemma 3.7. *Let V be an n-dimensional real vector space with an inner product h. Suppose a linear endomorphism S has the property that for each X with $h(X, X) \neq 0$ we have $SX = \lambda X$, where λ is a certain scalar possibly depending on X. Then $S = \lambda I$, where λ is a constant.*

Proof. Let $\{X_1, \ldots, X_n\}$ be an orthonormal basis in V relative to h. By assumption there exist scalars λ_i, $1 \leq i \leq n$, such that $SX_i = \lambda_i X_i$, $1 \leq i \leq n$. We are done if we show that all λ_i's are equal. Let $i \neq j$. Then $Z = X_i + 2X_j$ is non-null relative to h, that is, $h(Z, Z) \neq 0$. By assumption, we have $SZ = \lambda Z$ for some λ. On the other hand, we have

$$SZ = S(X_i + 2X_j) = \lambda_i X_i + 2\lambda_j X_j = \lambda(X_i + 2X_j).$$

From linear independence, we get $\lambda_i = \lambda_j$. This completes the proof of the lemma.

Remark 3.2. If x_0 is a point of an affine hypersphere, the geodesic with initial condition (x_0, X_0), where $X_0 \in T_{x_0}(M^n), X_0 \neq 0$, is contained in the plane Σ_0 (the so-called normal section) through x_0 and spanned by X_0 and ξ_0. We have also seen in Section 3 of Chapter I that for a centro-affine hypersurface every geodesic lies on a 2-plane. In this case, the geodesic with initial condition (x_0, X_0) is contained in the intersection of the hypersurface with the plane spanned by x_0 and X_0.

Definition 3.4. For a Blaschke hypersurface M, the *affine Gauss–Kronecker curvature* (also called the *affine Gaussian curvature*) K is defined to be $\det S$. The *affine mean curvature* H is defined to be $\frac{1}{n} \operatorname{tr} S$.

At this point, it will be instructive to go over several examples. Since the Blaschke structure of a nondegenerate hypersurface is determined as soon as we know how to get an affine normal field, we show how this can be done in concrete cases. Actually, the proof of Theorem 3.1 is constructive in the sense that starting with a tentative choice of ξ we can carry out necessary changes by concrete computation as follows:

Procedure for finding the affine normal field

(1) Choose a tentative transversal vector field ξ. Compute τ.

(2) Determine the affine fundamental form h for ξ by using the formula of Gauss : $D_X f_*(Y) = f_*(\nabla_X Y) + h(X, Y)\xi$. Verify that h is nondegenerate.

(3) Determine the induced volume element θ for ξ from

$$\theta(X_1, \ldots, X_n) = \det [f_* X_1, \ldots, f_* X_n, \xi].$$

(4) Choose a unimodular basis $\{X_1, \ldots, X_n\}$ with

$$\theta(X_1, \ldots, X_n) = 1,$$

set $h_{ij} = h(X_i, X_j)$ and compute $\det {}_\theta h = \det [h_{ij}]$.

(5) Take $\phi = |\det {}_\theta h|^{\frac{1}{n+2}}$ and set $\bar{\xi} = \phi \xi + Z$, where Z is to be determined by

$$\tau + \frac{1}{\phi} h(Z, \cdot) + d \log \phi = 0.$$

If $\tau = 0$, then this equation is simply $h(Z, X) = -X\phi$ for every X.

(6) Once we get the affine normal field $\bar{\xi}$, it is easy to compute the affine metric $\bar{h} = h/\phi$, the affine shape operator S, and the induced connection ∇.

We illustrate this method in the following examples.

Example 3.1. *Elliptic paraboloid.* See Figure 3(a). Let us consider the graph of the function $z = \frac{1}{2}(x^2 + y^2)$ as the immersion $f : M = \mathbf{R}^2 \to \mathbf{R}^3$ given by

(3.1) $$(x, y) \mapsto f(x, y) = (x, y, \frac{1}{2}(x^2 + y^2)) \in \mathbf{R}^3.$$

Using ∂_x, ∂_y for coordinate vector fields on $\mathbf{R}^2$ we have

$$f_*(\partial_x) = (1, 0, x),$$

$$f_*(\partial_y) = (0, 1, y).$$

Let us take $\xi = (0, 0, 1)$ as a tentative transversal vector field. For the corresponding volume element θ we have

$$\theta(\partial_x, \partial_y) = \det [f_*(\partial_x), f_*(\partial_y), \xi]$$

$$= \det \begin{bmatrix} 1 & 0 & 0 \\ 0 & 1 & 0 \\ x & y & 1 \end{bmatrix} = 1$$

so that $\{\partial_x, \partial_y\}$ is a unimodular basis. We now get

$$D_{\partial_x} f_*(\partial_x) = (0, 0, 1) = \xi,$$

$$D_{\partial_x} f_*(\partial_y) = (0, 0, 0) = D_{\partial_y} f_*(\partial_x),$$

$$D_{\partial_y} f_*(\partial_y) = (0, 0, 1) = \xi,$$

from which we have

$$\nabla_{\partial_x}\partial_x = \nabla_{\partial_x}\partial_y = \nabla_{\partial_y}\partial_x = \nabla_{\partial_y}\partial_y = 0$$

and

$$h(\partial_x, \partial_x) = 1, \quad h(\partial_x, \partial_y) = 0, \quad h(\partial_y, \partial_y) = 1.$$

Therefore

$$\det {}_\theta h = \det \begin{bmatrix} 1 & 0 \\ 0 & 1 \end{bmatrix} = 1$$

so that ξ itself is the affine normal field (note that $\phi = 1, Z = 0$ in the final step). Obviously, $S = 0$ (improper affine sphere). The induced connection ∇ is flat and $\{x, y\}$ is a flat coordinate system for ∇. The affine metric is positive-definite.

This example can easily be generalized to the n-dimensional elliptic paraboloid

$$(x^1, \ldots, x^n) \mapsto (x^1, \ldots, x^n, \frac{1}{2} \sum_{i=1}^{n} (x^i)^2).$$

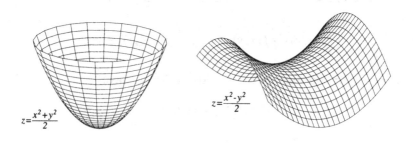

$$z = \frac{x^2 + y^2}{2}$$

$$z = \frac{x^2 - y^2}{2}$$

Figure 3(a)   Figure 3(b)

Example 3.2. *Hyperbolic paraboloid.* See Figure 3(b). This is given by $f : M = \mathbf{R}^2 \to \mathbf{R}^3$ defined by

(3.2) $$f(x, y) = (x, y, \frac{1}{2}(x^2 - y^2)).$$

By similar computation we find that it is an improper affine sphere whose affine metric has signature $(+, -)$. A generalization of this example to n dimensions depends on the quadratic function

$$F(x^1, \ldots, x^n) = \frac{1}{2} \left(\sum_{i=1}^{p} (x^i)^2 - \sum_{i=p+1}^{n} (x^i)^2 \right)$$

with resulting affine metric of signature $(p, n - p)$.

Example 3.3. *Graph of a function – general.* Let

$$x^{n+1} = F(x^1, \ldots, x^n)$$

be a differentiable function on a domain G in $\mathbf{R}^n$ and consider the immersion

$$(3.3) \qquad f : (x^1, \ldots, x^n) \in G \mapsto (x^1, \ldots, x^n, F(x^1, \ldots, x^n)) \in \mathbf{R}^{n+1}.$$

As in the case of Example 3.1, we start with a tentative choice of transversal field $\xi = (0, \ldots, 0, 1)$. Since $D_{\partial_i} \xi = 0$, we have $\tau = 0$. Denoting by ∂_j the coordinate vector field $\partial / \partial x^j$ we have

$$f_*(\partial_j) = (0, \ldots, 1, \ldots, 0, F_j),$$

where on the right-hand side 1 appears on the j-th component and $F_j = \partial F / \partial x^j$. Thus

$$D_{\partial_i} f_*(\partial_j) = F_{ij} \xi, \quad \text{where} \quad F_{ij} = \frac{\partial^2 F}{\partial x^i \partial x^j},$$

which implies

$$\nabla_{\partial_i}(\partial_j) = 0 \quad \text{and} \quad h(\partial_i, \partial_j) = F_{ij}.$$

Thus the immersion is nondegenerate if and only if $\det [F_{ij}] \neq 0$. We find $\det_\theta h$ as follows. We have

$$\theta(\partial_1, \ldots, \partial_n) = \det [f_*(\partial_1), \ldots, f_*(\partial_n), \xi] = 1,$$

and also $h(\partial_i, \partial_j) = F_{ij}$. Hence we take $\phi = |\det [F_{ij}]|^{\frac{1}{n+2}}$. Finally, we can find Z such that $\phi \xi + Z$ is the affine normal field by solving

$$\frac{\partial \phi}{\partial x^i} + \sum_{j=1}^{n} h_{ij} Z^j = 0$$

and finding

$$Z = \sum_{i=1}^{n} Z^i \partial_i.$$

Letting $[F^{ij}]$ be the inverse matrix of the matrix $[F_{ij}]$, we can find the affine normal field $\bar{\xi}$ in the form

$$(3.4) \qquad \bar{\xi} = -\sum_{k,i=1}^{n}(F^{ki}\frac{\partial\phi}{\partial x^i})f_*(\partial_k) + \phi\xi,$$

where we recall $\phi = |\det[F_{ij}]|^{\frac{1}{n+2}}$.

Remark 3.3. If $F(x_1,\ldots,x_n)$ is a differentiable function defined on the whole of $\mathbf{R}^n$ such that $\det[F_{ij}]$ is a positive constant, then F must be a quadratic function. This result is due to Jörgens. There is a generalization of this result from the point of view of affine differential geometry by Calabi [Cal1]. Jörgens' theorem has applications to the theory of surfaces (see, for example, [Sp, vol. IV, p. 165, p. 390]).

Example 3.4. *Euclidean surfaces.* Let M be a surface immersed in Euclidean 3-space E^3. Recall from Euclidean differential geometry that we have the first fundamental form, namely, the induced metric g on M, and the second fundamental form h, which is related to the Euclidean shape operator A by $h(X,Y) = g(AX,Y)$ for all $X,Y \in T_x(M)$. If N denotes a unit normal field, then we have

$$(3.5) \qquad D_X Y = \nabla_X^0 Y + h(X,Y)N,$$

where ∇^0 is the Levi-Civita connection of the metric g, and

$$(3.6) \qquad D_X N = -AX.$$

From such knowledge how can we get the equiaffine quantities for M? First of all, M is nondegenerate if and only if the Euclidean Gauss curvature $\bar{K} = \det A$ never vanishes. If $\{X_1,X_2\}$ is an orthonormal basis for g in $T_x(M)$, then the induced volume element θ (for N) has value $\theta(X_1,X_2) = \det[X_1\ X_2\ N] = 1$. Thus

$$(3.7) \qquad \det_\theta h = \det[h_{ij}] = \det A = \bar{K}.$$

Thus according to the *procedure for finding the affine normal*, we have

$$(3.8) \qquad \xi = \bar{K}^{\frac{1}{4}}N + Z,$$

where Z is a vector field on M such that $h(Z,X)/\phi + d\log\phi = 0$, that is,

$$(3.9) \qquad h(Z,X) = -X\phi, \text{ where } \phi = \bar{K}^{\frac{1}{4}}.$$

This may be written simply as $Z = -\text{grad } \bar{K}^{\frac{1}{4}}$, where grad is taken relative to the nondegenerate metric h. This formula is the same as that in [Bl, p. 166, (186)]. In the special cases, we obtain the following information.

First, suppose $\bar{K}$ is constant. Then from (3.8) we get $Z = 0$. This means that $\xi = \bar{K}^{\frac{1}{4}} N$ everywhere. Furthermore, the induced connection ∇ coincides with the Levi-Civita connection ∇^0 and the affine shape operator S equals $\bar{K}^{\frac{1}{4}} A$. Conversely, if ξ and N have the same direction everywhere, then $\bar{K}$ is constant. Second, suppose $\bar{K}$ has a critical point at x_0. Then from the computation above, we see that $Z = 0$ at x_0. Thus $\xi = \bar{K}^{\frac{1}{4}} N$ at x_0.

Example 3.5. *Quadrics with center in* $\mathbf{R}^3$. These surfaces are expressed as follows relative to a suitable affine coordinate system:

(1) $x^2 + y^2 + z^2 = 1$ (ellipsoid);

(2) $x^2 + y^2 - z^2 = 1$ (hyperboloid of one sheet);

(3) $x^2 + y^2 - z^2 = -1$ (hyperboloid of two sheets).

By changing to notation $x = (x^1, x^2, x^3)$ and $y = (y^1, y^2, y^3)$, we consider an inner product

$$g(x, y) = x^1 y^1 + x^2 y^2 + x^3 y^3$$

or

$$g(x, y) = x^1 y^1 + x^2 y^2 - x^3 y^3$$

or

$$g(x, y) = x^1 y^1 - x^2 y^2 - x^3 y^3.$$

Then the quadrics are expressed by $g(x, x) = 1$. In the first case, g is positive-definite and the surface is a Euclidean sphere of radius 1. Since its Gaussian curvature $\bar{K}$ is 1, we see from Example 3.4 that the affine normal field coincides with the usual Euclidean unit normal field. Hence S is the identity, the induced connection coincides with the Levi-Civita connection.

In the second case, g is a Lorentz–Minkowski metric with signature $(+, +, -)$. The surface (2) $g(x, x) = 1$ has induced metric with signature $(+, -)$. We can take $N = -x$ as spacelike unit normal vector field and find the metric shape operator A to be the identity. The second fundamental form $h(X, Y)$ coincides with $g(X, Y)$. Just as in Example 3.4, we may interpret all this from the affine point of view and conclude that the affine normal field coincides with N, and the affine metric has signature $(+, -)$.

In the third case, instead of the corresponding metric above, we may keep the metric $g(x, y) = x^1 y^1 + x^2 y^2 - x^3 y^3$ and consider the surface $g(x, x) = -1$, limited to one of the two sheets, say $x_3 > 0$. As is well-known, this is one of the standard models for hyperbolic plane H^2 as a surface with positive-definite metric of constant curvature -1 imbedded in the 3-dimensional Lorentz–Minkowski space. With timelike unit normal field $N = -x$, the shape operator A is the identity, the second fundamental form h equals the induced metric g, which is positive-definite, and the Gaussian curvature is -1. As nondegenerate surface, the affine metric has signature $(+, +)$, and the affine Gaussian curvature K is 1.

For detail on the surfaces, see [O'N, pp. 108–114].

4. Cubic forms

We continue with a nondegenerate hypersurface $f : M \to \mathbf{R}^{n+1}$ with its Blaschke structure. Our discussions here are mainly based on Theorems 3.1 and 3.3 in the preceding section.

From the Codazzi equation for h we see that the cubic form

$$C(X,Y,Z) = (\nabla_X h)(Y,Z)$$

is symmetric in X, Y, and Z. (It is simpler than (2.6), because $\tau = 0$.) Now in addition to the induced connection ∇ on M, we may consider the Levi-Civita connection $\hat{\nabla}$ for the affine metric h. We consider the *difference tensor* of type $(1,2)$

$$(4.1) \qquad\qquad K(X,Y) = \nabla_X Y - \hat{\nabla}_X Y.$$

Since both ∇ and $\hat{\nabla}$ have zero torsion, we have $K(X,Y) = K(Y,X)$. We shall also write

$$(4.1a) \qquad K_X Y = K(X,Y) \qquad \text{and} \qquad K_X = \nabla_X - \hat{\nabla}_X \, ;$$

thus for each $X \in T_x(M)$, K_X is a tensor of type $(1,1)$ that maps $Y \in T_x(M)$ to $K(X,Y) \in T_x(M)$. We can now relate the cubic form to the difference tensor.

Proposition 4.1. *We have*

$$(4.2) \qquad\qquad C(X,Y,Z) = -2h(K_X Y, Z).$$

Proof. We apply the derivation $\nabla_X = \hat{\nabla}_X + K_X$ to h; the $(1,1)$-tensor K_X acts on h as derivation by

$$(K_X h)(Y,Z) = -h(K_X Y, Z) - h(Y, K_X Z).$$

Since $\hat{\nabla}_X h = 0$, we get

$$(4.3) \qquad (\nabla_X h)(Y,Z) = (K_X h)(Y,Z) = -h(K_X Y, Z) - h(Y, K_X Z).$$

We know that $(\nabla_X h)(Y,Z)$ is symmetric in X, Y and Z and that $h(K_X Y, Z)$ is symmetric in X and Y. It follows that $h(Y, K_X Z)$ is symmetric in X, Y, as well as in X, Z, and therefore in X, Y, Z. From (4.3) we obtain

$$C(X,Y,Z) = (\nabla_X h)(Y,Z) = -2h(K_X Y, Z).$$

Remark 4.1. From (4.2) we have $h(K_X Y, Z) = h(Y, K_X Z)$, which says that the operator K_X is symmetric relative to h.

Corollary 4.2. *The induced connection* ∇ *and the Levi-Civita connection* $\hat{\nabla}$ *coincide with each other if and only if* $K = 0$, *that is, if and only if the cubic form* C *vanishes identically.*

We shall later see that $C = 0$ if and only if M is a quadratic surface (a classical result due to Pick and Berwald).

Theorem 4.3. *We have the apolarity condition*

(4.4) $$\operatorname{tr} K_X = 0 \quad \text{for all} \quad X \in T_x(M);$$

in index notation, $\sum_{j=1}^{n} K_{ij}^{j} = 0$ *for each fixed* i.

Proof. By applying the derivation $\nabla_X = \hat{\nabla}_X + K_X$ to the volume element $\theta = \omega_h$ we obtain
$$0 = \nabla_X \theta = (\hat{\nabla}_X + K_X)\theta = K_X \theta,$$

which implies (4.4).

Remark 4.2. In determining the affine normal field ξ for a nondegenerate hypersurface we require that ξ satisfies $\nabla \theta = 0$ (equivalent to $\tau = 0$) and that the volume element ω_h of h coincides with θ. Sometimes, it is convenient to relax the second condition to $\omega_h = c\theta$ for some positive constant c. Now this condition is equivalent to $\nabla \omega_h = 0$, since a ∇-parallel volume element is unique up to a positive scalar multiple. We may further note that $\nabla \omega_h = 0$ if and only if $\operatorname{tr} K_X = 0$ for each X. Hence we can say that apolarity is expressed by $\nabla \omega_h = 0$ as well.

Before we further rephrase the apolarity condition we recall pertinent facts concerning the notion of trace.

For a $(1,1)$-tensor A over a finite-dimensional vector space, the notion of trace is well-known. We may define it by expressing A by a matrix relative to any basis and taking the sum of the diagonal components of the matrix, this sum being independent of the choice of basis we use. Another way to define $\operatorname{tr} A$ is to take any nonzero skew-symmetric n-form θ. Then $\operatorname{tr} A$ is a unique number determined by

$$\sum_{i=1}^{n} \theta(X_1, \ldots, AX_i, \ldots, X_n) = (\operatorname{tr} A)\theta(X_1, \ldots, X_n),$$

where $\{X_1, \ldots, X_n\}$ is an arbitrary basis of V and the choice of θ is also arbitrary.

Based on this, we further define the following.

(1) Let h be an inner product on a real n-dimensional vector space. For a bilinear function $\alpha : V \times V \to \mathbf{R}$, the trace of α relative to h, denoted by $\operatorname{tr}_h \alpha$, is defined as follows. Let A be a $(1,1)$-tensor such that

$$\alpha(X, Y) = h(AX, Y) \quad \text{for all} \quad X, Y \in V.$$

Then we define

$$\operatorname{tr}_h \alpha = \operatorname{tr} A.$$

In index notation, let $h = (h_{ij}), \alpha = (\alpha_{ij})$. Then

$$\operatorname{tr}_h \alpha = \sum_{i,j=1}^{n} h^{ij} \alpha_{ij},$$

where $[h^{ij}]$ is the inverse of $[h_{ij}]$.

(2) Let h be an inner product on a real n-dimensional vector space, as before. Let β be a bilinear map $V \times V \to W$, where W is another finite-dimensional vector space. We define the trace of β relative to h, denoted by $\operatorname{tr}_h \beta$, as a unique element of W such that for an arbitrary linear function ϕ on W we have

$$\phi(\operatorname{tr}_h \beta) = \operatorname{tr}_h(\phi \circ \beta),$$

the right-hand side being defined as in (1).

Suppose $\dim W = m$. For index notation, take a basis $\{X_1, \dots, X_n\}$ in V and a basis $\{Y_1, \dots, Y_m\}$ in W. We write $\beta = (\beta_{ij}^r)$, where $1 \le i, j \le n$, $1 \le r \le m$, that is,

$$\beta(X_i, X_j) = \sum_{r=1}^{m} \beta_{ij}^r Y_r.$$

Then

$$\operatorname{tr}_h \beta = \sum_{i,j=1}^{n} \sum_{r=1}^{m} h^{ij} \beta_{ij}^r Y_r.$$

We now state the following.

Theorem 4.4. *The apolarity condition* (4.4) *is equivalent to each of the following conditions* (4.5), (4.5a) *and* (4.6):

(4.5) $\operatorname{trace}_h\{(Y, Z) \mapsto C(X, Y, Z)\} = 0$ *for all* $X \in T_x(M)$;

in index notation, $\sum_{j,k=1}^{n} h^{jk} C_{ijk} = 0$ *for each fixed* i;

(4.5a) $\operatorname{tr}_h(\nabla_X h) = 0$ *for all* $X \in T_x(M)$;

(4.6) $\operatorname{tr}_h K = 0$;

in index notation, $\sum_{j,k=1}^{n} h^{jk} K_{jk}^i = 0$ *for each fixed* i.

Proof. For (4.5), note that

$$\operatorname{trace}_h\{(Y, Z) \mapsto C(X, Y, Z)\} = \operatorname{trace}_h\{(Y, Z) \mapsto -2h(K_X Y, Z)\}$$

by Proposition 4.1. By the remark in (1), this is then equal to $-2\operatorname{tr}(K_X)$. Thus (4.5) is equivalent to (4.4).

As for (4.6), we have for any $X \in T_x(M)$

$$h(\operatorname{tr}_h K, X) = \operatorname{trace}_h\{(Y, Z) \mapsto h(K(Y, Z), X)\}$$
$$= \operatorname{trace}_h\{(Y, Z) \mapsto h(K_X Y, Z)\} = \operatorname{tr} K_X,$$

which shows that (4.4) and (4.6) are equivalent. This completes the proof of Theorem 4.4.

The following is an important classical theorem due to Maschke (for analytic surfaces), Pick (for surfaces) and Berwald (for hypersurfaces).

Theorem 4.5. *Let* $f : M \to \mathbf{R}^{n+1}$, $n \geq 2$, *be a nondegenerate hypersurface with Blaschke structure. If the cubic form* C *vanishes identically, then* $f(M)$ *is a hyperquadric in* $\mathbf{R}^{n+1}$.

Remark 4.3. Actually, this theorem and Lemma 4.6 below hold for a nondegenerate hypersurface with an equiaffine transversal vector field. The same proof is valid, since it does not depend on apolarity. Further generalizations of Theorem 4.5 will be given in Section 6 of Chapter IV.

Proof. Because of the importance of the theorem, we shall give two proofs; one is more geometric and the other more algebraic. They both depend on the following lemma.

Lemma 4.6. *If the cubic form of a Blaschke hypersurface* M *is identically* 0, *then* M *is an affine hypersphere.*

Proof. By assumption, $C = \nabla h = 0$. Since $\nabla_X h = 0$ for every $X \in \mathfrak{X}(M)$, the derivation of the algebra of tensor fields that extends the endomorphism $R(X, Y)$ maps h into 0. Hence $h(R(X, Y)Y, Y) = 0$. From the equation of Gauss, we obtain

$$h(Y, Y)h(SX, Y) = h(X, Y)h(SY, Y).$$

Let $\{X_1, \ldots, X_n\}$ be an orthonormal basis for h. Then

$$h(X_j, X_j)h(SX_i, X_j) = h(X_i, X_j)h(SX_j, X_j).$$

Hence $h(SX_i, X_j) = 0$ for $i \neq j$. It follows that there are scalars ρ_i such that $SX_i = \rho_i X_i$, for $1 \leq i \leq n$. By Lemma 3.7 we see that all ρ_i's are equal, that is, $S = \rho I$. This completes the proof of Lemma 4.6.

First Proof of Theorem 4.5.

We begin with a geometric observation that each point $x \in M$ determines a coordinate system in $\mathbf{R}^{n+1}$ by means of the tangent hyperplane $f_*(T_x(M))$ and the affine normal vector ξ_x as follows: an arbitrary point y in $\mathbf{R}^{n+1}$ can be represented in the form $y = f(x) + f_*(U) + \mu \xi_x$, where U is a certain vector in $T_x(M)$ and μ is a certain real number (see Figure 4). Now for each point x of M we define the *Lie quadric* $\mathfrak{F}_x$ as a hyperquadric in the affine space $\mathbf{R}^{n+1}$,

$$\mathfrak{F}_x = \{f(x) + f_*(U) + \mu\xi_x : U \in T_x(M), \mu \in \mathbf{R}, h(U,U) + \rho\mu^2 - 2\mu = 0\}.$$

For example, the point x itself (for which $U = 0$, $\mu = 0$) belongs to $\mathfrak{F}_x$. The idea of the first proof is to show that all these quadrics $\mathfrak{F}_x, x \in M$, coincide with each other and thus contain M.

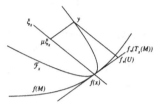

Figure 4

Let x_0 be an arbitrary fixed point in M and let y_0 be a point of $\mathfrak{F}_{x_0}$. For each $x \in M$ we can write

(4.7) $$y_0 = f(x) + f_*(U_x) + \mu\xi_x,$$

where U is a vector field on M and μ is a function. They are both differentiable. Under the assumption $C = 0$ we shall show that y_0 belongs to $\mathfrak{F}_x$ for every $x \in M$, that is, $\mathfrak{F}_{x_0} \subset \mathfrak{F}_x$, so actually, $\mathfrak{F}_{x_0} = \mathfrak{F}_x$.

Now if we take D_X of both sides of the equation (4.7), then we get

(4.8) $$0 = D_X y_0 = f_*(X) + f_*(\nabla_X U) + h(X,U)\xi + (X\mu)\xi + \mu(D_X\xi).$$

By Lemma 4.6 we have $S = \rho I$, where ρ is a constant. Thus $D_X\xi = -\rho X$. Using this in (4.8) we get

(4.9) $$\nabla_X U = (\mu\rho - 1)X$$

and

(4.10) $$X\mu + h(X,U) = 0.$$

Now we consider the function $\Phi = h(U,U) + \rho\mu^2 - 2\mu$ on M. We shall show that Φ is constant on M. At $x = x_0$, Φ is 0, because $y_0 \in \mathfrak{F}_{x_0}$. Hence $\Phi = 0$ at every point $x \in M$, that is, $y_0 \in \mathfrak{F}_x$ for every x.

In order to show that Φ is constant on M, it is sufficient to show that $X\Phi = 0$ for every vector field X on M. We have

$$\begin{aligned} X\Phi &= (\nabla_X h)(U,U) + 2h(\nabla_X U, U) + 2\rho\mu X\mu - 2X\mu \\ &= 2(\mu\rho - 1)[h(X,U) + X\mu] \quad \text{by (4.9)} \\ &= 0 \quad \text{by (4.10).} \end{aligned}$$

This completes the first proof.

Second Proof of Theorem 4.5.

We define a tensor field g of type $(0,2)$ along the immersion f in the following way. For each $x \in M$, g_x is a symmetric bilinear function on $T_{f(x_0)}(\mathbf{R}^{n+1})$ determined by

(4.11) $\qquad g(f_*X, f_*Y) = h(X,Y) \quad \text{for} \quad X, Y \in T_x(M),$

(4.12) $\qquad g(f_*X, \xi) = 0, \text{ where } \xi \text{ is the affine normal vector,}$

(4.13) $\qquad\qquad\qquad g(\xi, \xi) = \rho.$

Recall that ρ is a constant such that $S = \rho I$. We shall prove that g is parallel in $\mathbf{R}^{n+1}$, that is,

(4.14) $\qquad\qquad Xg(U,V) = g(D_X U, V) + g(U, D_X V)$

for each $X \in T_x(M)$ and for all vector fields U and V along f. We consider the three cases:

Case (i): $U = f_*(Y), V = f_*(Z)$, where Y, Z are vector fields on M. Then

$$Xg(U,V) = Xh(Y,Z) = h(\nabla_X Y, Z) + h(Y, \nabla_X Z),$$

$$g(D_X U, V) = g(D_X f_* Y, f_* Z) = g(f_*(\nabla_X Y) + h(X,Y)\xi, f_* Z) = h(\nabla_X Y, Z),$$

and

$$g(U, D_X V) = h(Y, \nabla_X Z);$$

hence (4.14) holds.

Case (ii): $U = f_*(Y), V = \xi$. Then $Xg(U,V) = 0$. We have also

$$g(D_X U, \xi) = g(f_*(\nabla_X Y) + h(X,Y)\xi, \xi) = h(X,Y)\rho$$

and

$$g(U, D_X \xi) = g(U, -f_*(SX)) = -g(f_*(Y), f_*(\rho X)) = -\rho h(Y,X)$$

so that (4.14) holds.

Case (iii): $U = V = \xi$. We have

$$Xg(U,V) = X\rho = 0 \text{ and } g(D_X \xi, \xi) = g(-f_*(SX), \xi) = 0.$$

Next we define a 1-form λ along $f : x \in M \mapsto \lambda_x \in T^*_{f(x)}(\mathbf{R}^{n+1}) = \mathbf{R}_{n+1}$ by setting

(4.15) $\qquad\qquad \lambda(f_*X) = g(f_*X, f(x)) \quad \text{for} \quad X \in T_x(M),$

(4.16) $$\lambda(\xi) = g(\xi, f(x)) + 1,$$

where $f(x)$ denotes the position vector for the point $f(x)$.

Again, we show that λ is parallel. For a vector field Y on M, we have

$$\begin{aligned}
X(\lambda(f_*Y)) &= X(g(f_*Y, f(x))) \\
&= g(D_X f_*(Y), f(x)) + g(f_*Y, f_*X) \\
&= g(f_*(\nabla_X Y), f(x)) + h(X, Y)g(\xi, f(x)) + h(X, Y),
\end{aligned}$$

$$\begin{aligned}
\lambda(D_X f_*Y) &= \lambda(f_*(\nabla_X Y) + h(X, Y)\xi) \\
&= g(f_*(\nabla_X Y), f(x)) + h(X, Y)\{g(\xi, f(x)) + 1\},
\end{aligned}$$

so that

$$(D_X \lambda)(f_*Y) = X(\lambda(f_*Y)) - \lambda(D_X f_*(Y)) = 0.$$

Similarly we have

$$\begin{aligned}
(D_X \lambda)(\xi) &= X(\lambda(\xi)) - \lambda(D_X \xi) \\
&= X(g(\xi, f(x)) + 1) - \lambda(-f_*(SX)) \\
&= g(D_X \xi, f(x)) + g(\xi, f_*X) + g(f_*(SX), f(x)) \\
&= -g(\rho f_*X, f(x)) + g(\rho f_*(X), f(x)) = 0.
\end{aligned}$$

Thus λ is parallel. This means that λ is given by a covector a (which belongs to the dual space $\mathbf{R}_{n+1}$ of $\mathbf{R}^{n+1}$), that is, $\lambda(U) = \langle U, a \rangle$ for each vector U in $\mathbf{R}^{n+1}$. We may find an affine function ψ on $\mathbf{R}^{n+1}$ such that $d\psi = \lambda$. We may also assume that $\psi(f(x_0)) = \phi(f(x_0))$ at a point $x_0 \in M$, where ϕ is defined by $\phi(p) = \frac{1}{2}g(p, p)$, where $p \in \mathbf{R}^{n+1}$. Now we have

$$\begin{aligned}
d(\phi \circ f)(X) &= \frac{1}{2}Xg(f(x), f(x)) = g(f_*X, f(x)) \\
&= \lambda(f_*X) = (d\psi)(f_*X)
\end{aligned}$$

so that

$$d(\phi \circ f) = d(\psi \circ f).$$

Hence $\phi \circ f = \psi \circ f$ on M. This means that $f(M)$ lies in a quadratic hypersurface.

Remark 4.4. For any affine coordinate system $\{u^1, \ldots, u^{n+1}\}$ we may write

$$\phi(u) = \sum_{i,j=1}^{n+1} a_{ij}u^i u^j, \quad \psi(u) = 2\sum_{i=1}^{n+1} a_i u^i + b,$$

so that $\phi = \psi$ is an equation for a quadratic hypersurface.

5. Conormal maps

Let $f : M \to \mathbf{R}^{n+1}$ be a nondegenerate hypersurface with an equiaffine transversal vector field ξ. Let $\mathbf{R}_{n+1}$ be the dual space of the underlying vector space for $\mathbf{R}^{n+1}$. If we regard $\mathbf{R}_{n+1}$ as an affine space, we call it the *dual affine space*. We now define a map $v : M \to \mathbf{R}_{n+1} - \{0\}$ as follows.

For each $x \in M$, let v_x be an element of $\mathbf{R}_{n+1}$ such that

(5.1) $\qquad v_x(\xi_x) = 1 \quad$ and $\quad v_x(f_*X) = 0 \quad$ for all $\quad X \in T_x(M)$.

Such v_x is uniquely determined and is called the *conormal vector* at x. If $\{X_1, \ldots, X_n\}$ is a unimodular basis in $T_x(M)$ relative to $\theta : \theta(X_1, \ldots, X_n) = 1$, then v can be expressed by

(5.2) $\qquad\qquad v(Y) = \det [f_*(X_1), \cdots, f_*(X_n), Y],$

where Y is an arbitrary vector in $\mathbf{R}^{n+1}$. The right-hand side of (5.2) is independent of the choice of unimodular basis $\{X_1, \ldots, X_n\}$. From (5.2) we can check that $x \mapsto v_x$ is differentiable as follows. Take a differentiable unimodular frame field $\{X_1, \ldots, X_n\}$. Then (5.2) shows that v_x depends differentiably on $x \in M$.

We have thus a differentiable map $v : M \to \mathbf{R}_{n+1} - \{0\}$, called the *conormal map*. Denoting by D the usual flat affine connection on $\mathbf{R}_{n+1}$, we have

Proposition 5.1. *We have*

(5.3) $\qquad v_*(Y)(\xi) = 0 \quad$ and $\quad v_*(Y)(f_*(X)) = -h(Y, X)$
$$\text{for all} \quad X, Y \in T_x(M).$$

The conormal map v is an immersion.

Proof. We first note that $v_*(Y) = D_Y v$, where D_Y denotes covariant differentiation in $\mathbf{R}_{n+1}$ for the flat affine connection with respect to Y. By differentiating $v(\xi) = 1$ relative to Y we get

$$\begin{aligned} 0 = Y(v(\xi)) &= (D_Y v)(\xi) + v(D_Y \xi) \\ &= (v_* Y)(\xi) + v(-f_*(SY)) = (v_* Y)(\xi), \end{aligned}$$

proving the first equation.

To prove the second equation, we start with $v(f_*(X)) = 0$ and get

$$\begin{aligned} 0 = Y(v(f_*X)) &= (D_Y v)(f_*X) + v(D_Y f_*X) \\ &= (v_*(Y))(f_*X) + v(f_*(\nabla_Y X) + h(Y, X)\xi), \end{aligned}$$

from which we get

$$v_*(Y)(f_*(X)) = -h(Y, X).$$

To show that v is an immersion, assume that $v_*(Y) = 0$ for $Y \in T_x(M)$. Then by (5.3) we get $h(Y, X) = 0$ for every $X \in T_x(M)$. Since h is nondegenerate, we conclude that $Y = 0$.

Now for the immersion $v : M \to \mathbf{R}_{n+1} - \{0\}$, we observe that for each $x \in M$ the conormal vector v_x is transversal to $v(M)$, because $v_x(\xi_x) = 1$ but $v_*(X)\xi = (D_X v)(\xi) = 0$ and therefore v_x and $v_*(X)$ are linearly independent for each $X \neq 0$. With this remark in mind, we can state the following:

Corollary 5.2. *A nondegenerate hypersurface with an equiaffine transversal vector field is determined uniquely up to translation by the conormal map and the affine fundamental form.*

Proof. Let f^1 and f^2 be two nondegenerate immersions with the same conormal map v and the affine fundamental form h. Then the equations (5.1) and (5.3) imply

$$v(f_*^1 X) = v(f_*^2 Y) \quad \text{and} \quad v_*(Y)(f_*^1 X) = v_*(Y)(f_*^2 X) \quad \text{for any } X \text{ and } Y.$$

Hence we have $f_*^1 X = f_*^2 X$ for all X, which implies that f^1 differs from f^2 by a constant vector.

By the same remark, we can consider $v : M \to \mathbf{R}_{n+1} - \{0\}$ as a centro-affine hypersurface by taking $-v$ as transversal vector field so that

$$T_{v(x)}(\mathbf{R}_{n+1}) = v_*(T_x(M)) + \text{Span}\,\{v_x\}.$$

Using this we write the formula of Gauss for $v : M \to \mathbf{R}_{n+1} - \{0\}$

(5.4) $$D_X(v_* Y) = v_*(\bar{\nabla}_X Y) + \bar{h}(X, Y)(-v),$$

where $\bar{\nabla}$ is the induced affine connection on M by v and $\bar{h}$ is the affine fundamental form for v. Note that $\bar{h}$ may be degenerate. From Section 3 of Chapter I we know that the induced connection $\bar{\nabla}$ is projectively flat. The curvature tensor $\bar{R}$ can be expressed by

$$\bar{R}(X, Y)Z = \bar{h}(Y, Z)X - \bar{h}(X, Z)Y$$

and the normalized Ricci tensor $\bar{\gamma}$ for $\bar{\nabla}$ is equal to $\bar{h}$. We have

Proposition 5.3. *The connection $\bar{\nabla}$ on M induced by the conormal immersion $v : M \to \mathbf{R}_{n+1} - \{0\}$ as centro-affine hypersurface is projectively flat and its normalized Ricci tensor γ is equal to the fundamental form $\bar{h}$.*

Proposition 5.4. *We have*

(5.5) $$\bar{h}(X, Y) = h(SX, Y) \quad \text{for all} \quad X, Y \in T_x(M)$$

(5.6) $$Xh(Y, Z) = h(\nabla_X Y, Z) + h(Y, \bar{\nabla}_X Z)$$
$$\text{for all } Y, Z \in \mathfrak{X}(M) \text{ and } X \in T_x(M).$$

Proof. From $(v_* Y)(\xi) = 0$ we obtain

$$
\begin{aligned}
0 = X((v_* Y)(\xi)) &= D_X(v_* Y)(\xi) + (v_* Y)(D_X \xi) \\
&= [v_*(\bar{\nabla}_X Y) - \bar{h}(X,Y)v](\xi) + (v_* Y)(D_X \xi) \\
&= -\bar{h}(X,Y) + (v_* Y)(-f_*(SX)) \\
&= -\bar{h}(X,Y) + h(SX,Y)
\end{aligned}
$$

by virtue of (5.3). Hence we have (5.5). To prove (5.6), we start with $v_*(Y)(f_* Z) = -h(Y,Z)$ as in (5.3). We get

$$
X(v_*(Y)(f_* Z)) = (D_X v_*(Y))(f_* Z) + v_*(Y)(D_X(f_* Z)).
$$

Here

$$
(D_X v_*(Y))(f_* Z) = v_*(\bar{\nabla}_X Y)(f_* Z) + \bar{h}(X,Y)(-v)(f_* Z) = -h(\bar{\nabla}_X Y, Z)
$$

and

$$
v_*(Y)(D_X(f_* Z)) = v_*(Y)(f_*(\nabla_X Z) + h(X,Z)\xi) = -h(Y, \nabla_X Z).
$$

From these equations we obtain $Xh(Y,Z) = h(\bar{\nabla}_X Y, Z) + h(Y, \nabla_X Z)$. Interchanging Y and Z, we get (5.6). This completes the proof of Proposition 5.4.

In view of (5.6) we see that the connection $\bar{\nabla}$ induced on M by the conormal map v is conjugate to the connection ∇. The reader should refer to Section 4 of Chapter I for the definition of conjugate connections and related facts. In [Schr] ∇ is called the first connection and $\bar{\nabla}$ the second connection. The Levi-Civita connection $\hat{\nabla}$ is equal to $\frac{1}{2}(\nabla + \bar{\nabla})$, as was stated in Corollary 4.4 of Chapter I. It follows that ∇ and $\bar{\nabla}$ coincide if and only if ∇ and $\hat{\nabla}$ coincide. From Theorem 4.5 we get

Proposition 5.5. *Let M be a nondegenerate hypersurface in $\mathbf{R}^{n+1}$ with an equiaffine transversal vector field. Then $f(M)$ is a hyperquadric if and only if ∇ and $\bar{\nabla}$ coincide.*

Actually, the following generalization is known, see [Si1]; refer also to [Wi].

Theorem 5.6. *Let M be a nondegenerate hypersurface in $\mathbf{R}^{n+1}$ with an equiaffine transversal vector field. Then $f(M)$ is a hyperquadric if and only if any two of the connections ∇, $\bar{\nabla}$, and $\hat{\nabla}$ are projectively equivalent.*

Proof. If any two of the connections are projectively equivalent, the third one is projectively equivalent to them; this fact follows from $\nabla_X - \hat{\nabla}_X =$

$\hat{\nabla}_X - \bar{\nabla}_X = \frac{1}{2}(\nabla_X - \bar{\nabla}_X)$. Thus it is sufficient to show that if ∇ and $\bar{\nabla}$ are projectively equivalent, then actually they coincide. By assumption, we have

$$\bar{\nabla}_X Z = \nabla_X Z + \rho(X)Z + \rho(Z)X,$$

where ρ is a 1-form. Substituting this in equation (5.6) we get

$$Xh(Y,Z) = h(\nabla_X Y, Z) + h(Y, \nabla_X Z) + \rho(X)h(Y,Z) + \rho(Z)h(X,Y),$$

that is,

$$(\nabla_X h)(Y,Z) = \rho(X)h(Y,Z) + \rho(Z)h(X,Y).$$

Since $(\nabla_X h)(Y,Z)$ is symmetric in X and Y by Codazzi's equation, it follows that $\rho(X)h(Y,Z)$ is symmetric in X and Y. Since h is nondegenerate, we get $\rho(X)Y = \rho(Y)X$ for all X and Y. This obviously implies that $\rho = 0$. Hence ∇ and $\bar{\nabla}$ coincide.

In the remainder of this section, we shall discuss a geometric application of the conormal map to the question of shadow boundaries. For simplicity, we shall treat only surfaces in $\mathbf{R}^3$.

Let M be a nondegenerate surface imbedded in $\mathbf{R}^3$. A curve x_t on M is said to be a *shadow boundary* for parallel lighting in the direction of a vector a if the line through each point x_t in the direction of a is tangent to M at x_t. In this case, the cylinder through the curve x_t with generators parallel to a is tangent to M along x_t. We now prove

Theorem 5.7. *Let M be a nondegenerate surface imbedded in $\mathbf{R}^3$ with an equiaffine transversal vector field ξ. A curve x_t on M is a shadow boundary if and only if it is a pregeodesic relative to the connection $\bar{\nabla}$.*

Proof. Recall that x_t is a pregeodesic for $\bar{\nabla}$ if $\bar{\nabla}_t \dot{x}_t = \phi(t)\dot{x}_t$ (see Section 3 of Chapter I). Now using the conormal vector v we set $v_t = v(x_t)$. Thus $v_*(\dot{x}_t) = dv_t/dt$. Hence we get

$$\frac{d}{dt}\frac{dv_t}{dt} = v_*(\bar{\nabla}_t \dot{x}_t) - \bar{h}(\dot{x}_t, \dot{x}_t)v_t = \phi_t \frac{dv_t}{dt} + \psi_t v_t,$$

where $\psi_t = -\bar{h}(\dot{x}_t, \dot{x}_t)$. Hence we get a second-order linear differential equation

$$(5.7) \qquad\qquad \frac{d^2 v_t}{dt^2} = \phi_t \frac{dv_t}{dt} + \psi_t v_t.$$

It follows that

$$v_t = \sigma_t \alpha + \rho_t \beta,$$

where α and β are certain constant covectors and σ_t and ρ_t are functions of t. Take a vector $a \neq 0$ such that $\alpha(a) = \beta(a) = 0$. Since $v_t(a) = 0$ for each t, we see that a is tangent to M at x_t.

Conversely, suppose x_t is a shadow boundary for parallel lighting in the direction of a. Then for $v_t = v(x_t)$ we have $v_t(a) = 0$, since a is tangent to M at x_t. Then $dv_t/dt(a) = d^2 v_t/dt^2(a) = 0$. Hence the covectors v_t, dv_t/dt and $d^2 v_t/dt^2$ are linearly dependent. Thus we have an equation of the form (5.7), which implies that $\bar{\nabla}_t \dot{x}_t = \phi_t \dot{x}_t$, that is, x_t is a pregeodesic for $\bar{\nabla}$. This completes the proof of Theorem 5.7.

The following gives another characterization of quadrics among the Blaschke surfaces in $\mathbf{R}^3$; the proof depends on apolarity.

Theorem 5.8. *Let M be a Blaschke surface imbedded in $\mathbf{R}^3$. If every shadow boundary is a plane curve, then M is a quadric.*

Proof. By Theorem 5.7, the assumption means that every $\bar{\nabla}$-geodesic is a plane curve. Let x_t be a $\bar{\nabla}$-geodesic and write X for $\dot{x}_t$. Then using $\nabla_t X = \bar{\nabla}_t X + 2K_X X = 2K_X X$ we get

$$D_t X = \nabla_t X + h(X, X)\xi_t$$
$$= 2K(X, X) + h(X, X)\xi_t.$$

From this we can compute

$$D_t^2 X = 2(\nabla_X K)(X, X) + 8K(K(X, X), X) - h(X, X)SX - 2C(X, X, X)\xi$$

by noting $2h(K(X, X), X) = -C(X, X, X)$ by virtue of Proposition 4.1.

By assumption, the curve x_t lies on a certain 2-plane. Thus the determinant of the matrix with column vectors X, $D_t X$, and $D_t^2 X$ must be 0. Using the expressions above we have then

$$\det [X \quad 2K(X, X) + h(X, X)\xi \quad L(X) - 2C(X, X, X)\xi] = 0,$$

where we write $L(X) = L(X, X, X)$ with $L(X, Y, Z) = 2(\nabla_X K)(Y, Z) + 8K(K(X, Y), Z) - h(X, Y)SZ$. Expanding the determinant we get

$$-4C(X, X, X)\det [X \quad K(X, X) \quad \xi] - h(X, X)\det [X \quad L(X) \quad \xi] = 0.$$

The tensors $h, K, S, C, \nabla K, L$ over the real vector space $T_x(M^2)$ can be extended to tensors over the complexification of $T_x(M^2)$. Since the last equation holds for all $X \in T_x(M^2)$, it holds in the complexification. If we take a complex vector X such that $h(X, X) = 0$, then we obtain

$$C(X, X, X)\det [X \quad K(X, X) \quad \xi] = 0.$$

Now if $\det [X \quad K(X, X) \quad \xi] \neq 0$, then $C(X, X, X) = 0$. If $\det [X \quad K(X, X) \quad \xi] = 0$, then this implies that $K(X, X) = aX$ so that $C(X, X, X) = -2h(K(X, X), X) = -2ah(X, X) = 0$. In any case, we have seen that $C(X, X, X) = 0$, provided $h(X, X) = 0$.

If h is positive-definite, we take an orthonormal basis $\{e_1, e_2\}$, namely, $h(e_1, e_1) = h(e_2, e_2) = 1, h(e_1, e_2) = 0$. In the complexification, we may take $X = e_1 + ie_2$ so that $h(X, X) = 0$. Hence $C(X, X, X) = 0$, which implies

$$C_{111} - 3C_{122} = 0, \qquad C_{222} - 3C_{112} = 0.$$

Since apolarity gives

$$C_{111} + C_{122} = 0, \qquad C_{112} + C_{222} = 0,$$

we conclude that all $C_{ijk} = 0$, that is, $C = 0$.

If h is indefinite, the argument is similar and simpler in that we need not go to the complexification of $T_x(M^2)$ in order to produce a nonzero vector X such that $h(X, X) = 0$. In any case, $C = 0$ and thus the surface is a quadric by Theorem 4.5.

The notion of affine distance is closely related to the conormal vector. For a Blaschke hypersurface M imbedded in $\mathbf{R}^{n+1}$, we identify $x \in M$ with its position vector x. Pick a point p in $\mathbf{R}^{n+1}$ and write

$$(5.8) \qquad\qquad x - p = Z_x + \rho(x)\xi_x,$$

where Z_x is in $T_x(M)$. Using the conormal vector v we get from (5.8)

$$(5.9) \qquad\qquad \rho(x) = v(x - p).$$

This number $\rho(x)$ is defined as the *affine distance* from p to x. The function ρ is the *affine distance function* (or *affine support function*) from p.

We have two results related to affine distance.

Proposition 5.9. *For a given point p in $\mathbf{R}^{n+1}$, the function $v(x - p)$ on M has an extremum at $y \in M$ if and only if the vector $\overrightarrow{yp}$ is in the direction of the affine normal vector ξ_y.*

Proof. From (5.8) we get for any $X \in T_x(M)$

$$X = D_X(x - p) = D_X Z + (X\rho)\xi + \rho D_X \xi$$
$$= \nabla_X Z + h(X, Z)\xi + (X\rho)\xi - \rho S X$$

and hence

$$(5.10) \qquad -X - \rho S X + \nabla_X Z = 0 \quad \text{and} \quad h(X, Z) + X\rho = 0.$$

Now assume that $\rho = v(x - p)$ has an extremum at $x = y$. Then $X\rho = 0$ for every $X \in T_x(M)$. From (5.10) we get $h(X, Z) = 0$. This implies that $Z = 0$ and hence $y - p = \rho\xi$, that is, $\overrightarrow{yp}$ is in the direction of ξ. Conversely, assume that $\overrightarrow{yp}$ is in the direction of ξ. From (5.8) we get $Z_y = 0$. From (5.10) we get $X\rho = 0$ for every $X \in T_y(M)$. Thus $v(x - p)$ has an extremum at $x = y$.

Proposition 5.10. *Let $p \in M$. The affine distance function from p is constant on M if and only if M is a proper affine hypersphere (with p as center).*

Proof. Use (5.8) and (5.10).

Let $f : M \to \mathbf{R}^{n+1}$ be a nondegenerate immersion. We identify $f(x)$ with x for simplicity. Relative to any Euclidean inner product on $\mathbf{R}^{n+1}$, let N denote the Euclidean unit normal. The Euclidean support function $\bar{\rho}$ is defined by

$$\bar{\rho} = \langle x, N \rangle$$

where $\langle \, , \, \rangle$ is the Euclidean inner product. Let $\bar{K}$ denote the Euclidean Gauss–Kronecker curvature function. Then we have the following result going back to [Tz].

Theorem 5.11. *The immersion f is an affine hypersphere with the origin as its center if and only if $\bar{K}/\bar{\rho}^{n+2}$ is constant.*

Proof. The Euclidean structure equation is written as

$$D_X Y = \nabla_X^0 Y + h^0(X, Y)N,$$
$$D_X N = -AX.$$

Using the *procedure for finding the affine normal field ξ* in Section 3, we have

$$\xi = \alpha N + Z,$$

where $\alpha = \bar{K}^{\frac{1}{n+2}}$ and Z is determined by

$$X\alpha + h^0(X, Z) = 0 \quad \text{for all} \quad X \in T(M).$$

Let v^0 be the dual vector of N: $v^0(y) = \langle y, N \rangle$. Since $v^0(N) = 1$ and $v^0(X) = 0$ for $X \in T(M)$, we have

$$v^0 = \alpha v,$$

where v is the affine conormal vector. Hence, $\bar{\rho} = v^0(x) = \alpha v(x) = \alpha \rho$. This implies

$$\frac{\bar{K}}{\bar{\rho}^{n+2}} = \frac{1}{\rho^{n+2}}.$$

Therefore, the left-hand side is constant if and only if the function ρ is constant and f is an affine hypersphere by Proposition 5.10.

Remark 5.1. Tzitzéica [Tz] introduced surfaces with the property that the Euclidean Gauss curvature is proportional to the fourth power of the support function from the origin and called them S-surfaces. This property amounts to the property that ρ is constant as in the proof above; thus S-surfaces are proper affine spheres with center at the origin. Tzitzéica's property in terms of Euclidean structure is invariant under a centro-affine transformation, as he directly observed. This work is regarded as the source of affine differential geometry of surfaces.

6. Laplacian for the affine metric

In this section we shall compute the Laplacian for the affine metric applied to various objects. We first recall the general setting for the Laplacian. Let $f : M \to \tilde{M}$ be a differentiable map from an n-dimensional manifold into an m-dimensional manifold $\tilde{M}$. We assume that M is provided with a nondegenerate metric, say h, and its Levi-Civita connection $\hat{\nabla}$ and $\tilde{M}$ with a torsion-free affine connection $\tilde{\nabla}$. In this situation, we can define the tension field $\tau(f)$ of f as a vector field defined along f, namely, a section $M \to f^{-1}(T(\tilde{M}))$. When $\tilde{M}$ is a vector space V, $\tau(f)$ reduces to a function $M \to V$, which we call the Laplacian $\triangle f$. When $\tilde{M}$ is an affine space $\mathbf{R}^{n+1}$, we may also regard $\tau(f)$ as a function on M with values in the associated vector space $\mathbf{R}^{n+1}$ and call it the Laplacian $\triangle f$.

Now for a map $f : M \to \tilde{M}$, we define the *Hessian* of f denoted by Hess $_f$ as follows: For any $X, Y \in \mathfrak{X}(M)$ we set

$$(6.1) \qquad \text{Hess}_f(X, Y) = \tilde{\nabla}_X f_*(Y) - f_*(\hat{\nabla}_X Y),$$

which can easily be verified to be tensorial, that is, the value of (6.1) at $x \in M$ that lies in $T_{f(x)}(\tilde{M})$ depends only on X_x and Y_x. Now we define the *tension field* $\tau(f)$ by

$$(6.2) \qquad \tau(f) = \text{tr}_h \text{Hess}_f.$$

The notion of trace here is more general than what we had previously, as the value of this trace at $x \in M$ is in $T_{f(x)}(\tilde{M})$. In terms of an orthonormal basis $\{X_1, \ldots, X_n\}$ in $T_x(M)$ for the metric h, $h(X_i, X_j) = \epsilon_i \delta_{ij}$ with $\epsilon_i = \pm 1$, we have

$$(6.3) \qquad \tau(f)(x) = \sum_{i=1}^{n} \epsilon_i \text{Hess}_f(X_i, X_i) \in T_{f(x)}(\tilde{M}),$$

independent of the choice of orthonormal basis.

If $\tau(f)$ is identically 0, we say that the map f is *harmonic*.

When $\tilde{M}$ is a real vector space of finite dimension V, the tension field $\tau(f)$ for a function $f : M \to V$ can be viewed as a function $M \to V$. It is called the *Laplacian* of f and denoted by $\triangle f$. In this case, the Hessian can be written in the form

$$(6.4) \qquad \text{Hess}_f(X, Y) = X(Yf) - (\hat{\nabla}_X Y)f,$$

where we note that Yf, for example, is the directional derivative of the vector-valued function. The Hessian defined in this way is an extension of the usual Hessian that is defined only at critical points of f. The Laplacian is expressed by

$$(6.5) \qquad \Delta f = \sum_{i=1}^{n} \epsilon_i [X_i^2 f - (\hat{\nabla}_{X_i} X_i)f].$$

In particular, when $V = \mathbf{R}$, this is the classical Laplacian for a real-valued function f relative to the metric h on M (and relative to the usual flat connection on $\mathbf{R}$). There is another way to obtain the Laplacian Δf in this case. Using the metric h we define the *gradient* of f, grad f, to be the vector field such that

$$(6.6) \qquad h(\operatorname{grad} f, Y) = df(Y) = Yf \quad \text{for all} \quad Y \in \mathfrak{X}(M).$$

Furthermore we define the *divergence* of W, div W, for any vector field W by

$$(6.7) \qquad \operatorname{div} W = \operatorname{trace} \{X \in T_x(M) \mapsto \hat{\nabla}_X W \in T_x(M)\}.$$

(For a Blaschke hypersurface M, however, $\hat{\nabla}_X$ can be replaced by ∇_X because $\nabla_X W = \hat{\nabla}_X W + K_X W = \hat{\nabla}_X W + K_W X$ and $\operatorname{tr} K_W = 0$ by apolarity.) Now we can get

$$(6.8) \qquad \Delta f = \operatorname{div}(\operatorname{grad} f).$$

For an expression for Δ in coordinates, see Appendix 2.

We prove

Proposition 6.1. *Assume the affine metric h is positive-definite. Then, at a point y where the affine distance function ρ has a minimal value, the affine shape operator S is definite. In particular, if furthermore ρ is negative at y, then S is positive-definite at y.*

Proof. Recall the notation in (5.8) and (5.9). Assume ρ takes a minimal value at y. Then $Z_y = 0$ as shown in the proof of Proposition 5.9. On the other hand, we have

$$\begin{aligned}
\operatorname{Hess}_\rho(X, Y) &= X(Y(\rho)) - (\hat{\nabla}_X Y)\rho \\
&= -X(h(Y, Z)) + h(\hat{\nabla}_X Y, Z) && \text{by (5.10)} \\
&= -(\nabla_X h)(Y, Z) - h(Y, \nabla_X Z) - h(K_X Y, Z) \\
&= -\frac{1}{2} C(X, Y, Z) - h(Y, (I + \rho S)X) && \text{by (5.10)}.
\end{aligned}$$

That is, we have

$$(6.9) \qquad \operatorname{Hess}_\rho(X, Y) = -\frac{1}{2} C(X, Y, Z) - h(Y, (I + \rho S)X).$$

Hence, at y, we have

$$h(Y,(I+\rho S)X) = -\text{Hess}_\rho(X,Y).$$

Since ρ has a local minimum at y, we see

$$h(X,(I+\rho S)X) \le 0 \quad \text{for all} \quad X \in T_y(M),$$

which implies

$$-\rho h(SX,X) \ge h(X,X) > 0 \quad \text{for all} \quad X \ne 0 \in T_y(M),$$

because h is positive-definite. This proves Proposition 6.1.

We now have

Proposition 6.2. *For the affine distance function ρ, we have*

$$(6.10) \qquad\qquad \triangle\rho = -n(1+H\rho),$$

where H is the affine mean curvature.

Proof. This follows from (6.9) and apolarity (4.5). Or we can use (5.10) and (6.8).

Remark 6.1. If we define an $(n-1)$-form β by

$$\beta(X_1,\ldots,X_{n-1}) = \frac{1}{n}\theta(Z,X_1,\ldots,X_{n-1}) = \frac{1}{n}\omega(Z,X_1,\ldots,X_{n-1},\xi),$$

where Z is the vector field given in (5.8), then we have

$$(6.11) \qquad\qquad d\beta = (1+H\rho)\theta.$$

We shall next compute the Laplacian $\triangle v$ of the conormal map $v : M \to \mathbf{R}_{n+1}$ as follows.

Theorem 6.3.

$$(6.12) \qquad\qquad \triangle v = v_*(\text{tr}_h K) - (\text{tr } S)v.$$

Proof. We recall that $\nabla_X - \hat\nabla_X = K_X = \hat V_X - \bar\nabla_X$ from Section 4. Using this we compute

$$
\begin{aligned}
\text{Hess}_v(X,Y) &= D_X(v_* Y) - v_*(\hat\nabla_X Y) \\
&= D_X(v_* Y) - v_*(\bar\nabla_X Y + K_X Y) \\
&= \bar h(X,Y)(-v) - v_*(K(X,Y)) \\
&= h(SX,Y)(-v) - v_*(K(X,Y))
\end{aligned}
$$

by virtue of (5.4) and (5.5). To take the trace of Hess $_v$, let $\{X_1, \ldots, X_n\}$ be any basis in $T_x(M)$ and let $[h^{ij}] = [h_{ij}]^{-1}$, where $h_{ij} = h(X_i, X_j)$. Then by writing $SX_i = \sum_k S_i^k X_k$ we have

$$\sum_{i,j} h^{ij} h(SX_i, X_j)v = \sum_{i,j,k} h^{ij} S_i^k h_{kj} v$$

$$= \sum_{i,k} \delta_k^i S_i^k v$$

$$= \sum_i S_i^i v = (\text{tr } S)v.$$

Thus we get (6.12).

Corollary 6.4. *For a Blaschke hypersurface, the conormal map is harmonic if and only if the affine mean curvature is 0.*

Proof. By (4.6) we have $\text{tr }_h K = 0$. Hence we get

(6.13) $$\triangle v = -(\text{tr } S)v,$$

which implies the result.

Remark 6.2. The Laplacian for the normal map $x \mapsto \xi_x$ is discussed in Note 3.

To conclude this section, we shall compute the Laplacian of the immersion $f : M \to \mathbf{R}^{n+1}$ of a Blaschke hypersurface itself. Thus for $X, Y \in \mathfrak{X}(M)$ we have

$$D_X f_*(Y) = f_*(\nabla_X Y) + h(X, Y)\xi,$$

where ξ is the affine normal field, ∇ the induced connection, and h the affine metric. We may rewrite it in the form

$$D_X f_*(Y) - f_*(\hat{\nabla}_X Y) = f_*(K(X, Y)) + h(X, Y)\xi,$$

that is,

$$\text{Hess}_f(X, Y) = f_*(K(X, Y)) + h(X, Y)\xi.$$

By taking the trace of Hess $_f$ we get $\tau(f) = n\xi$ by $\text{tr }_h K = 0$ again. Hence we can state

Theorem 6.5. *For a Blaschke hypersurface $f : M \to \mathbf{R}^{n+1}$, the Laplacian $\triangle f$ divided by n is equal to the affine normal field.*

According to our treatment in Section 3, the definition of the Blaschke structure (∇, h, S), particularly that of the affine metric h, depended on the construction of the affine normal field ξ in Theorem 3.1. Now Theorem 6.5 gives a way of determining ξ from the affine metric h. So a natural question is how we can determine the affine metric of a nondegenerate hypersurface before we know its affine normal field. We can do this as follows. Let $\bar{\xi}$ be a tentative choice of transversal vector field with associated affine fundamental form $\bar{h}$ and associated volume element $\bar{\theta}$. By using Lemma 3.1 we can easily verify that

$$(6.14) \qquad \hat{h} = \bar{h}/|\det_{\bar{\theta}} \bar{h}|^{\frac{1}{n+2}}$$

is independent of the choice of $\bar{\xi}$. For the affine normal field ξ the corresponding affine fundamental form, namely the affine metric h, satisfies $|\det_{\theta} h| = 1$, because its induced volume element θ coincides with the volume element for h. This means that in (6.14) we have $\hat{h} = h$. Summarizing these discussions we have

Corollary 6.6. *For a nondegenerate hypersurface* $f : M \rightarrow \mathbf{R}^{n+1}$, *we can find the affine metric from* (6.14), *where* $\bar{h}$ *is the affine fundamental form for an arbitrary transversal vector field* $\bar{\xi}$. *From the affine metric we can find the affine normal field by means of Theorem 6.5.*

7. Lelieuvre's formula

In Section 5, we have defined the conormal map $v : M \rightarrow \mathbf{R}_{n+1} - \{0\}$ to a given nondegenerate immersion $f : M \rightarrow \mathbf{R}^{n+1}$ with an equiaffine transversal vector field ξ. Corollary 5.2 states that the immersion f is in turn determined by v and the affine fundamental form h. In this section, we shall obtain Lelieuvre's formula, which gives the immersion f explictly in terms of h and v. Note that we may regard the mapping f_* as an $\mathbf{R}^{n+1}$-valued 1-form on M and write it by $df: df(X) = f_*(X)$. It is determined by the property

$$v(df(X)) = 0 \quad \text{and} \quad D_Y v(df(X)) = -h(Y, X).$$

Now assume that we have a nondegenerate metric h on an n-dimensional manifold M and a centro-affine immersion $v : M \rightarrow \mathbf{R}_{n+1}$, namely, an immersion v such that, for each $x \in M$, $v(x)$ and $v_*(X)$ are linearly independent for each $X \in T_x(M)$. We define an $\mathbf{R}^{n+1}$-valued 1-form π by

$$(7.1) \qquad v(\pi(X)) = 0,$$
$$(7.2) \qquad (D_Y v)(\pi(X)) = -h(Y, X).$$

Since the vector v_x is transversal to $v(M)$, this form is well-defined. Assume for the moment that the 1-form π is closed. Then we have locally an $\mathbf{R}^{n+1}$-valued map, say f, such that $df = \pi$ and the map f is an immersion in view of (7.2). Next define a vector field ξ by the property

$$(7.3) \qquad v_x(\xi_x) = 1 \quad \text{and} \quad v_*(Y)(\xi) = 0.$$

Then ξ is, of course, transversal to the map f. Since $v(D_Y \xi) = Y(v(\xi)) - (D_Y v)(\xi) = 0$, the vector $D_Y \xi$ is tangent to the map f; that is ξ is equiaffine and v turns out to be the conormal vector of the map f with transversal field ξ. We remark that ξ is not necessarily Blaschke. We summarize the argument as follows:

Theorem 7.1. *Let h be a nondegenerate metric on a simply connected n-manifold and $v : M \to \mathbf{R}_{n+1}$ a centro-affine immersion. If the $\mathbf{R}^{n+1}$-valued 1-form π defined by (7.1) and (7.2) is closed, then there exists a nondegenerate immersion $f : M \to \mathbf{R}^{n+1}$ together with an equiaffine transversal field ξ such that the conormal vector field of f coincides with v and the affine fundamental form coincides with h.*

Let us look for a local expression of the form π. We fix a local basis of vector fields $\{X_i\}$, $1 \le i \le n$, and set $h(X_i, X_j) = h_{ij}$. We determine a set of 1-forms α^i by $\alpha^i(X_j) = \delta^i_j$. We set furthermore $v_i = D_{X_i} v$ and define a 1-form τ with values in $\overset{n}{\wedge}\mathbf{R}_{n+1}$ by

$$\tau = \sum_{i,j}(h_{ij}v_1 \wedge \cdots \wedge v_{j-1} \wedge v \wedge v_{j+1} \wedge \cdots \wedge v_n \, \alpha^i).$$

Recall here that Det denotes the determinant function of $\mathbf{R}^{n+1}$. The dual determinant function Det^* on $\mathbf{R}_{n+1}$ is defined by the property

$$\text{Det}\,(a_1, \ldots, a_{n+1})\,\text{Det}^*(v_1, \ldots, v_{n+1}) = \det\,[v_i(a_j)].$$

By using Det^*, the vector space $\overset{n}{\wedge}\mathbf{R}_{n+1}$ is identified with $\mathbf{R}^{n+1}$ by the rule

$$v(v_1 \wedge \cdots \wedge v_n) = \text{Det}^*(v, v_1, \ldots, v_n)$$

where $v, v_1, \ldots, v_n \in \mathbf{R}_{n+1}$. With this identification, the form τ can be regarded as a 1-form with values in $\mathbf{R}^{n+1}$. We easily see

$$\tau(X_j) = \sum_i (h_{ij}v_1 \wedge \cdots \wedge v_{i-1} \wedge v \wedge v_{i+1} \wedge \cdots \wedge v_n)$$

and

$$v(\tau(X_j)) = 0,$$
$$v_i(\tau(X_j)) = -h_{ij}\text{Det}^*(v, v_1, \ldots, v_n).$$

Therefore

$$\pi = \text{Det}^*(v, v_1, \ldots, v_n)^{-1}\,\tau.$$

The associated affine immersion f is given by

$$(7.4) \quad f(x) = \int_{x_0}^x \text{Det}^*(v, v_1, \ldots, v_n)^{-1} \sum_{i,j}(h_{ij}v_1 \wedge \cdots \wedge v_{j-1} \wedge v \wedge v_{j+1} \wedge \cdots \wedge v_n)\,\alpha^i$$

relative to a base point x_0. If we started with a nondegenerate immersion f with equiaffine transversal field ξ, then the map f would be recovered by (7.4) up to translation. This integral is called *Lelieuvre's formula*.

Remark 7.1. In the conclusion of Theorem 7.1, an equiaffine transversal field ξ coincides with the affine normal field if and only if

$$(7.5) \qquad |\text{Det}^*(v, v_1, \ldots, v_n)| = |\det [h_{ij}]|^{\frac{1}{2}}.$$

This follows from the formula

$$\text{Det}(X_1, \ldots, X_n, \xi)\,\text{Det}^*(v, v_1, \ldots, v_n) = \det [h_{ij}]$$

and from the fact that ξ is the affine normal field if and only if

$$\text{Det}(X_1, \ldots, X_n, \xi) = |\det [h_{ij}]|^{\frac{1}{2}},$$

which is the expression of the volume condition in Theorem 3.3.

We shall now examine the condition that the 1-form π is closed. The fundamental equation for the centro-affine immersion v relative to the transversal vector field $-v$ is written as follows:

$$(7.6) \qquad D_X(v_* Y) = v_*(\bar{\nabla}_X Y) + \bar{h}(X, Y)(-v).$$

Then we have

Lemma 7.2. *The 1-form τ is closed if and only if the metric h satisfies the equation*

$$(7.7) \qquad (\bar{\nabla}_X h)(Y, Z) = (\bar{\nabla}_Y h)(X, Z)$$

for all tangent vectors X, Y, Z.

Proof. Differentiating (7.1) relative to X we obtain

$$0 = (D_Y v)(\pi(X)) + v(D_Y(\pi(X))),$$

thus

$$v(D_Y(\pi(X))) = h(Y, X),$$

which implies

$$v(D_Y(\pi(X))) = v(D_X(\pi(Y))).$$

On the other hand, we have $v([X, Y]) = 0$. Hence

$$(7.8) \qquad v(d\pi(X, Y)) = 0.$$

Next differentiate (7.2) relative to Z and obtain

$$-Z(h(X,Y)) = D_Z(v_* Y)(\pi(X)) + v_* Y(D_Z(\pi(X)))$$
$$= v_*(\bar{\nabla}_Z Y)(\pi(X)) - \bar{h}(Z,Y)v(\pi(X)) + v_* Y(D_Z(\pi(X)))$$
$$= -h(\bar{\nabla}_Z Y, X) + v_* Y(D_Z(\pi(X))),$$

that is,

$$v_* Y(D_Z \pi(X)) = -Z(h(X,Y)) + h(\bar{\nabla}_Z Y, X).$$

We have a similar equation with X and Z interchanged as well as

$$v_* Y(\pi([X,Z])) = h(\bar{\nabla}_Z X, Y) - h(\bar{\nabla}_X Z, Y),$$

because $\bar{\nabla}$ is torsion-free, $[X,Z] = \bar{\nabla}_X Z - \bar{\nabla}_Z X$, and by virtue of (7.2). From these three equations we get

$$(7.9) \qquad (D_Y v)(d\pi(X,Z)) = -(\bar{\nabla}_X h)(Z,Y) + (\bar{\nabla}_Z h)(X,Y).$$

From (7.8) and (7.9) we have Lemma 7.2.

Remark 7.2. The equation (7.7) for the tensor h is called the Codazzi equation relative to the affine connection $\bar{\nabla}$. Its role and the meaning of Lemma 7.2 will be further clarified in Section 8.

We have

Corollary 7.3. *Let h be a nondegenerate metric on a simply connected n-manifold and $v : M \to \mathbf{R}_{n+1}$ a centro-affine immersion. Suppose the Codazzi equation (7.7) is satisfied. Then a nondegenerate immersion f with equiaffine transversal vector field ξ is uniquely (up to parallel translation) determined by h and v as in Theorem 7.1. Moreover, ξ coincides with the affine normal field for f up to a constant scalar factor if and only if the Laplacian $\triangle v$ relative to h is proportional to v at each point.*

Proof. By Theorem 7.1 and Lemma 7.2 we get f with ξ. The computation of $\triangle v$ in the proof of Theorem 6.3 (without assuming that the hypersurface is Blaschke) shows that

$$(7.10) \qquad \triangle v = v_*(\mathrm{tr}\,_h K) - (\mathrm{tr}\, S)v.$$

The conclusion of the theorem follows, since $\mathrm{tr}\,_h K = 0$ is a necessary and sufficient condition for equiaffine ξ to coincide with the affine normal field up to a constant scalar multiple.

We may now provide the following refinement of Theorem 7.1. Let g be a nondegenerate metric on M. We define a 1-form ω with values in $\mathbf{R}^{n+1}$ (like the form π),

$$(7.11) \qquad \omega_g = |\mathrm{Det}\,^*(v, v_1, \ldots, v_n)|^{\frac{2}{n}-1} |\det [g_{ij}]|^{-\frac{1}{n}} \mu,$$

where

$$\mu = \sum_{i,j}(g_{ij}v_1 \wedge \cdots \wedge v_{j-1} \wedge v \wedge v_{j+1} \wedge \cdots \wedge v_n \, \alpha^i),$$

which is similar to τ we had before.

We may easily verify that the form ω_g is invariant by a conformal change of the metric g. If we take

$$h = |\mathrm{Det}^*(v, v_1, \ldots, v_n)|^{\frac{2}{n}} |\det [g_{ij}]|^{-\frac{1}{n}} g,$$

we get

(7.12) $$|\det [h_{ij}]|^{\frac{1}{2}} = |\mathrm{Det}^*(v, v_1, \ldots, v_n)|$$

and thus $\omega_h = \omega_g$ coincides with the form π we defined for h before. Thus we conclude that in the conformal class of the given metric g, there is a unique metric h satisfying (7.12) for which the form π coincides with ω_g. Now in view of Theorem 7.1 and Remark 7.1, these cosiderations lead to

Theorem 7.4. *Let g be a nondegenerate metric on a simply connected n-manifold M and $v : M \to \mathbf{R}^{n+1}$ a centro-affine immersion. If the form ω_g, which depends only on the conformal class of g, is closed, then there is a Blaschke immersion $f : M \to \mathbf{R}^{n+1}$ whose affine metric is conformal to g and whose conormal map coincides with v. The affine metric is uniquely determined by the conformal class of g and the map v.*

The results presented here are found in [LNW]. We shall now explain the 2-dimensional version originally found in [Bl]. Assume now that $n = 2$ and that the affine metric of a Blaschke immersion f is indefinite. We choose an *asymptotic coordinate system* $\{x^1, x^2\}$ with respect to h (see Appendix 2) so that $h = 2F\,dx^1 dx^2$. Set $X_1 = \partial/\partial x^1$ and $X_2 = \partial/\partial x^2$. By definition, we have

$$\mathrm{Det}\,(f_* X_1, f_* X_2, \xi)\,\mathrm{Det}^*(v, v_1, v_2) = -F^2$$

and

$$\mathrm{Det}\,(f_* X_1, f_* X_2, \xi) = \theta(X_1, X_2) = F.$$

Hence

$$\pi = v \wedge v_1 \, dx^1 - v \wedge v_2 \, dx^2;$$

that is

(7.13) $$f(x) = f(x_0) + \int_{x_0}^{x}(v \wedge v_1 \, dx^1 - v \wedge v_2 \, dx^2).$$

This formula is found in [Bl, pp.140–141]. In his approach, given $v = v(x^1, x^2)$, where (x^1, x^2) are going to be asymptotic coordinates, we want to find an immersion $(x^1, x^2) \to f(x^1, x^2)$ such that

$$\frac{\partial f}{\partial x^1} = v \wedge v_1, \qquad \frac{\partial f}{\partial x^2} = v_2 \wedge v.$$

Its integrability condition

$$(v \wedge v_1)_2 = (v_2 \wedge v)_1$$

reduces to

(7.14) $$v_{12} \wedge v = 0.$$

Note that this condition also says that the Laplacian Δv relative to any metric with asymptotic coordinates (x^1, x^2) is proportional to v at each point (see (A2.12) in Appendix 2). Thus under this condition, not only is the integrability condition satisfied but also the resulting immersion f with equiaffine ξ is a Blaschke surface. (This is a peculiarity for dimension 2.) Within the conformal class of metrics that have (x^1, x^2) as asymptotic coordinates, the affine metric h of a unique Blaschke surface with v as conormal map is given by $2F dx^1 dx^2$, where $F = -\text{Det}^*(v, v_1, v_2)$.

When the affine metric is definite, we choose an *isothermal coordinate system* $\{x^1, x^2\}$ (see Appendix 2). Then we get

(7.15) $$f(x) = f(x_0) + \int_{x_0}^{x} (v \wedge v_2 \, dx^1 - v \wedge v_1 \, dx^2).$$

An interpretation similar to the indefinite case is possible.

8. Fundamental theorem

The fundamental theorem for surfaces in Euclidean 3-space says that if a 2-dimensional manifold M is given a positive-definite Riemannian metric g and a symmetric bilinear form h that together satisfy the equations of Gauss and Codazzi, then there is locally an immersion of M into E^3 with g and h as the first and second fundamental forms. Moreover, two such local immersions are unique up to Euclidean congruence. If M is simply connected, we can find a global isometric immersion. This type of fundamental theorem can be established in the case of a hypersurface in Euclidean space and, more generally, in the case of isometric immersion of a pseudo-Riemannian manifold (M, g) into a pseudo-Riemannian manifold $(\tilde{M}, \tilde{g})$ of constant sectional curvature, where the dimensions of M and $\tilde{M}$ can be arbitrary.

Actually, one unifying method for proving such existence and uniqueness theorems is to appeal to the theory of connections in an appropriate vector bundle over M. This is what we do to obtain the *fundamental theorem in affine differential geometry*. We thus begin with the *preliminary existence theorem* based on the idea that appears in the general theory for differentiable immersions ([Wet]).

Theorem 8.1. *Let M be a simply connected differentiable n-manifold with a torsion-free affine connection ∇, a symmetric $(0,2)$-tensor field h, and a $(1,1)$-tensor field S that satisfy the equation of Gauss (2.1), the equations of Codazzi*

(2.2*),(2.3*), and the equation of Ricci (2.4*) (see Theorem 2.4). Then there exist a ∇-parallel volume element θ on M and a global equiaffine immersion $f : (M, \nabla, \theta) \to \mathbf{R}^{n+1}$ with h and S as affine fundamental form and shape operator. Such an immersion is uniquely determined up to affine transformation of $\mathbf{R}^{n+1}$. If, moreover, h is nondegenerate and the given ∇ and h satisfy the apolarity condition (4.5), then there is a parallel volume element ω in $\mathbf{R}^{n+1}$ such that f is a Blaschke immersion.

Proof. Let $B = TM \oplus N$ be the Whitney sum (or direct sum) of the tangent bundle TM and the trivial line bundle $N = M \times \mathbf{R}$ over M. Let ξ be a nonvanishing section of N. We introduce a connection D in B by

$$D_X Y = \nabla_X Y + h(X, Y)\xi \quad \text{for all} \quad Y \in \mathfrak{X}(M),$$

$$D_X \xi = -SX.$$

It is easy to check that D is a linear connection in the vector bundle B. By the kind of computation we did in the proof of Theorem 2.1 (and indeed assuming $\tau = 0$ in the current case) we can easily check that D is flat, that is, the curvature tensor of D is 0. Since M is simply connected, it follows that there is a connection-preserving bundle isomorphism ϕ from B onto the trivial bundle $M \times \mathbf{R}^{n+1}$ with flat linear connection, which we also denote by D. Let π be the projection of $M \times \mathbf{R}^{n+1}$ onto $\mathbf{R}^{n+1}$. We define an $\mathbf{R}^{n+1}$-valued 1-form α on M by $\alpha_p(X) = \pi(\phi(X))$ for all $p \in M$ and $X \in T_p(M)$. If X, Y are vector fields on M, then we have

$$(d\alpha)(X, Y) = X\alpha(Y) - Y\alpha(X) - \alpha([X, Y])$$
$$= \pi(D_X\phi(Y) - D_Y\phi(X) - \phi([X, Y])).$$

Since ϕ preserves the connection, we have

$$D_X\phi(Y) - D_Y\phi(X) - \phi([X, Y]) = \phi(\nabla_X Y - \nabla_Y X - [X, Y]) = 0.$$

Hence $d\alpha = 0$. Thus there exists a function $f : M \to \mathbf{R}^{n+1}$ such that $df = \alpha$, that is, $f_*(X) = \phi(X)$, showing that f is an immersion. Moreover, we have

$$D_X(f_* Y) = D_X\phi(Y) = \phi(\nabla_X Y + h(X, Y)\xi) = f_*(\nabla_X Y) + h(X, Y)\phi(\xi),$$

where $\xi' = \phi(\xi)$ may be considered a transversal vector field to $f(M)$. It now follows that f is an affine immersion and that S is the affine shape operator.

Now using any parallel volume element, say ω_1, on $\mathbf{R}^{n+1}$, we get the induced volume element θ_1 by $\theta_1(X_1, \ldots, X_n) = \omega_1(X_1, \ldots, X_n, \xi')$. Since $D_X\xi'$ is tangent to $f(M)$, θ_1 is parallel relative to ∇. Hence $f : (M, \nabla, \theta_1) \to \mathbf{R}^{n+1}$ is an equiaffine immersion.

If we assume that h is nondegenerate and that apolarity holds, that is, $\nabla \omega_h = 0$, we have $\omega_h = c\theta_1$ with a certain positive constant c. Replacing ω_1 by $\omega = c\,\omega_1$, we can get the volume element θ induced by ω to coincide with

ω_h. This means that f is a Blaschke immersion. This completes the proof of Theorem 8.1.

Now for a Blaschke hypersurface $f : M \to \mathbf{R}^{n+1}$ we have the induced connection ∇, the affine metric h, the affine shape operator S, and the cubic form $C = \nabla h$ that satisfy the fundamental equations in Theorem 3.3 and the apolarity condition in Theorem 4.4. In 1918, Radon proved for a 2-dimensional manifold M that a nondegenerate metric h and a cubic form C satisfying the apolarity condition (4.5) can be realized as the affine metric and the cubic form for a Blaschke immersion, provided a certain integrability condition is fulfilled. We shall now present a new formulation of *Radon's theorem*. The basic data we take is the pair (∇, h) of a torsion-free affine connection ∇ and a nondegenerate metric h satisfying the compatibility condition (4.6) of Chapter I and the apolarity condition $\nabla \omega_h = 0$. Actually, there is a one-to-one correspondence between the set of all such pairs (∇, h) and the set of all pairs (h, C) of a nondegenerate metric h and a cubic form C satisfying the apolarity condition (4.5) of Chapter II.

Theorem 8.2. *Let M be a simply connected manifold with a torsion-free affine connection ∇ and a nondegenerate metric h. We assume that*

(1) (∇, h) is compatible, that is, $(\nabla_X h)(Y, Z) = (\nabla_Y h)(X, Z)$ for all $X, Y, Z \in \mathfrak{X}(M)$;

(2) The apolarity $\nabla \omega_h = 0$ is satisfied, where ω_h is the volume element for the metric h.

Then there is a Blaschke immersion $f : M \to \mathbf{R}^{n+1}$, relative to a parallel volume element ω in $\mathbf{R}^{n+1}$, such that ∇ is the induced connection and h the affine metric if and only if the conjugate connection $\bar{\nabla}$ is projectively flat.

Proof. Assume that ∇ and h are realized by a Blaschke immersion. Then, as we know from Proposition 5.4, the conjugate connection of ∇ relative to h coincides with the connection $\bar{\nabla}$ induced by the conormal immersion ν. Since ν is centro-affine, it follows from Section 3 of Chapter I that $\bar{\nabla}$ is projectively flat. We shall now prove the converse.

Assume that the conjugate connection $\bar{\nabla}$ is projectively flat. Since $\nabla \omega_h = 0$, it follows from Corollary 4.4 (2) of Chapter I, that $\bar{\nabla} \omega_h = 0$. Thus the Ricci tensor of $\bar{\nabla}$ is symmetric. We denote by $\bar{\gamma}$ the normalized Ricci tensor for $\bar{\nabla}$. Since h is nondegenerate, we can define a tensor field S of type $(1,1)$ by

$$h(SX, Y) = \bar{\gamma}(X, Y).$$

Since $\bar{\gamma}$ is symmetric, it follows that $h(SX, Y) = h(X, SY)$, that is, S satisfies the Ricci equation.

Now, $\bar{\nabla}$ is projectively flat by assumption. Hence

$$\bar{R}(X, Y)Z = \bar{\gamma}(Y, Z)X - \bar{\gamma}(X, Z)Y = h(SY, Z)X - h(SX, Z)Y$$

by virtue of Theorem 3.3 of Chapter I. By using Proposition 4.6 of Chapter I, we have

$$\begin{aligned} h(R(X,Y)W,Z) &= -h(W,\bar{R}(X,Y)Z) \\ &= -h(W,h(SY,Z)X - h(SX,Z)Y) \\ &= -h(Z,h(X,W)SY - h(Y,W)SX), \end{aligned}$$

which implies

$$R(X,Y)W = h(Y,W)SX - h(X,W)SY.$$

Thus S satisfies the equation of Gauss. Since (∇,h) is compatible, the Codazzi equation for h is satisfied. The apolarity condition $\nabla\omega_h = 0$ is also satisfied.

In order to appeal to Theorem 8.1, it remains to show that the Codazzi equation for S is satisfied. If $n > 2$, we know from Proposition 2.3 that the equation is a consequence of the Gauss equation and the Codazzi equation for h. For $n = 2$, the projective flatness of $\bar{\nabla}$ implies

$$(\bar{\nabla}_X\bar{\gamma})(Y,Z) = (\bar{\nabla}_Y\bar{\gamma})(X,Z).$$

But then

$$\begin{aligned} (\bar{\nabla}_X\bar{\gamma})(Y,Z) &= X\gamma(Y,Z) - \bar{\gamma}(\bar{\nabla}_XY,Z) - \bar{\gamma}(Y,\bar{\nabla}_XZ) \\ &= Xh(SY,Z) - h(SY,\bar{\nabla}_XZ) - h(SZ,\bar{\nabla}_XY) \\ &= h(\nabla_X(SY),Z) - h(SZ,\bar{\nabla}_XY). \end{aligned}$$

By assumption, (∇,h) is compatible and thus $\bar{\nabla}$ has torsion 0. We obtain

$$\begin{aligned} (\bar{\nabla}_X\bar{\gamma})(Y,Z) &- (\bar{\nabla}_Y\bar{\gamma})(X,Z) \\ &= h(\nabla_X SY,Z) - h(\nabla_Y SX,Z) - h(SZ,\bar{\nabla}_XY) + h(SZ,\bar{\nabla}_YX) \\ &= h(\nabla_X SY,Z) - h(\nabla_Y SX,Z) - h(SZ,\nabla_XY - \nabla_YX) \\ &= h((\nabla_X S)Y - (\nabla_Y S)X,Z). \end{aligned}$$

Since h is nondegenerate, we get $(\nabla_X S)Y = (\nabla_Y S)X$, as desired. This completes the proof of Theorem 8.2.

Remark 8.1. In Theorem 8.2, the condition that $\bar{\nabla}$ is projectively flat may be replaced as follows (see [DNV]).

(i) For $n = 2$:

$$\text{trace}_h\{(X,W) \to (\nabla_X R)(Y,Z)W\} = 0 \quad \text{for all } Y \text{ and } Z;$$

or (as in [Schr, p. 139])

$$\text{trace}_h\{(Y,Z) \to (\nabla_Y \text{Ric})(Z,X)\} = 0 \quad \text{for all } X.$$

(ii) For $n \geq 3$:
$R(X, Y)Z = 0$ whenever X, Y, and Z are orthogonal to each other relative to h.

For another variation and a generalization to the case where h is degenerate, see [O1].

9. Some more formulas

We shall further investigate the relationship between the curvature tensor R of the induced connection ∇ and the curvature tensor $\hat{R}$ of the Levi-Civita connection $\hat{\nabla}$ for the affine metric of a Blaschke hypersurface.

Proposition 9.1. *We have*

$$(9.1) \qquad R(X, Y) = \hat{R}(X, Y) + (\hat{\nabla}_X K)_Y - (\hat{\nabla}_Y K)_X + [K_X, K_Y]$$
$$= \hat{R}(X, Y) + (\nabla_X K)_Y - (\nabla_Y K)_X - [K_X, K_Y],$$

$$(9.2) \qquad R(X, Y)Z = \frac{1}{2}\{h(Y, Z)SX - h(X, Z)SY$$
$$+ h(SY, Z)X - h(SX, Z)Y\} - [K_X, K_Y]Z.$$

Proof. To obtain (9.1) we must compute $R(X, Y) = [\nabla_X, \nabla_Y] - \nabla_{[X,Y]}$ by using $\nabla_X = \hat{\nabla}_X + K_X$ from (4.2) and by noting

$$(\hat{\nabla}_X K)_Y = \hat{\nabla}_X K_Y - K_Y \hat{\nabla}_X - K_{\hat{\nabla}_X Y}$$

and

$$(\nabla_X K)_Y = (\hat{\nabla}_X K)_Y + (K_X \cdot K)_Y$$
$$= (\hat{\nabla}_X K)_Y + [K_X, K_Y] - K_{K_X Y}.$$

Next, from (9.1) and the equation of Gauss we have

$$(*) \qquad h(\hat{R}(X, Y)Z, W) = h(Y, Z)h(SX, W) - h(X, Z)h(SY, W)$$
$$+ h((\hat{\nabla}_Y K)_X Z, W) - h((\hat{\nabla}_X K)_Y Z, W)$$
$$- h([K_X, K_Y]Z, W).$$

Interchange Z and W in this equation and subtract. Since K_X and $(\hat{\nabla}_Y K)_X = \hat{\nabla}_Y(K_X) - K_{\hat{\nabla}_Y X}$ are symmetric operators relative to h, we obtain

$$2h(\hat{R}(X, Y)Z, W) = h(Y, Z)h(SX, W) - h(X, Z)h(SY, W)$$
$$+ h(X, W)h(SY, Z) - h(Y, W)h(SX, Z)$$
$$- 2h([K_X, K_Y]Z, W).$$

This leads to (9.2).

Proposition 9.2. *The Ricci tensor of the affine metric h is given by*

$$(9.3) \qquad \widehat{\mathrm{Ric}}\,(Y,Z) = \frac{1}{2}\{h(Y,Z)\mathrm{tr}\,S + (n-2)h(SY,Z)\} + \mathrm{tr}\,(K_Y K_Z),$$

where $\mathrm{tr}\,(K_Y K_Z) = h(K_Y, K_Z)$ *(inner product extending h to the tensor space of symmetric endomorphisms, such as* K_Y *).*

Proof. Take the trace of the linear map $X \mapsto \hat{R}(X,Y)Z$ using (9.2) and noting the following:

$$\mathrm{trace}\,\{X \mapsto h(X,Z)SY\} = h(SY,Z),$$

$$\mathrm{trace}\,\{X \mapsto h(SX,Z)Y = h(X,SZ)Y\} = h(Y,SZ) = h(SY,Z),$$

$$\mathrm{trace}\,\{X \mapsto [K_Y,K_X]Z\}$$
$$= \mathrm{trace}\,\{X \mapsto K_Y K_X Z\} - \mathrm{trace}\,\{X \mapsto K_X K_Y Z\}$$
$$= \mathrm{trace}\,\{X \mapsto K_Y K_Z X\}$$
$$= \mathrm{tr}\,(K_Y K_Z),$$

where we have used

$$K_X K_Y Z = K_{K_Y Z} X \quad \text{and} \quad \mathrm{tr}\, K_{K_Y Z} = 0 \text{ (apolarity)}.$$

Remark 9.1. (9.2) is the same as equation (13) in [Schr, p.136], and (9.3) is the same as (2.22) in Schneider's paper [Schn1] and formula (3.18) as in Calabi's paper [Cal4].

From (9.3) we immediately get

Proposition 9.3. *The scalar curvature* $\hat{\rho} = \frac{1}{n(n-1)}(\sum_{i,j} h^{ij}\hat{R}_{ij})$ *of the affine metric can be expressed by*

$$(9.4) \qquad\qquad\qquad \hat{\rho} = H + J,$$

where $H = \frac{1}{n}\mathrm{tr}\,S$ *is the affine mean curvature defined before and*

$$(9.5) \qquad\qquad\qquad J = \frac{1}{n(n-1)}h(K,K)$$

is called the Pick invariant.

Here h is extended as inner product in the tensor space of type $(1,2)$; if $K = (K^i_{jk})$ in index notation, then

$$h(K,K) = \sum_{i,j,k,p,q,r} h_{ip}h^{jq}h^{kr}K^i_{jk}K^p_{qr}.$$

For the cubic form $C = (C_{ijk})$ we have the inner product

$$h(C,C) = \sum_{i,j,k,p,q,r} h^{ip} h^{jq} h^{kr} C_{ijk} C_{pqr}.$$

From (4.2) we have

$$C_{ijk} = -2 \sum_r h_{kr} K_{ij}^r$$

so that

$$h(C,C) = 4h(K,K).$$

For $n = 2$, (9.4) is the same as (14), p. 136 of [Schr], and also [Bl, p. 158]. Since $\widehat{\mathrm{Ric}}(Y,Z) = \hat{\rho} h(Y,Z)$, we obtain from (9.3) and (9.4)

$$(9.6) \qquad \mathrm{tr}(K_Y K_Z) = J h(Y,Z) \quad \text{for all} \quad Y, Z.$$

This formula appears as (150) in [Bl, p. 157].

Going back to the general case, we shall define two more tensors. Let $\hat{L}$ be the symmetric bilinear form defined by

$$(9.7) \qquad \hat{L}(X,Z) = \mathrm{trace}\{Y \mapsto (\hat{\nabla}_Y K)(X,Z)\}.$$

We shall prove
Proposition 9.4.

$$(9.8) \qquad \hat{L}(X,Z) = \frac{n}{2}\{h(X,Z)H - h(SX,Z)\} = -\frac{n}{2}h(S^0 X,Z),$$

where $S^0 = S - HI$ *with* $H = \frac{1}{n}\mathrm{tr}\,S$.
Proof. Go back to the equation (*) in the proof of Proposition 9.1. Adding this equation and the equation obtained by interchanging Z and W we obtain

$$
\begin{aligned}
(9.9) \qquad 0 = {} & 2h((\hat{\nabla}_Y K)_X Z, W) - 2h((\hat{\nabla}_X K)_Y Z, W) \\
& + h(Y,Z)h(SX,W) - h(X,Z)h(SY,W) \\
& + h(Y,W)h(SX,Z) - h(X,W)h(SY,Z).
\end{aligned}
$$

Eliminating W from this we write

$$
\begin{aligned}
(9.10) \qquad 2(\hat{\nabla}_Y K)_X Z & + h(Y,Z)SX + h(SX,Z)Y \\
& = 2(\hat{\nabla}_X K)_Y Z + h(X,Z)SY + h(SY,Z)X.
\end{aligned}
$$

By taking the trace of the linear map that takes Y to each side of the equation above, we obtain (9.8) by virtue of

$$\mathrm{trace}\{Y \mapsto (\hat{\nabla}_X K)_Y Z\} = 0,$$

which can be established as follows.

From $K_Y Z = K_Z Y$ we have $(\hat{\nabla}_X K)_Y Z = (\hat{\nabla}_X K)_Z Y$. Hence we have

$$\text{trace}\{Y \mapsto (\hat{\nabla}_X K)_Y Z\} = \text{tr}\,(\hat{\nabla}_X K)_Z.$$

But this is equal to

$$\text{tr}\,(\hat{\nabla}_X K)_Z = \text{tr}\,(\hat{\nabla}_X(K_Z)) - \text{tr}\,K_{\hat{\nabla}_X Z} = X(\text{tr}\,K_Z) = 0,$$

where we use apolarity twice.

Corollary 9.5. *A Blaschke hypersurface is an affine hypersphere if and only if the tensor $\hat{L}$ is identically zero.*

Proof. Immediate from Proposition 9.4.

Corollary 9.6. *For a Blaschke hypersurface, the affine metric h and the cubic form C determine the shape operator S uniquely.*

Proof. We know from Proposition 4.1 that (h, C) determines K and hence J. From Proposition 9.3, H is determined. By Proposition 9.4, S is determined.

Remark 9.2. (9.8) is the same as (2.24) in [Schn1]. The tensor $\hat{L}$ is the same as the tensor $c_{ij} = \nabla_s T_{ij}^s$ in (15), p.136 in [Schr].

We go back to (9.9). By using Proposition 4.1, we can rewrite (9.9) in the form

(9.11)
$$(\hat{\nabla}_X C)(Y, Z, W) - (\hat{\nabla}_Y C)(X, Z, W)$$
$$= h(X, Z)h(SY, W) - h(Y, Z)h(SX, W)$$
$$+ h(X, W)h(SY, Z) - h(Y, W)h(SX, Z).$$

By using $\nabla_X = \hat{\nabla}_X + K_X$ it is easy to see that

$$(\hat{\nabla}_X C)(Y, Z, W) - (\hat{\nabla}_Y C)(X, Z, W) = (\nabla_X C)(Y, Z, W) - (\nabla_Y C)(X, Z, W),$$

which implies that ∇C is totally symmetric (that is, symmetric in all of its four variables) if and only if $\hat{\nabla} C$ is totally symmetric. In this case, we get

$$h(X, Z)h(SY, W) - h(Y, Z)h(SX, W)$$
$$+ h(X, W)h(SY, Z) - h(Y, W)h(SX, Z) = 0.$$

By taking the trace of the linear map taking X to the left-hand side of the equation above, we easily obtain $SY = \frac{1}{n}\text{tr}\,S\,Y$, that is, M^n is an affine hypersphere. Thus we get

Proposition 9.7. *For a Blaschke hypersurface M^n, the following conditions are equivalent:*

(i) *∇C is totally symmetric;*

(ii) *$\hat{\nabla} C$ is totally symmetric;*

(iii) M^n is an affine hypersphere.

We also consider the second covariant differentials of C:

(9.12) $(\nabla^2 C)(Y,Z,W;X,U) = (\nabla_U \nabla_X C)(Y,Z,W) - (\nabla_{\nabla_U X} C)(Y,Z,W)$

and

(9.13) $(\hat{\nabla}^2 C)(Y,Z,W;X,U) = (\hat{\nabla}_U \hat{\nabla}_X C)(Y,Z,W) - (\hat{\nabla}_{\hat{\nabla}_U X} C)(Y,Z,W).$

By definition of the curvature tensors we have

(9.14)
$$(\nabla^2 C)(Y,Z,W;X,U) - (\nabla^2 C)(Y,Z,W;U,X)$$
$$= (R(U,X)C)(Y,Z,W),$$

(9.15)
$$(\hat{\nabla}^2 C)(Y,Z,W;X,U) - (\hat{\nabla}^2 C)(Y,Z,W;U,X)$$
$$= (\hat{R}(U,X)C)(Y,Z,W).$$

We can now prove

Theorem 9.8. *For a Blaschke hypersurface, ∇C and $\nabla^2 C$ are both totally symmetric if and only if $C = 0$ or $S = 0$.*

Proof. If $C = 0$ or $S = 0$, it is easy to see that ∇C and $\nabla^2 C$ are totally symmetric. To prove the converse, assume ∇C and $\nabla^2 C$ are both totally symmetric. Then we know $S = HI$. If $H = 0$, then $S = 0$. Assume $H \neq 0$. The Gauss equation gives

$$R(X,Y)Z = H[h(Y,Z)X - h(X,Z)Y].$$

Using this expression we evaluate $R(X,Y)C = 0$. We get

$$h(Y,U)C(X,V,W) - h(X,U)C(Y,V,W)$$
$$+ h(Y,V)C(U,X,W) - h(X,V)C(U,Y,W)$$
$$+ h(Y,W)C(U,V,X) - h(X,W)C(Y,V,Y) = 0.$$

Taking the trace of the bilinear map that sends (X,W) to the left-hand side of the equation above we obtain, by using apolarity,

$$(n+1)C(U,V,Y) = 0, \text{ that is, } C = 0.$$

Remark 9.3. Proposition 9.7 and Theorem 9.8 appear in [BNS]. Formulas (9.2), (9.11), (9.15), etc. will be used in the next section for the computation of the Laplacian of the Pick invariant.

We shall complete this section by defining an analogue of the tensor $\hat{L}$. We set

(9.16) $$L(X,Z) = \text{trace}\,\{Y \mapsto (\nabla_Y K)((X,Z)\}.$$

Then:

Proposition 9.9. *We have*

(9.17) $\hat{L}(X,Z) - L(X,Z) = 2\operatorname{tr}(K_X K_Z),$

(9.18) $L(X,Z) = -\dfrac{n}{2}h(S^0 X, Z) - 2\operatorname{tr}(K_X K_Z).$

Proof. Using $\nabla_Y - \hat{\nabla}_Y = K_Y$ we get

$$\nabla_Y K = \hat{\nabla}_Y K + K_Y \cdot K,$$

$$(K_Y \cdot K)_X Z = K_Y K_X Z - K_{K_Y X} Z - K_X K_Y Z$$
$$= K_{K_X Z} Y - K_Z K_X Y - K_X K_Z Y.$$

Since $\operatorname{tr} K_{K_X Z} = 0$, we obtain (9.17). Now (9.18) follows from (9.8) and (9.17).

Remark 9.4. We check the signs of all the quantities. If we change an affine normal field ξ to $-\xi$, then h changes to $-h$, but ∇ and $\bar{\nabla}$ remain the same. $C, S, H, J, \hat{\rho}$ change to $-C, -S, -H, -J, -\hat{\rho}$.

10. Laplacian of the Pick invariant

For a Blaschke hypersurface $f : M^n \to \mathbf{R}^{n+1}$, let J be the Pick invariant defined by (9.5). We want to compute and estimate $\triangle J$ when M^n is an affine hypersphere. For the sake of convenience, we set

(10.1) $u = h(C,C) = 4n(n-1)J.$

We first prove

Lemma 10.1. *For an affine hypersphere, we have*

(10.2) $\triangle u = 2h(\triangle C, C) + 2h(\hat{\nabla} C, \hat{\nabla} C).$

Proof. We introduce the notation

$$P(U, X; Y, Z, W) = (\hat{\nabla}_U \hat{\nabla}_X C)(Y, Z, W) - (\hat{\nabla}_{\hat{\nabla}_U X} C)(Y, Z, W),$$

namely, $\hat{\nabla}^2 C(Y, Z, W; X, U)$. We have

(10.3) $P(U, X; Y, Z, W) - P(X, U; Y, Z, W) = (\hat{R}(U, X)C)(Y, Z, W).$

Covariant differentiation of (9.11) gives

(10.4) $P(U, X; Y, Z, W) - P(X, U; Y, Z, W)$
$$= h(X,Z)h((\hat{\nabla}_U S)Y, W) - h(Y,Z)h((\hat{\nabla}_U S)X, W)$$
$$+ h(X,W)h((\hat{\nabla}_U S)Y, Z) - h(Y,W)h((\hat{\nabla}_U S)X, Z).$$

Assume now that the hypersurface is an affine hypersphere:$S = HI$. Then $\hat{\nabla}S = 0$ and

$$P(U,X;Y,Z,W) = P(U,Y;X,Z,W) \quad \text{by (10.4)}$$
$$= P(Y,U;X,Z,W) + (\hat{R}(U,Y)C)(X,Z,W) \quad \text{by (10.3)}$$
$$= P(Y,Z;X,U,W) + (\hat{R}(U,Y)C)(X,Z,W) \quad \text{by (10.4)}.$$

Therefore

$$(\triangle C)(Y,Z,W) = \text{trace}_h\{(U,X) \mapsto P(U,X;Y,Z,W)\}$$
$$= \text{trace}_h\{(U,X) \mapsto P(Y,Z;X,U,W)\}$$
$$+ \text{trace}_h\{(U,X) \mapsto (\hat{R}(U,Y)C)(X,Z,W)\}$$
$$= \text{trace}_h\{(U,X) \mapsto (\hat{R}(U,Y)C)(X,Z,W)\} \quad \text{by apolarity}.$$

From (9.2), we see that

$$\hat{R}(U,Y)Z = H(h(Y,Z)U - h(U,Z)Y) - [K_U, K_Y]Z,$$

from which we get

$$\text{trace}_h\{(U,X) \mapsto C(\hat{R}(U,Y)X,Z,W)\}$$
$$= -(n-1)HC(Y,Z,W) - \text{trace}_h\{(U,X) \mapsto C([K_U,K_Y]X,Z,W)\},$$
$$\text{trace}_h\{(U,X) \mapsto C(X, \hat{R}(U,Y)Z,W)\},$$
$$= -HC(Y,Z,W) - \text{trace}_h\{(U,X) \mapsto C(X,[K_U,K_Y]Z,W)\}.$$

Hence,

$$(10.5) \quad (\triangle C)(Y,Z,W)$$
$$= (n+1)HC(Y,Z,W) + \text{trace}_h\{(U,X) \mapsto C([K_U,K_Y]X,Z,W)$$
$$+ C(X,[K_U,K_Y]Z,W) + C(X,Z,[K_U,K_Y]W)\}.$$

Now the definition of the Hessian shows that

$$\text{Hess}_u(U,X) = UXu - (\hat{\nabla}_U X)u$$
$$= 2h(\hat{\nabla}_U \hat{\nabla}_X C - \hat{\nabla}_{\hat{\nabla}_U X}C, C) + 2h(\hat{\nabla}_X C, \hat{\nabla}_U C);$$

whence

$$\triangle u = 2h(\triangle C, C) + 2h(\hat{\nabla}C, \hat{\nabla}C).$$

This completes the proof of Lemma 10.1.

Proposition 10.2. *Assume the hypersurface is a locally strictly convex affine hypersphere. Then the squared norm u of the cubic form C satisfies the differential inequality*

$$\triangle u \geq 2(n+1)Hu + \frac{n+1}{2n(n-1)}u^2.$$

Proof. We evaluate the term $2h(\triangle C, C)$ in (10.2). We use an h-orthonormal basis $\{X_i\}$: $h(X_i, X_j) = \delta_{ij}$. Put $C_{ijk} = C(X_i, X_j, X_k)$. Then $K_{X_i X_j} = -\frac{1}{2}\sum C_{ijk}X_k$. We have by (10.5)

$$(\triangle C)(X_i, X_j, X_k) = (n+1)HC_{ijk} + \frac{1}{4}\sum_{\ell,m,p}\{C_{i\ell m}C_{\ell mp}C_{pjk}$$

$$+ (C_{ijm}C_{\ell mp} - C_{\ell jm}C_{imp})C_{\ell pk} + (C_{ikm}C_{\ell mp} - C_{\ell km}C_{imp})C_{\ell pj}\}.$$

Therefore,

$$4h(\triangle C, C) = 4(n+1)H\sum(C_{ijk})^2$$

$$+ \sum C_{i\ell m}C_{\ell mp}C_{pjk}C_{ijk}$$

$$+ \sum(C_{ijm}C_{\ell mp} - C_{\ell jm}C_{imp})C_{\ell pk}C_{ijk}$$

$$+ \sum(C_{ikm}C_{\ell mp} - C_{\ell km}C_{imp})C_{\ell pj}C_{ijk}.$$

We set here

$$a_{ij} = \sum_{k,\ell}C_{ik\ell}C_{jk\ell},$$

$$b_{ij;k\ell} = \sum_m C_{ijm}C_{k\ell m} - C_{jkm}C_{i\ell m}.$$

Then it is easy to see

$$4h(\triangle C, C)$$

$$= 4(n+1)Hu + \sum(a_{ij})^2 + \sum(b_{ij;k\ell})^2$$

$$\geq 4(n+1)Hu + \sum(a_{ii})^2 + \sum_{i\neq j}(b_{ij;ji})^2 + \sum_{i\neq j}(b_{ii;jj})^2$$

$$\geq 4(n+1)Hu + \frac{1}{n}(\sum a_{ii})^2 + \frac{1}{n(n-1)}\left\{(\sum_{i\neq j}b_{ij;ji})^2 + (\sum_{i\neq j}b_{ii;jj})^2\right\}$$

$$= 4(n+1)Hu + \frac{1}{n}(\sum a_{ii})^2 + \frac{1}{n(n-1)}\left\{(\sum_{i,j}b_{ij;ji})^2 + (\sum_{i,j}b_{ii;jj})^2\right\}$$

(because $b_{ii;ii} = 0$),

whereas we have

$$\sum a_{ii} = u, \quad \sum b_{ij;ji} = u, \quad \sum b_{ii;jj} = -u.$$

Hence,

$$4h(\triangle C, C) \geq 4(n+1)Hu + \frac{(n+1)}{n(n-1)}u^2.$$

Using this in (10.2) completes the proof of Proposition 10.2.

Remark 10.1. According to [Cal2], a more precise estimate is possible by introducing the function

$$\psi = (h(C,C))^{\frac{1}{2}}.$$

At any point where $\psi \neq 0$, ψ is differentiable and we can see

$$\psi \operatorname{Hess}_\psi(U,X)$$
$$= \{h(C,C)h(\hat{\nabla}_X C, \hat{\nabla}_U C) - h(\hat{\nabla}_X C, C)h(\hat{\nabla}_U C, C)\}/h(C,C)$$
$$+ h(\hat{\nabla}_U \hat{\nabla}_X C - \hat{\nabla}_{\hat{\nabla}_U X} C, C),$$

which implies

$$\psi \triangle \psi \geq h(\triangle C, C).$$

In the rest of this section, we shall write down the equality (10.2) more concretely in the case $n = 2$.

Proposition 10.3. *For an affine sphere M^2, we have*

(10.6) $$\triangle J = 6J(H+J) + \frac{1}{4}h(\hat{\nabla}C, \hat{\nabla}C),$$

(10.7) $$\triangle \log|J| = 6(H+J) \quad \text{where} \quad J \neq 0.$$

Proof. Since $u = 8J$ when $n = 2$, it is enough to see $h(\triangle C, C) = 24J(J+H)$ for (10.6). Let $\{X_i\}$ be an h-orthonormal basis: $h(X_i, X_j) = \epsilon_i \delta_{ij}$ where $\epsilon_1 = 1$ and $\epsilon_2 = \epsilon = \pm 1$. Set

$$C_{ijk} = C(X_i, X_j, X_k)$$

and

$$C_{ijk\ell} = (\hat{\nabla}_{X_\ell} C)(X_i, X_j, X_k).$$

Notice that $\hat{\nabla}C$ is totally symmetric by Proposition 9.7. By the apolarity we see

$$C_{112} = -C_{222}, \qquad C_{122} = -C_{111},$$
$$C_{1112} = -C_{1222}, \qquad C_{1122} = -C_{1111}, \qquad C_{2222} = C_{1111}.$$

If we set now

$$a_{ij} = \sum_{k,\ell} \epsilon_k \epsilon_\ell C_{ik\ell} C_{jk\ell},$$

$$b_{ij;k\ell} = \sum_m \epsilon_m C_{ijm} C_{k\ell m} - \epsilon_m C_{jkm} C_{i\ell m},$$

then we have

$$4h(\triangle C, C) = 4(n+1)Hu + \sum \epsilon_i \epsilon_j (a_{ij})^2 + \sum \epsilon_i \epsilon_j \epsilon_k \epsilon_\ell (b_{ij;k\ell})^2.$$

Using the apolarity, we see

$$a_{ij} = \frac{1}{2} h(C, C) \epsilon_i \delta_{ij},$$

$$b_{11;22} = b_{22;11} = -\frac{1}{2} \epsilon h(C, C),$$

$$b_{12;21} = b_{21;12} = \frac{1}{2} \epsilon h(C, C),$$

and other b's vanish. Hence,

$$\sum \epsilon_i \epsilon_j (a_{ij})^2 = \frac{1}{2} h(C, C)^2, \qquad \sum \epsilon_i \epsilon_j \epsilon_k \epsilon_\ell (b_{ij;k\ell})^2 = h(C, C)^2.$$

Therefore

$$h(\triangle C, C) = 3Hh(C, C) + \frac{3}{8} h(C, C)^2$$

$$= 24J(J + H),$$

completing the proof of (10.6). From the identity

$$\text{Hess}_{\log |J|}(X, Y) = \frac{\text{Hess}_J(X, Y)}{J} - \frac{XJ \cdot YJ}{J^2},$$

which holds for $J \neq 0$, we have

$$\triangle \log |J| = \frac{\triangle J}{J} - \frac{4 \text{trace}_h\{(X, Y) \mapsto h(\hat{\nabla}_X C, C) h(\hat{\nabla}_Y C, C)\}}{h(C, C)^2}$$

$$= 6(H + J) + A$$

where

$$A = \frac{2\text{trace}_h\{(X, Y) \mapsto h(\hat{\nabla}_X C, \hat{\nabla}_Y C) h(C, C) - 2h(\hat{\nabla}_X C, C) h(\hat{\nabla}_Y C, C)\}}{h(C, C)^2}.$$

Since

$$\sum \epsilon_i \epsilon_j \epsilon_k C_{ijk} C_{ijk1}$$

$$= C_{111}C_{1111} + 3\epsilon C_{112}C_{1121} + 3C_{122}C_{1221} + \epsilon C_{222}C_{2221}$$

$$= C_{111}C_{1111} + 3\epsilon(-C_{222})(-C_{1222}) + 3(-C_{111})(-C_{1111}) + \epsilon C_{222}C_{2221}$$

$$= 4(C_{111}C_{1111} + \epsilon C_{222}C_{1222})$$

and

$$\sum \epsilon_i \epsilon_j \epsilon_k C_{ijk} C_{ijk2}$$
$$= C_{111}C_{1112} + 3\epsilon C_{112}C_{1122} + 3C_{122}C_{1222} + \epsilon C_{222}C_{2222}$$
$$= C_{111}(-\epsilon C_{1222}) + 3\epsilon(-C_{222})(-C_{1111}) + 3\epsilon(-C_{111})C_{1222} + \epsilon C_{222}C_{1111}$$
$$= 4\epsilon(C_{222}C_{1111} - C_{111}C_{1222}),$$

we have

$$\text{trace}_h\{(X,Y) \mapsto h(\hat{\nabla}_X C, C)h(\hat{\nabla}_Y C, C)\}$$
$$= \sum \epsilon_i \epsilon_j \epsilon_k \epsilon_\ell \epsilon_p \epsilon_q \epsilon_r C_{ijk} C_{ijk\ell} C_{pqr} C_{pqr\ell}$$
$$= 16\{(C_{111}C_{1111})^2 + (C_{222}C_{1222})^2 + \epsilon(C_{222}C_{1111})^2 + \epsilon(C_{111}C_{1222})^2\}$$
$$= 16((C_{111})^2 + \epsilon(C_{222})^2)((C_{1111})^2 + \epsilon(C_{1222})^2).$$

On the other hand,

$$h(C,C) = \sum \epsilon_i \epsilon_j \epsilon_k C_{ijk} C_{ijk} = 4((C_{111})^2 + \epsilon(C_{222})^2)$$

and

$$\sum \epsilon_i \epsilon_j \epsilon_k \epsilon_\ell C_{ijk\ell} C_{ijk\ell} = 8((C_{1111})^2 + \epsilon(C_{1222})^2);$$

hence

$$h(\hat{\nabla}C, \hat{\nabla}C)h(C,C) = 32((C_{111})^2 + \epsilon(C_{222})^2)((C_{1111})^2 + \epsilon(C_{1222})^2).$$

That is, we see

$$A = 0,$$

concluding the proof of (10.7).

11. Behavior of the cubic form on surfaces

We first study the behavior of the cubic form in the case of a nondegenerate affine surface. Assume that C is the cubic form of a nondegenerate affine surface M and that $C_x \neq 0$ at $x \in M$. By a *null direction* of C, we mean the direction of a vector $X \neq 0$ such that $C(X,X,X) = 0$.

Proposition 11.1. *If h is positive-definite, then C has three distinct null directions.*

Proof. Take a basis $\{X_1, X_2\}$ such that $h_{11} = h_{22} = 1$, $h_{12} = 0$. By apolarity we have $C_{111} + C_{221} = 0$ and $C_{112} + C_{222} = 0$. Setting $a = C_{111}$ and $b = C_{112}$, we have $C(X,X,X) = a(x^1)^3 + 3b(x^1)^2 x^2 - 3ax^1(x^2)^2 - b(x^2)^3$ for $X = x^1 X_1 + x^2 X_2$.

Case where $b = 0$:

$$C(X, X, X) = ax^1[(x^1)^2 - 3(x^2)^2]$$

so that $(0, 1), (\sqrt{3}, 1)$, and $(\sqrt{3}, -1)$ give three distinct null directions.
   Case where $b \neq 0$: If we write $t = x^2/x^1$ and $c = a/b$, solving the equation $C(X, X, X) = 0$ is reduced to solving

$$f(t) = t^3 + 3ct^2 - 3t - c = 0.$$

One can show that this equation has three distinct roots by checking the values $f(c_1) > 0$ and $f(c_2) < 0$, where $c_1 = -c - \sqrt{c^2 + 1}$ and $c_2 = -c + \sqrt{c^2 + 1}$ are the two critical points of f such that $c_1 < c_2$.

Proposition 11.2. *If h is indefinite, then C has*
either

   (a) *one null direction of multiplicity 1, in which case C can be written in the form $C(X, X, X) = \mu(X)g(X, X)$, where μ is a 1-form and g is a definite inner product on $T_x(M)$, each unique up to a scalar multiple;*
or

   (b) *a null direction of multiplicity 3, in which case there is a nonzero X in $T_x(M)$ such that*

$$(11.1) \qquad h(X, X) = 0, \quad \text{and} \quad C(X, U, V) = 0 \quad \text{for all} \quad U, V \in T_x(M).$$

Moreover, we can find $Y \in T_x(M)$ such that

$$(11.2) \qquad h(Y, Y) = 0, \quad h(X, Y) = 1, \quad \text{and} \quad C(Y, Y, Y) = 1.$$

A pair of vectors $\{X, Y\}$ satisfying conditions (11.1) and (11.2) is uniquely determined. The direction of X satisfying (11.1) is unique.
   We have case (b) if and only if the Pick invariant J is 0.
Proof. We use a null basis $\{X_1, X_2\}$ so that $h_{11} = h_{22} = 0$ and $h_{12} = 1$. From apolarity we get $C_{112} = C_{122} = 0$. Thus

$$C(X, X, X) = C_{111}(x^1)^3 + C_{222}(x^2)^3.$$

Case (a). If $C_{111} \neq 0$ and $C_{222} \neq 0$, let (α, β) be their real cubic root. Then

$$C(X, X, X) = (\alpha x^1 + \beta x^2)(\alpha^2(x^1)^2 - \alpha\beta x^1 x^2 + \beta^2(x^2)^2).$$

We may define a 1-form μ by

$$\mu(X) = \alpha x^1 + \beta x^2$$

and an inner product g by

$$g(X, Y) = \alpha^2 x^1 y^1 - 2\alpha\beta(x^1 y^2 + x^2 y^1) + \beta^2 x^2 y^2.$$

Clearly, g is positive-definite and C has only one null direction. The uniqueness assertion is also obvious.

Case (b). If $C_{111} = 0$, then $X = (1,0)$ is a null direction of multiplicity 3. Since $C(X, X_i, X_j) = C_{1ij} = 0$ for all i, j, we see that X is in the kernel of C, that is,

$$C(X, U, V) = 0 \quad \text{for all} \quad U, V.$$

Obviously, $h(X, X) = 0$. If $C_{222} = 0$, then $X = (0, 1)$ is the vector. The uniqueness assertions are easy to verify. The additional statement in Proposition 11.2 follows from

$$8J = h(C, C) = \sum_{i,j,k,p,q,r} h_{ip} h_{jq} h_{kr} C_{ijk} C_{pqr} = 2C_{111} C_{222}$$

in terms of the same null basis $\{X_1, X_2\}$. This completes the proof of Proposition 11.2.

Remark 11.1. Each of the two cases in Proposition 11.2 is actually possible at a point of an affine surface. For example, for the graph of $z = xy + \frac{1}{6}(x^3 + y^3)$ at the point $(0, 0, 0)$ the vector $\partial/\partial x - \partial/\partial y$ gives the only null direction of the cubic form. On the other hand, for the Cayley surface, namely, for the graph of $z = xy - \frac{1}{3}x^3$, the cubic form has a null direction of multiplicity 3 at every point as we shall see in Chapter III.

As an application of Proposition 11.2 we show that a nondegenerate surface is a ruled surface if and only if the affine metric h is hyperbolic and the Pick invariant J is 0 everywhere. Theorems 11.3 and 11.4 give more precise statements. We first give

Definition 11.1. A surface M in $\mathbf{R}^3$ is said to be *ruled* if every point of M is on (an open segment of) a line in $\mathbf{R}^3$ that lies in M.

Theorem 11.3. *If a nondegenerate surface M is ruled, then its affine metric is hyperbolic and the Pick invariant is 0 at every point.*

Proof. Let x_0 be an arbitrary point of M. By assumption, there is an open segment x_t with affine parameter $t, -\epsilon < t < \epsilon$, through x_0 that lies on the surface. Then

$$D_t(\dot{x}_t) = \nabla_t \dot{x}_t + h(\dot{x}_t, \dot{x}_t)\xi = 0,$$

which implies $\nabla_t \dot{x}_t = 0$ and $h(\dot{x}_t, \dot{x}_t) = 0$. Thus we have

$$C(\dot{x}_t, \dot{x}_t, \dot{x}_t) = (\nabla_t h)(\dot{x}_t, \dot{x}_t) = (d/dt)h(\dot{x}_t, \dot{x}_t) - 2h(\nabla_t \dot{x}_t, \dot{x}_t) = 0.$$

At x_0, $X_1 = \dot{x}_0$ satisfies $h_{11} = h(X_1, X_1) = 0$, $C_{111} = C(X_1, X_1, X_1) = 0$. By taking a vector X_2 such that X_1, X_2 form a local null basis we see from apolarity that $C_{112} = C_{122} = 0$ and $J = \frac{1}{8}C_{111}C_{222} = 0$.

We wish to state the converse of Theorem 11.3 in the following precise form. By a line field $[X]$ on a manifold M we mean a 1-dimensional differentiable distribution (so that locally it is defined as the span of a vector field X).

Theorem 11.4. *For a nondegenerate surface M assume that the Pick invariant J is identically zero but the cubic form C is never 0. Then M admits a line field $[X]$ such that for each point x of M the integral curve of $[X]$ through x is an open segment of a line in $\mathbf{R}^3$ that lies on M. Indeed, we can find a nonvanishing vector field X that induces the line field.*

Proof. From the assumption, the affine metric h is hyperbolic. From Proposition 11.2, each point x admits a vector $X \in T_x(M)$ such that $h(X, X) = 0$ and $C(X, U, V) = 0$ for all $U, V \in T_x(M)$. Actually, by taking a differentiable local null basis $\{X_1, X_2\}$ and going through the proof, we see that we may find a local vector field X satisfying the conditions stated above. Let Y be a local vector field such that $h(Y, Y) = 0$, $h(X, Y) = 1$, and $C(Y, Y, Y) = 1$, by adjusting X if necessary. We know that such a pair $\{X, Y\}$ is uniquely determined by (11.1) and (11.2) so that they are defined globally on M. From $h(X, X) = 0$ we get $h(\hat{\nabla}_X X, X) = 0$, hence $\hat{\nabla}_X X = \lambda X$, where λ is a scalar function. Since $h(K_X X, X) = C(X, X, X) = 0$, $h(K_X X, Y) = C(X, X, Y) = 0$, we see that $K_X X = 0$ and hence $\nabla_X X = \hat{\nabla}_X X + K_X X = \lambda X$. From $D_X X = \nabla_X X + h(X, X)\xi$ we get $D_X X = \nabla_X X = \lambda X$. This implies that the integral curve of X is a pregeodesic in (M, ∇) as well as in the affine space $\mathbf{R}^3$. Thus it is part of a straight line. This proves Theorem 11.4.

Remark 11.2. A result similar to Theorem 11.4 is usually stated as follows: A nondegenerate surface with hyperbolic affine metric and $J = 0$ is a ruled surface. In this form, the proof requires analyticity. In our formulation, if we assume that the induced connection ∇ is complete, then every line determined by the line field is an entire line in $\mathbf{R}^3$.

III

Models with remarkable properties

In this chapter we discuss many of the important models and properties of nondegenerate hypersurfaces that were developed during the periods I and II described in the Introduction. In addition, we present more recent results and new approaches. Sections 1 and 5 deal with affine spheres, both proper and improper, that are also ruled surfaces. We relate them to newer results on affine spheres whose affine metrics have constant curvature. In Sections 2, 3, and 4 we emphasize the equiaffinely homogeneous surfaces, including a new model that was discovered in 1991. Section 6 gives a characterization of Cayley surfaces by the property that the cubic form C is not 0 but parallel: $\nabla C = 0$. In Section 7, we discuss global properties of ovaloids and prove a well-known theorem that a compact affine hypersphere must be an ellipsoid. Section 8 is devoted to some new characterizations of ellipsoids in the 2-dimensional case. Section 9 is concerned with the classical Minkowski formulas and their applications to other known characterizations of ellipsoids. In Section 10 we prove a theorem due to Blaschke and Schneider on a certain elliptic differential equation on a hypersurface, which provides a different proof for the theorem on compact affine hyperspheres. In Section 11, we deal with a variational formula for affine volume and prove a characterization of an elliptic paraboloid among affine minimal surfaces originally due to Calabi.

1. Ruled affine spheres

We discuss in this section affine spheres that are ruled (recall Definition 11.1 of the previous chapter). We first consider the graph of $z = xy + \Phi(x)$, where $\Phi(x)$ is an arbitrary smooth function defined on **R**. We show that this nondegenerate surface is an improper affine sphere that is ruled. Among these surfaces there are two particularly important surfaces, namely,

(1) the graph of $z = xy - \frac{1}{3}x^3$, called the Cayley surface,
(2) the graph of $z = xy + \log x$, $x > 0$.

Each of these two surfaces is equiaffinely homogeneous, that is, there is a Lie subgroup of the group of all equiaffine transformations of $\mathbf{R}^3$ that acts transitively on the surface.

Let us now consider the graph $f : (x, y) \mapsto (x, y, xy + \Phi(x))$. Using coordinate vector fields $\partial_x = \partial/\partial x, \partial_y = \partial/\partial y$ we can proceed in the same manner as in Example 3.1 of Chapter II. We get

$$f_*(\partial_x) = (1, 0, y + \Phi'), \quad f_*(\partial_y) = (0, 1, x),$$

$$\theta(\partial_x, \partial_y) = 1,$$

$$D_x f_*(\partial x) = (0, 0, \Phi'') = \Phi'' \xi,$$

$$D_x f_*(\partial_y) = (0, 0, 1) = \xi, \quad D_y f_*(\partial_y) = (0, 0, 0),$$

where $\xi = (0, 0, 1)$ is a tentative transversal vector field. Thus

$$h(\partial_x, \partial_x) = \Phi'', \quad h(\partial_x, \partial_y) = 1, \quad h(\partial_y, \partial_y) = 0$$

and

$$\nabla_x \partial_x = \nabla_x \partial_y = \nabla_y \partial_x = \nabla_y \partial_y = 0.$$

Since

$$\det {}_\theta h = \det \begin{bmatrix} \Phi'' & 1 \\ 1 & 0 \end{bmatrix} = -1,$$

we see that ξ is the affine normal field. Thus $S = 0$ and the induced connection ∇ is flat; indeed, $\{x, y\}$ is a flat coordinate system for ∇.

For each point $(x, y, xy + \Phi(x))$, the line $(x, 0, \Phi(x)) + t(0, 1, x)$ goes through the point and lies on the surface. In fact, if we take the curve $p(x) = (x, 0, \Phi(x))$ in the xz-plane and the vector field $q(x) = (0, 1, x)$, then $p(x) + tq(x)$ represents a ruling of the surface.

Continuing with the previous notation we see that the cubic form $C = \nabla h$ is given by

$$C(\partial_x, \partial_x, \partial_x) = \Phi^{(3)}, \quad C(\partial_x, \partial_x, \partial_y) = 0,$$

$$C(\partial_x, \partial_y, \partial_y) = 0, \quad C(\partial_y, \partial_y, \partial_y) = 0.$$

In index notation with $(x^1, x^2) = (x, y)$ this means that C_{111} is the only nonzero component of C. On the other hand, $h_{11} = \Phi''$, $h_{12} = 1$, $h_{22} = 0$, from which we get $h^{11} = 0$, $h^{12} = 1$, $h^{22} = -\Phi''$. It follows that the Pick invariant $J = \frac{1}{8} h(C, C)$ (defined in Proposition 9.3 of Chapter II) is identically 0. Our surface (with $\Phi^{(3)} \neq 0$) provides an example of a surface with $J = 0$ but $C \neq 0$. Finally, we remark that the affine metric h of our surface is flat (the curvature of h is 0), as follows from Proposition 9.3 of Chapter II.

To sum up, we have

Theorem 1.1. *For any smooth function $\Phi(x)$, the graph of $z = xy + \Phi(x)$ is a nondegenerate surface and a ruled improper affine sphere whose affine metric is flat.*

In Section 5 we shall prove that, conversely, every ruled improper affine sphere is locally the graph of a function $z = xy + \Phi(x)$.

There are two special cases.

Example 1.1. *The graph of* $z = xy - x^3/3$. (See Figure 5.)
This surface, called the *Cayley surface*, has the property that $C \neq 0, \nabla C = 0$, as we can see from the computation above. The surface is equiaffinely homogeneous; we take the group consisting of all matrices of the form

$$\begin{bmatrix} 1 & b & a & ab - \frac{1}{3}b^3 \\ 0 & 1 & b & a \\ 0 & 0 & 1 & b \\ 0 & 0 & 0 & 1 \end{bmatrix} \quad \text{acting on} \quad \begin{bmatrix} z \\ y \\ x \\ 1 \end{bmatrix}.$$

The surface is the orbit of the origin $(0, 0, 0)$ under this 2-dimensional abelian Lie group.

Later, in Section 6, we shall characterize the Cayley surface as a non-degenerate surface in $\mathbf{R}^3$ with the property $C \neq 0, \nabla C = 0$; see Section 6.

We now consider

Example 1.2. *The graph of* $z = xy + \log x, x > 0$. (See Figure 6.)
This surface is also equiaffinely homogeneous with the group

$$\begin{bmatrix} 1 & b/a & 0 & -\log a \\ 0 & 1/a & 0 & 0 \\ 0 & 0 & a & b \\ 0 & 0 & 0 & 1 \end{bmatrix} \quad \text{acting on} \quad \begin{bmatrix} z \\ x \\ y \\ 1 \end{bmatrix},$$

where $a > 0$. The 2-dimensional group is isomorphic to the group of matrices $\begin{bmatrix} a & b \\ 0 & 1 \end{bmatrix}$ (thus solvable but non-abelian). This surface, which was missing in the classification list of equiaffinely homogeneous surfaces in the book of Guggenheimer [Gu], has been discovered in [NS2]. We shall give more detail on the classification result in Section 5.

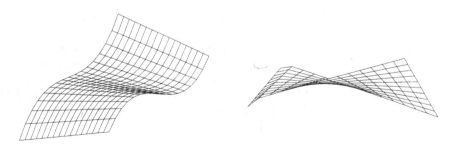

Figure 5   Figure 6

Next we consider the surface of the form

(1.1) $$f(u,v) = uA(v) + A'(v),$$

where $A(v)$ is a space curve in $\mathbf{R}^3$ parametrized by v and $A'(v) = dA/dv$. We assume that

(1.2) $$\det [A(v)\ A'(v)\ A''(v)] = a \text{ (nonzero constant)}.$$

We shall show that this nondegenerate surface is ruled and is a proper affine sphere. In order to simplify computation we reduce the question to the case where the constant a is equal to 1. For this purpose, we observe that if we take $B(v) = A(v)/a^{\frac{1}{3}}$, then (1.1) is of the form

$$f(u,v) = a^{\frac{1}{3}}[uB(v) + B'(v)],$$

where $\det [B(v)\ B'(v)\ B''(v)] = 1$. In the study of properties of the surface, the constant factor $a^{\frac{1}{3}}$ can obviously be dropped. This means we can go back to (1.1) and assume $a = 1$ in (1.2). This is clearly a ruled surface. We want to show that it is a proper affine sphere.

We have

$$f_u = A, \qquad f_v = uA' + A'',$$

$$f_{uu} = 0, \quad f_{uv} = A', \quad f_{vv} = uA'' + A'''.$$

The coefficients of the affine metric,

$$E = h(\partial_u, \partial_u), \quad F = h(\partial_u, \partial_v), \quad G = h(\partial_v, \partial_v),$$

can be obtained as follows: Let

$$L = \det [f_u\ f_v\ f_{uu}], \quad M = \det [f_u\ f_v\ f_{uv}], \quad N = \det [f_u\ f_v\ f_{vv}].$$

Then, for (1.1), we have

$$L = 0, \quad M = \det [A\ uA' + A''\ A'] = -1, \quad N = u^2 + \det [A\ A''\ A'''],$$

by noting (1.2) with $a = 1$ and its consequence $\det [A\ A'\ A'''] = 0$.

Thus $LN - M^2 = -1$, which means that the surface is nondegenerate and the coefficients E, F, and G are equal to L, M, and N, respectively, because $|LN - M^2| = 1$; see (6.14) of Chapter II. We may find the affine normal field ξ by using Theorem 6.5 of Chapter II, that is, by computing $\pm\frac{1}{2}\Delta f$. By the formula in Appendix 2 we find

$$\xi = \pm\frac{1}{2}[-(Gf_u + f_v)_u - (f_u)_v] = \pm(uA + A').$$

We take

(1.3) $$\xi = -(uA + A'),$$

which is nothing but the negative of the position vector, that is, a centro-affine transversal field. Hence we have the shape operator $S = I$. We have thus shown that the surface is a proper affine sphere.

To sum up, we have

Theorem 1.2. *A surface of the form* (1.1) *with the condition* (1.2) *is a ruled proper affine sphere.*

In Section 5, we shall prove the converse: every ruled proper affine sphere is of the form (1.1) with the condition (1.2). Here we shall consider one special case.

Example 1.3. Let $A(v) = (\sin v, -\cos v, 1)$ and let

(1.4) $$f(u, v) = uA(v) + A'(v) = (\cos v + u \sin v, \sin v - u \cos v, u).$$

The surface is the one-sheeted hyperboloid $x^2 + y^2 - z^2 = 1$ with one of the usual two rulings. The affine mean curvature is equal to 1, and so is the Gauss curvature of the affine metric.

2. Some more homogeneous surfaces

We shall discuss four more nondegenerate surfaces in $\mathbf{R}^3$ that are equiaffinely homogeneous. To emphasize the group-theoretic aspect, we shall deal with the general description of the orbit hypersurface under an n-dimensional Lie group of affine transformations of $\mathbf{R}^{n+1}$.

Let $\mathbf{SA}(n+1)$ be the group of all equiaffine transformations of $\mathbf{R}^{n+1}$. Let $\phi : G \to \mathbf{SA}(n+1)$ be an isomorphism of an n-dimensional Lie group G into $\mathbf{SA}(n+1)$. We assume that there exists a point $x_0 \in \mathbf{R}^{n+1}$ such that $f(g) = \phi(g)x_0 \in \mathbf{R}^{n+1}$ is an immersion of G into $\mathbf{R}^{n+1}$. In other words, the orbit $f(G) = \phi(G)x_0$ is a hypersurface. Note that for $a, g \in G$ we have

(2.1) $$f(ag) = \phi(a)f(g),$$

that is

(2.2) $$f \circ L_a = \phi(a) \circ f$$

where L_a denotes the left translation of G by a. The Lie algebra $\mathfrak{g}$ of left-invariant vector fields can be identified with the tangent space $T_e(G)$.

We assume that $\phi(a)x_0 = x_0$ only for the identity element e. Suppose we have a transversal vector field ξ for $f : G \to \mathbf{R}^{n+1}$ that is equivariant, i.e.

$\xi_{ag} = \phi(a)\xi_g$. Then ξ is equiaffine if and only if $\phi_*(X)\xi_e$ lies in $f_*(T_e(G))$ for every $X \in \mathfrak{g}$. This follows from

$$(D_X\xi)_e = \left(\frac{d}{dt}\right)_{t=0} \xi_{a_t}$$

$$= \left(\frac{d}{dt}\right)_{t=0} \phi(a_t)\xi_e = \phi_*(X)\xi_e,$$

where a_t is the 1-parameter subgroup $\exp tX$, and from

$$(D_X\xi)_a = D_{f_*(X_a)}\xi = D_{\phi_*(a)X_e}\xi = \phi_*(a)(D_{X_e}\xi),$$

since ξ is equivariant and since $\phi(a)$ preserves the flat connection D in $\mathbf{R}^{n+1}$. Let X and Y denote left-invariant vector fields on G. We write

$$D_X f_*(Y) = f_*(\nabla_X Y) + h(X, Y)\xi,$$

thereby getting the induced connection ∇ and the affine fundamental form h. We can verify that ∇ and h are left-invariant:

$$(D_X f_*(Y))_{ag} = [D_{L_a X} f_*(L_a Y)]_{ag}$$
$$= [D_{L_a X}\phi(a)f_*(Y)]_{ag}$$
$$= \phi(a)[D_X f_*(Y)]_g$$
$$= \phi(a)[f_*(\nabla_X Y)_g + h_g(X, Y)\xi_g]$$
$$= f_*(L_a \nabla_X Y)_g + h_g(X, Y)\phi(a)\xi_{ag}.$$

This means that ∇ and h can be expressed in terms of the Lie algebra $\mathfrak{g}$, that is, ∇ is a bilinear map

$$(X, Y) \in \mathfrak{g} \times \mathfrak{g} \mapsto \nabla_X Y \in \mathfrak{g},$$

and h is a symmetric bilinear function

$$(X, Y) \in \mathfrak{g} \times \mathfrak{g} \mapsto h(X, Y) \in \mathbf{R}.$$

The affine shape operator S also has the same kind of invariance property. Now, because of the invariance of h as well as of the induced volume element θ, it follows that the equiaffine, equivariant transversal field ξ is actually the affine normal field, up to a constant scalar. The connection ∇ is the Blaschke connection.

   The following method of computation is quite useful in practical applications. Let $X \in \mathfrak{g}$ and $a_t = \exp tX$ be the 1-parameter subgroup generated by X. For $Y \in \mathfrak{g}$ consider the vector field $Y_t = L_{a_t} Y_e$, with $Y_e = Y \in T_e(G)$, along the curve a_t. Then using

$$f_*(Y_t) = \phi(a_t)f_*(Y_e) \quad \text{and} \quad f_*(Y_e) = \phi_*(Y) \cdot x_0$$

we get

$$(D_X f_*(Y))_e = \left(\frac{d}{dt}\right)_{t=0} [\phi(a_t)f_*(Y_e)]$$
$$= \phi_*(X)f_*(Y_e) = \phi_*(X)\phi_*(Y) \cdot x_0.$$

Therefore

$$\phi_*(X)\phi_*(Y) \cdot x_0 = f_*(\nabla_X Y) + h(X,Y)\xi_e,$$

that is,

(2.3) $$\phi_*(X)\phi_*(Y) \cdot x_0 = \phi_*(\nabla_X Y) \cdot x_0 + h(X,Y)\xi_e$$

gives the expressions for $\nabla_X Y$ and $h(X,Y)$.

To find the shape operator S, we let $\xi_t = \phi(a_t) \cdot \xi_e$. Then

$$D_X \xi_t = \left(\frac{d}{dt}\right)_{t=0} [\phi(a_t)] \cdot \xi_e = \phi_*(X) \cdot \xi_e$$

so that

$$\phi_*(X) \cdot \xi_e = -f_*(SX),$$

that is,

(2.4) $$\phi_*(X) \cdot \xi_e = -\phi_*(SX) \cdot x_0$$

gives the expression for S.

We shall now apply this method of computation. For simplicity, we shall write

$$X \cdot x_0, \quad XX \cdot x_0, \quad X \cdot \xi_e,$$

instead of

$$\phi_*(X) \cdot x_0, \quad \phi_*(X)\phi_*(X) \cdot x_0, \quad \phi_*(X) \cdot \xi_e.$$

Example 2.1. *The graph of* $z = 1/xy$. (See Figure 7.)

This surface comes from the group

$$G = \{g = \begin{bmatrix} a & 0 \\ 0 & b \end{bmatrix} : a, b > 0\}$$

with representation

$$\phi(g) = \begin{bmatrix} a & 0 & 0 \\ 0 & b & 0 \\ 0 & 0 & 1/ab \end{bmatrix}$$

and

$$x_0 = \begin{bmatrix} 1 \\ 1 \\ 1 \end{bmatrix} \quad \text{and} \quad \xi = \begin{bmatrix} 1 \\ 1 \\ 1 \end{bmatrix}.$$

Then

$$f(g) = \begin{bmatrix} a \\ b \\ 1/ab \end{bmatrix}.$$

Let

$$X = \begin{bmatrix} 1 & 0 \\ 0 & 0 \end{bmatrix} \quad \text{and} \quad Y = \begin{bmatrix} 0 & 0 \\ 0 & 1 \end{bmatrix},$$

where $[X, Y] = 0$; we get

$$\phi_*(X) = \begin{bmatrix} 1 & 0 & 0 \\ 0 & 0 & 0 \\ 0 & 0 & -1 \end{bmatrix}, \quad X \cdot x_0 = \begin{bmatrix} 1 \\ 0 \\ -1 \end{bmatrix}$$

and

$$\phi_*(Y) = \begin{bmatrix} 0 & 0 & 0 \\ 0 & 1 & 0 \\ 0 & 0 & -1 \end{bmatrix}, \quad Y \cdot x_0 = \begin{bmatrix} 0 \\ 1 \\ -1 \end{bmatrix}.$$

Hence, from

$$XX \cdot x_0 = \begin{bmatrix} 1 \\ 0 \\ 1 \end{bmatrix} = \frac{1}{3} \begin{bmatrix} 1 \\ 0 \\ -1 \end{bmatrix} - \frac{2}{3} \begin{bmatrix} 0 \\ 1 \\ -1 \end{bmatrix} + \frac{2}{3} \begin{bmatrix} 1 \\ 1 \\ 1 \end{bmatrix},$$

$$XY \cdot x_0 = \begin{bmatrix} 0 \\ 0 \\ 1 \end{bmatrix} = -\frac{1}{3} \begin{bmatrix} 1 \\ 0 \\ -1 \end{bmatrix} - \frac{1}{3} \begin{bmatrix} 0 \\ 1 \\ -1 \end{bmatrix} + \frac{1}{3} \begin{bmatrix} 1 \\ 1 \\ 1 \end{bmatrix},$$

$$YY \cdot x_0 = \begin{bmatrix} 0 \\ 1 \\ 1 \end{bmatrix} = -\frac{2}{3} \begin{bmatrix} 1 \\ 0 \\ -1 \end{bmatrix} + \frac{1}{3} \begin{bmatrix} 0 \\ 1 \\ -1 \end{bmatrix} + \frac{2}{3} \begin{bmatrix} 1 \\ 1 \\ 1 \end{bmatrix},$$

we obtain

$$\nabla_X X = \frac{1}{3} X - \frac{2}{3} Y, \qquad h(X, X) = \frac{2}{3},$$

$$\nabla_X Y = -\frac{1}{3} X - \frac{1}{3} Y = \nabla_Y X, \quad h(X, Y) = \frac{1}{3},$$

$$\nabla_Y Y = -\frac{2}{3} X + \frac{1}{3} Y, \qquad h(Y, Y) = \frac{2}{3}.$$

The form h is positive-definite. From these we get

$$R(X, Y)Y = \frac{-2}{3} X + \frac{1}{3} Y,$$

$$R(X, Y)X = \frac{-1}{3} X + \frac{2}{3} Y,$$

$$(\nabla_X R)(X, Y)Y = \frac{-2}{3}X + \frac{2}{3}Y,$$

$$(\nabla_Y R)(X, Y)X = \frac{-2}{3}X + \frac{2}{3}Y,$$

and so on. For the shape operator, our general formula gives $S = -I$. This can also be seen from our equivariant choice of ξ that coincides with the position vector, that is, our surface is regarded as centro-affine.

More generally, the hypersurface $x^1 x^2 \cdots x^{n+1} = 1$ is equiaffinely homogeneous relative to the group of all diagonal matrices with diagonal components $a^1, a^2, \ldots, a^{n+1}$ such that the product is equal to 1. This hypersurface can be handled in the same group-theoretic way.

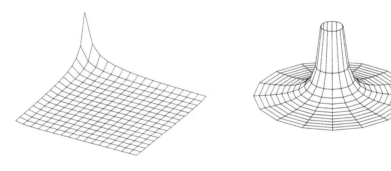

Figure 7   Figure 8

Example 2.2. *The graph of $z = 1/(x^2 + y^2)$.* (See Figure 8.)

This surface is equiaffinely homogeneous relative to the group

$$G = \left\{ g = \begin{bmatrix} c\cos t & -c\sin t & 0 \\ c\sin t & c\cos t & 0 \\ 0 & 0 & c^{-2} \end{bmatrix} : c > 0, t \in \mathbf{R} \right\}$$

with inclusion ϕ into $SA(3)$. For $x_0 = \begin{bmatrix} 1 \\ 0 \\ 1 \end{bmatrix}$, we have $f(g) = \phi(g)x_0 = \begin{bmatrix} c\cos t \\ c\sin t \\ c^{-2} \end{bmatrix}$. Let

$$X = \begin{bmatrix} 1 & 0 & 0 \\ 0 & 1 & 0 \\ 0 & 0 & -2 \end{bmatrix} \quad \text{and} \quad Y = \begin{bmatrix} 0 & -1 & 0 \\ 1 & 0 & 0 \\ 0 & 0 & 0 \end{bmatrix},$$

where $[X, Y] = 0$. Let $\xi = \begin{bmatrix} 1 \\ 0 \\ 1 \end{bmatrix}$ at x_0. Since

$$f_*(X) = \begin{bmatrix} 1 \\ 0 \\ -2 \end{bmatrix}, \quad f_*(Y) = \begin{bmatrix} 0 \\ 1 \\ 0 \end{bmatrix},$$

we get

$$XX \cdot x_0 = -X + 2\xi, \quad XY \cdot x_0 = Y, \quad YY \cdot x_0 = -\frac{1}{3}X - \frac{2}{3}\xi,$$

so that

$$\nabla_X X = -X, \quad \nabla_X Y = Y = \nabla_Y X, \quad \nabla_Y Y = -\frac{1}{3}X,$$

and

$$h(X, X) = 2, \quad h(X, Y) = 0, \quad h(Y, Y) = -\frac{2}{3}.$$

It follows that h is indefinite. We also have

$$R(X, Y)Y = \frac{2}{3}X, \quad R(X, Y)X = 2Y,$$

and

$$(\nabla_X R)(X, Y)Y = -\frac{4}{3}X, \quad (\nabla_Y R)(X, Y)Y = \frac{4}{3}Y,$$

$$(\nabla_X R)(X, Y)X = 4Y, \quad (\nabla_Y R)(X, Y)X = -\frac{4}{3}X.$$

Thus $\nabla R \neq 0$.

Example 2.3. *The surface $z^2(y - x^2)^3 = 1$.* (See Figure 9.)

We can parametrize the surface by $x = b$, $y = a^2 + b^2$, $z = a^{-3}$. We take the group G of all matrices of the form

$$\begin{bmatrix} a^{-3} & 0 & 0 & 0 \\ 0 & a^2 & 2ab & b^2 \\ 0 & 0 & a & b \\ 0 & 0 & 0 & 1 \end{bmatrix} \quad \text{acting on} \quad \begin{bmatrix} z \\ y \\ x \\ 1 \end{bmatrix}$$

and a basis $\{X, Y\}$ of its Lie algebra:

$$X = \begin{bmatrix} -3 & 0 & 0 & 0 \\ 0 & 2 & 0 & 0 \\ 0 & 0 & 1 & 0 \\ 0 & 0 & 0 & 0 \end{bmatrix} \quad \text{and} \quad Y = \begin{bmatrix} 0 & 0 & 0 & 0 \\ 0 & 0 & 2 & 0 \\ 0 & 0 & 0 & 1 \\ 0 & 0 & 0 & 0 \end{bmatrix}.$$

We take

$$x_0 = \begin{bmatrix} 1 \\ 1 \\ 0 \\ 1 \end{bmatrix} \quad \text{and} \quad \xi_e = \begin{bmatrix} 1 \\ 1 \\ 0 \\ 0 \end{bmatrix}.$$

We have then

$$X \cdot x_0 = \begin{bmatrix} -3 \\ 2 \\ 0 \\ 0 \end{bmatrix}, \quad Y \cdot x_0 = \begin{bmatrix} 0 \\ 0 \\ 1 \\ 0 \end{bmatrix},$$

$$XX \cdot x_0 = -f_*(X) + 6\xi, \quad XY \cdot x_0 = f_*(Y),$$

$$YX \cdot x_0 = 0, \quad YY \cdot x_0 = \frac{2}{5}f_*(X) + \frac{6}{5}\xi.$$

Hence we have

$$\nabla_X X = -X, \quad \nabla_X Y = Y, \quad \nabla_Y X = 0, \quad \nabla_Y Y = \frac{2}{5}X,$$

$$h(X, X) = 6, \quad h(X, Y) = 0, \quad h(Y, Y) = \frac{6}{5},$$

and

$$R(X, Y)Y = -\frac{6}{5}X, \quad R(X, Y)X = 0, \quad \nabla_Y R = 0, \quad \nabla_X R = -2R.$$

The affine metric is positive-definite. We note that the curvature tensor is recurrent. (A tensor field K is said to be *recurrent* if there is a 1-form σ such that $\nabla_X K = \sigma(X)K$ for each $X \in T_x(M)$.) As for the shape operator,

$$X \cdot \xi = \begin{bmatrix} -3 \\ 2 \\ 0 \\ 0 \end{bmatrix} \quad \text{and} \quad Y \cdot \xi = 0$$

lead to $SX = -X$ and $SY = 0$. Thus the surface is not an affine sphere. Because of homogeneity, the affine principal curvatures are -1 and 0 at every point. The affine Gauss–Kronecker curvature K is identically 0; the affine mean curvature is identically equal to $-\frac{1}{2}$.

Example 2.4. *The surface* $z^2(y - x^2)^3 = -1$. (See Figure 10.)

We can parametrize the surface by $x = b$, $y = -a^2 + b^2$, $z = a^{-3}$. We take the same group G as in the preceding example but different vectors

$$x_0 = \begin{bmatrix} 1 \\ -1 \\ 0 \\ 1 \end{bmatrix} \quad \text{and} \quad \xi = \begin{bmatrix} 1 \\ 1 \\ 0 \\ 0 \end{bmatrix}.$$

By a computation similar to that in Example 2.3 we can conclude that the affine metric is indefinite, that the shape operator has two constant eigenvalues, one of which is 0, and that the curvature tensor is recurrent.

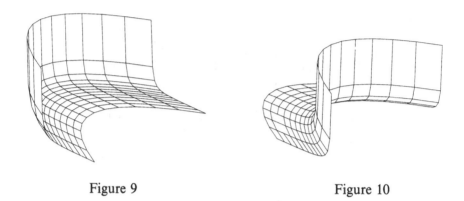

Figure 9   Figure 10

3. Classification of equiaffinely homogeneous surfaces

Having seen all the surfaces that will constitute the list of equiaffinely homogeneous surfaces, it is time to state the classification result.

Theorem 3.1. *Any nondegenerate surface in* $\mathbf{R}^3$ *that is homogeneous under equiaffine transformations is a quadric or is affinely congruent to one of the following surfaces:*

(I)	$xyz = 1,$
(II)	$(x^2 + y^2)z = 1,$
(III)	$x^2(z - y^2)^3 = 1,$
(IV)	$x^2(z - y^2)^3 = -1,$
(V)	$z = xy - \dfrac{1}{3}x^3,$
(VI)	$z = xy + \log x.$

The classification is up to affine congruence, and this means the following. If a nondegenerate surface M in $\mathbf{R}^3$ is equiaffinely homogeneous, then for any affine transformation ϕ of $\mathbf{R}^3$ onto itself the image $\phi(M^2)$ is again a nondegenerate equiaffinely homogeneous surface. We shall use a method different from that of Guggenheimer. Actually, we prove that we get the first four surfaces when the Pick invariant J is not 0. For the case $J = 0$ we get the last two surfaces – we refer to [NS2] for this part.

We now assume that $J \neq 0$. By changing the affine normal ξ to ξ/λ we may assume that the constant Pick invariant is equal to 2. Such ξ is uniquely determined.

We shall find local absolute invariants as follows. Let $\{X_1, X_2\}$ be a local basis of tangent vector fields satisfying

$$h(X_i, X_j) = \epsilon_i \delta_{ij}, \qquad \epsilon_1 = 1, \quad \epsilon_2 = \epsilon = \pm 1.$$

Put

$$C_{ijk} = C(X_i, X_j, X_k).$$

Lemma 3.2. *We may rechoose X_i as above in such a way that*

(3.1) $$C_{222} = 0, \quad C_{111} = -2, \quad C_{112} = 0, \quad C_{122} = 2\epsilon.$$

There are a finite number of possibilities for the choice of such a basis $\{X_1, X_2\}$.

Proof. If $\epsilon = 1$, then Proposition 11.1, Chapter II, implies the existence of $\{X_1, X_2\}$ such that $C_{222} = 0$. If $\epsilon = -1$, then arguments similar to Proposition 11.2, Chapter II, lead to $\{X_1, X_2\}$ with $C_{222} = 0$. From apolarity we get $C_{112} = 0$ and $C_{111} = -\epsilon C_{122}$. From $h(C,C) = C_{111}^2 + 3C_{122}^2 = 16$, we get $C_{111} = \pm 2$. If $C_{111} = 2$, we may perform an orthogonal transformation on $\{X_1, X_2\}$ relative to h (definite or indefinite) and rechoose $\{X_1, X_2\}$ so that $C_{111} = -2$, which then implies $C_{122} = 2\epsilon$. This argument also shows that there are a finite number of possibilities for choosing a desired basis $\{X_1, X_2\}$.

Let $\nabla_{X_i} X_j = \sum_k \Gamma_{ij}^k X_k$; then the identities (3.1) show that we can put

$$\Gamma_{11}^2 = a, \qquad \Gamma_{22}^1 = \epsilon(b-1),$$

and that

(3.2) $$\begin{aligned} \nabla_{X_1} X_1 &= X_1 + aX_2, & \nabla_{X_1} X_2 &= -\epsilon a X_1 - X_2, \\ \nabla_{X_2} X_1 &= -(b+1)X_2, & \nabla_{X_2} X_2 &= \epsilon(b-1)X_1. \end{aligned}$$

Hence

(3.3) $$[X_1, X_2] = -\epsilon a X_1 + bX_2.$$

We also express the shape operator by

$$S = \begin{pmatrix} p & \epsilon q \\ q & r \end{pmatrix}.$$

Assume here that the surface is equiaffinely homogeneous. Then the scalars a, b, p, q, and r are all constant. This may be seen by using a section from a neighborhood of a point of the surface into the group of equiaffine transformations and by observing that, because of the finiteness in the choice of $\{X_1, X_2\}$, the adapted frame field must be locally invariant under the local transformations that keep the surface invariant, and by finally noting that these transformations also leave the connection ∇ and the shape operator S invariant. Now the Codazzi equation for S implies

$$2a\epsilon q + (b+1)(p-r) = 0,$$
$$- a(p-r) + 2q(b-1) = 0.$$

The Gauss equation gives

$$3a + q = 0,$$
$$r + \epsilon a^2 + (b+1)(b+2) = 0,$$
$$p + \epsilon a^2 + (b-1)(b-2) = 0.$$

These two sets of equations show that

$$a = 0 \quad \text{and} \quad b(b+1) = 0$$

or

$$b = \frac{1}{2} \quad \text{and} \quad a^2\epsilon = \frac{3}{4}.$$

That is, we have three cases to consider:

A : $a = b = 0$ and $S = \begin{pmatrix} -2 & 0 \\ 0 & -2 \end{pmatrix}$;

B : $a = 0, \quad b = -1$ and $S = \begin{pmatrix} -6 & 0 \\ 0 & 0 \end{pmatrix}$;

C : $a^2 = \dfrac{3}{4}, \quad b = \dfrac{1}{2}, \quad \epsilon = 1.$

Note here that Case C can be reduced to Case B by a rotation of the tangent space by the angle $\frac{2}{3}\pi$.

Integration of Case A.

In Case A, we have

$$\begin{array}{ll} D_{X_1}X_1 = X_1 + \xi, & D_{X_1}X_2 = -X_2, \\ D_{X_2}X_1 = -X_2, & D_{X_2}X_2 = -\epsilon X_1 + \epsilon\xi, \\ D_{X_1}\xi = 2X_1, & D_{X_2}\xi = 2X_2. \end{array}$$

Since $[X_1, X_2] = 0$, we can put locally

$$X_1 = \frac{\partial}{\partial u} \quad \text{and} \quad X_2 = \frac{\partial}{\partial v}.$$

Relative to these coordinates, the map f satisfies

$$f_{uu} = f_u + \xi, \qquad f_{uv} = -f_v, \qquad \epsilon f_{vv} = -f_u + \xi,$$
$$\xi_u = 2f_u, \qquad \xi_v = 2f_v.$$

From the last two equations, $\xi = 2f + c$ for some constant vector c. Putting $w = f + \frac{1}{2}c$, we see

$$w_{uu} = w_u + 2w, \qquad w_{uv} = -w_v, \qquad w_{vv} = -\epsilon w_u + 2\epsilon w.$$

Let $w = e^{ku}$. It is a solution when $k = 2$. Similarly, $w = e^{ku+\ell v}$ is a solution if $k^2 = k + 2$, $k\ell = -\ell$, and $\ell^2 = -\epsilon k + 2\epsilon$; that is, when $k = -1$ and $\ell^2 = 3\epsilon$. Hence the map f is given as follows relative to some affine coordinates:

$$f = (e^{-u+\sqrt{3}v}, \quad e^{-u-\sqrt{3}v}, \quad e^{2u}) \qquad (\epsilon = 1)$$

or

$$f = (e^{-u}\cos\sqrt{3}v, \quad e^{-u}\sin\sqrt{3}v, \quad e^{2u}) \qquad (\epsilon = -1).$$

The corresponding equations of the immersed surfaces are

(I) $$xyz = 1$$

and

(II) $$(x^2 + y^2)z = 1.$$

Integration of Case B.

In this case, we have

$$D_{X_1}X_1 = X_1 + \xi, \qquad D_{X_1}X_2 = -X_2,$$
$$D_{X_2}X_1 = 0, \qquad D_{X_2}X_2 = -2\epsilon X_1 + \epsilon\xi,$$
$$D_{X_1}\xi = 6X_1, \qquad D_{X_2}\xi = 0.$$

Since $[X_1, X_2] = -X_2$, we can put

$$X_1 = \frac{\partial}{\partial u} \qquad \text{and} \qquad X_2 = e^{-u}\frac{\partial}{\partial v}.$$

Relative to these coordinates, the map f satisfies

$$f_{uu} = f_u + \xi, \qquad f_{uv} = 0, \qquad \epsilon e^{-2u}f_{vv} = -2f_u + \xi,$$
$$\xi_u = 6f_u, \qquad \xi_v = 0.$$

So we can put $f = \phi(u) + \psi(v)$. From the fourth equation, $\xi = 6f + \chi(v)$. The last equation implies $6\psi + \chi = \text{const}$. By rechoosing ϕ we get $\xi = 6\phi$. Then the third equation becomes $\psi_{vv} = e^{2u}(-2\epsilon\phi_u + 6\epsilon\phi)$. Hence both sides must be constant. Put $\psi_{vv} = A$. Then $\psi = Av^2 + Bv + C$ and $\phi = \frac{1}{5}\epsilon A e^{-2u} + De^{3u}$. That is, we have

$$f = A(v^2 + \frac{1}{5}\epsilon e^{-2u}) + Bv + De^{3u} + C.$$

Hence, for certain affine coordinates (x, y, z), we have

$$x = 5^{\frac{3}{2}}e^{3u}, \quad y = v, \quad z = v^2 + \frac{1}{5}\epsilon e^{-2u}.$$

The immersed surface is written by an equation

$$z = y^2 + \epsilon x^{-\frac{2}{3}}.$$

Or, separately depending on ϵ, we have

(III) $x^2(z - y^2)^3 = 1$ $(\epsilon = 1)$

and

(IV) $x^2(z - y^2)^3 = -1$ $(\epsilon = -1)$.

4. SL(n, **R**) and SL(n, **R**)/SO(n)

In this section, we first deal with the special linear group $SL(n, \mathbf{R})$ as a centro-affine hypersurface in an affine space. We then discuss the standard affine immersion of the homogeneous space $SL(n, \mathbf{R})/SO(n)$ into an affine space.

We consider the set $\mathfrak{gl}(n, \mathbf{R})$ of all $n \times n$ real matrices as a vector space (and thus as an affine space) of dimension n^2. Each matrix $x = [x_{ij}]$ corresponds to the vector in $\mathbf{R}^{n^2}$ with components

$$(x_{11}, x_{21}, \ldots, x_{n1}; \ldots; x_{1n}, x_{2n}, \ldots, x_{nn}),$$

which sets up a natural isomorphism of $\mathfrak{gl}(n, \mathbf{R})$ to $\mathbf{R}^{n^2}$. We may regard

$$SL(n, \mathbf{R}) = \{a \in \mathfrak{gl}(n, \mathbf{R}) : \det a = 1\}$$

as an $(n^2 - 1)$-dimensional hypersurface imbedded in $\mathfrak{gl}(n, \mathbf{R})$. We may also consider $SL(n, \mathbf{R})$ as a group of transformations acting on $\mathfrak{gl}(n, \mathbf{R})$ on the left, namely,

(4.1) $a \in SL(n, \mathbf{R}) : x \mapsto ax$ for $x \in \mathfrak{gl}(n, \mathbf{R})$.

We denote by ω the volume element in $\mathfrak{gl}(n, \mathbf{R})$ that corresponds to the standard determinant function det in $\mathbf{R}^{n^2}$. We want to show that the action of $SL(n, \mathbf{R})$ on $\mathfrak{gl}(n, \mathbf{R})$ preserves this volume element. This follows from

Lemma 4.1. *Denoting by l_a the action $x \mapsto ax$ in $\mathfrak{gl}(n, \mathbf{R})$ we have:*

(1) *Relative to the basis $\{e_{1,1}, e_{2,1}, \ldots, e_{n,1}; \ldots; e_{1,n}, e_{2,n}, \ldots, e_{n,n}\}$ in $\mathfrak{gl}(n, \mathbf{R})$, where $e_{i,j}$ denotes the $n \times n$ matrix with all components equal to 0 except the (i, j) component equal to 1, l_a can be expressed by the $n^2 \times n^2$ matrix with n blocks of a on the diagonal with all other components equal to 0;*

(2) $\det l_a = (\det a)^n$; *thus $\det l_a = 1$ for $a \in SL(n, \mathbf{R})$.*

Proof. Since $e_{i,j} e_{p,q} = \delta_{j,p} e_{i,q}$, we have

$$ae_{i,j} = \sum_{p=1}^{n} a_{pi} e_{p,j},$$

where a_{ij} denotes the (i, j)-component of a, which implies (1) and thus (2).

The hypersurface $SL(n, \mathbf{R})$ can be regarded as the orbit of the identity matrix e under the group $G = SL(n, \mathbf{R})$ acting on $\mathfrak{gl}(n, \mathbf{R})$. At e, the tangent space $T_e(G)$ is equal to

$$\mathfrak{sl}(n, \mathbf{R}) = \{a \in \mathfrak{gl}(n, \mathbf{R}) : \operatorname{tr} a = 0\},$$

which can be identified also with the Lie algebra of all left-invariant vector fields on $SL(n, \mathbf{R})$. We observe that the position vector e (from the origin $=$ the zero matrix) is transversal to $T_e(G)$, as $\operatorname{tr} e = n$. Since l_a maps the hypersurface $SL(n, \mathbf{R})$ onto itself, we may regard $SL(n, \mathbf{R})$ as a centro-affine hypersurface: at each point x we take the transversal vector ξ to be the opposite of the position vector x. For any vector fields X and Y on $SL(n, \mathbf{R})$ we may write

(4.2) $$D_X Y = \nabla_X Y - h(X, Y)x.$$

Lemma 4.2. *If X and Y are left-invariant vector fields on $SL(n, \mathbf{R})$, then for any $a \in SL(n, \mathbf{R})$ we have*

(1) $(D_X Y)_a = (l_a)_*(D_X Y)_e$; $(D_X Y)_e = XY$;

(2) $(\nabla_X Y)_a = (l_a)_*(\nabla_X Y)_e$;

that is, $\nabla_X Y$ is a left-invariant vector field on $SL(n, \mathbf{R})$;

(3) $[h(X, Y)]_a = [h(X, Y)]_e$,

that is, $h(X, Y)$ is constant on $SL(n, \mathbf{R})$.

Proof. (1) Let $a_t = \exp tX$ be the 1-parameter subgroup of $SL(n, \mathbf{R})$ generated by X so that $(da_t/dt)_{t=0} = X_e$. Then aa_t is a curve starting at a where the tangent vector is equal to $a(da/dt)_{t=0} = aX_e = X_a$. Along this curve aa_t, Y is given by $l_{aa_t} Y_e$. By definition of $D_X Y$ we have

$$(D_X Y)_a = \left(\frac{d}{dt}\right)_{t=0} [aa_t, Y_e] = aX_e Y_e.$$

This equation together with the equation at $a = e$ leads to (1).

(2) and (3) follow from (1) and the invariance of the transversal field under $SL(n, \mathbf{R})$.

From (2) in Lemma 4.2 we see that the induced connection ∇ is a left-invariant affine connection on $SL(n, R)$. As such, it is expressed algebraically in terms of the Lie algebra $\mathfrak{g} = \mathfrak{sl}(n, \mathbf{R})$ by a bilinear mapping

$$\nabla : \mathfrak{g} \times \mathfrak{g} \to \mathfrak{g}.$$

In view of (1) for $a = e$ in Lemma 4.2, we know that $\nabla_X Y$ is the trace zero part of the matrix product XY, namely,

$$(4.3) \qquad \nabla_X Y = XY - \frac{1}{n} \operatorname{tr}(XY)I.$$

From (3) in Lemma 4.2 we see that the affine normal field coincides with the centro-affine transversal field up to a scalar. Thus the connection ∇ coincides with the Blaschke connection on $SL(n, \mathbf{R})$. It also follows that h is expressed in terms of $\mathfrak{g}$ by a bilinear function

$$(4.4) \qquad h(X, Y) = -\frac{1}{n} \operatorname{tr}(XY).$$

We note that h coincides, up to a scalar, with the Killing form of the Lie algebra $\mathfrak{sl}(n, \mathbf{R})$. It is easily seen that h is nondegenerate. Since ∇ is induced by a centro-affine transversal field, it is projectively flat. Since $S = I$, the Gauss equation gives

$$R(X, Y)Z = -\frac{1}{n} \operatorname{tr}(YZ)X + \frac{1}{n} \operatorname{tr}(XZ)Y$$

and

$$\operatorname{Ric}(Y, Z) = \frac{2 - n^2}{n} \operatorname{tr}(YZ).$$

We note that ∇ is also right-invariant, as we can see by verifying that ∇ in (4.3) is invariant under $\operatorname{Ad}(a)$, $a \in SL(n, \mathbf{R})$.

To sum up, we have

Theorem 4.3. *For the natural imbedding of $SL(n, \mathbf{R})$ in $\mathfrak{gl}(n, \mathbf{R})$ with centro-affine transversal vector field, the induced connection coincides with the*

Blaschke connection. It is given by (4.3) and is bi-invariant and projectively flat.

The case $n = 2$ deserves special mention. In this case, (4.3) and (4.4) give

$$(4.5) \qquad \nabla_X Y = \frac{1}{2}[X, Y],$$

$$(4.6) \qquad h(X, Y) = -\frac{1}{2}\operatorname{tr}(XY).$$

We find that ∇ is the so-called (0)-connection on $SL(2, \mathbf{R})$ (see [KN2, p. 199]). The (0)-connection on any Lie group G is the bi-invariant torsion-free connection defined by (4.5) and is known to be locally symmetric, that is, $\nabla R = 0$. In our case of $SL(2, \mathbf{R})$, it coincides with the Levi-Civita connection of the metric h. By Theorem 4.5 of Chapter II, we know that $SL(2, \mathbf{R})$ lies in $\mathfrak{gl}(2, \mathbf{R})$ as a quadratic hypersurface. Indeed, the defining equation $\det x = 1$ for $SL(2, R)$ is a quadratic equation on $\mathfrak{gl}(n, \mathbf{R})$. We may also verify that ∇ is complete.

Remark 4.1. $SL(2, \mathbf{R})$ is a 3-dimensional Blaschke hypersurface in $\mathbf{R}^4$ with locally symmetric connection. It happens to be a hyperquadric. For the problem of deciding which hypersurfaces have locally symmetric induced connections, see Section 8 of this chapter (case of compact surfaces), Section 4 of Chapter IV (hypersurfaces of dimension ≥ 3), and Note 8 (surfaces).

Remark 4.2. Theorem 4.3 suggests the problem of determining which Lie groups admit bi-invariant, projectively flat affine connections. See [NP1] for this and other related problems.

Our next aim is to discuss a natural imbedding of $SL(n, \mathbf{R})/SO(n)$ into an affine space. In order to have a clear picture of the general situation we consider a homogeneous space $M = G/H$ of a connected Lie group G over a closed subgroup H. Let $n = \dim(G/H)$ and let $f : G/H \to \mathbf{R}^{n+1}$ be an immersion of M into an affine space $\mathbf{R}^{n+1}$. Further, assume that there is a representation $\phi : G \to \mathbf{SA}(n + 1)$, where $\mathbf{SA}(n + 1)$ is the group of unimodular affine transformations of $\mathbf{R}^{n+1}$, such that

$$(4.7) \qquad f(ax) = \phi(a)f(x) \quad \text{for all} \quad a \in G, \ x \in M,$$

where ax denotes the natural action of $a \in G$ on M. We say that (f, ϕ) is an *equivariant immersion* of (M, G) into $(\mathbf{R}^{n+1}, \mathbf{SA}(n + 1))$. If we write $x_0 = \pi(e)$, where $\pi : G \to G/H$ is the natural projection, then $f(ax_0) = \phi(a)f(x_0)$; thus $f(M)$ is the orbit of $p_0 = f(x_0)$ by $\phi(G)$.

Furthermore, let ξ be an equiaffine transversal vector field for the immersion f that is equivariant, namely,

$$(4.8) \qquad \xi_{ax} = \phi(a)\xi_x \quad \text{for all} \quad a \in G, \ x \in M.$$

Lemma 4.4.

(1) *If ξ is an equivariant transversal vector field, then $\xi_0 = \xi|_e$ is a vector at $p_0 = f(x_0)$ transversal to $f_*(T_{x_0}(M))$ and invariant under $\phi(H)$. Conversely, given any vector ξ_0 transversal to $f_*(T_{x_0}(M))$ and invariant under $\phi(H)$, we can get an equivariant transversal vector field to f.*

(2) *In the correspondence $\xi \rightleftharpoons \xi_0$, ξ is equiaffine if and only if ξ_0 satisfies the following condition:*

(*) *For any X in the Lie algebra $\mathfrak{g}$ of G, $\phi_*(X_e) \cdot \xi_0$ is tangent to $f(M)$, that is, it lies in $f_*(T_{x_0}(M))$.*

Proof. (1) The first assertion is obvious. For the converse, given ξ_0 invariant under $\phi(H)$, which fixes x_0, we may define ξ_x to be $\phi(a)\xi_0$ by taking $a \in G$ such that $a(x_0) = x$ and by noting that the result is independent of the choice of a.

(2) Assume that ξ is equiaffine. For any $X_0 \in T_{x_0}(M)$ we take $X \in \mathfrak{g}$ such that $\pi_*(X) = X_0$. Let $a_s = \exp sX$ be the 1-parameter subgroup of G generated by X and set $x_s = \pi(a_s)$. Since ξ is equivariant, we have $\xi_{x_s} = \phi(a_s)\xi_0$. Thus we get

$$(4.9) \qquad D_{X_0}\xi = \left(\frac{d}{ds}\right)_{s=0}[\phi(a_s)\xi_0] = \left[\phi_*\left(\frac{da_s}{ds}\right)_{s=0}\right]\xi_0 = \phi_*(X)\xi_0,$$

showing that ξ is equiaffine at x_0 if and only if (*) is satisfied. Now observe that ξ, being equivariant, is equiaffine if it is so at x_0.

Suppose we take an equivariant transversal vector field ξ for an equivariant immersion f and consider the induced structure (∇, h, S, θ).

Proposition 4.5. *The induced connection ∇ on $M = G/H$ is an invariant affine connection. The affine fundamental form h, the induced volume element θ, and the shape operator S are invariant under G.*

Proof. If X and Y are vector fields on M, then for any $a \in G$ we have

$$(4.10) \qquad \begin{aligned} D_{aX}f_*(aY) &= D_{\phi(a)f_*(X)}\phi(a)f_*(Y) \\ &= D_{f_*X}f_*Y = D_X f_* Y. \end{aligned}$$

Since the decomposition $T_{f(x)}(\mathbf{R}^{n+1}) = f_*(T_x(M)) + \mathbf{R}\{\xi_x\}$ is equivariant, it follows from (4.10) that

$$\nabla_{aX}aY = a\nabla_X Y,$$

that is, ∇ is an invariant connection on G/H. It also follows that h is G-invariant. The invariance of θ and S is easy to show.

We shall now apply the framework above to the case where $G/H = SL(n, \mathbf{R})/SO(n)$ and the immersion f defined as follows.

We denote by $\mathfrak{s}(n)$ the vector space of all real symmetric matrices of degree n, which is identified with $\mathbf{R}^m$ with $m = \frac{1}{2}n(n+1)$. The mapping

$$(4.11) \qquad f : SL(n, \mathbf{R}) \to \mathfrak{s}(n) \quad \text{by} \quad f(a) = a^t a$$

induces an imbedding

$$f : SL(n, \mathbf{R})/SO(n) \to \mathfrak{s}(n).$$

We have also a representation ϕ of $SL(n, \mathbf{R})$ on $\mathfrak{s}(n)$ by

(4.12) $$\phi(a)X = aX^t a \quad \text{for} \quad X \in \mathfrak{s}(n).$$

The mapping f is equivariant in the sense that

(4.13) $$f(ab) = \phi(a)f(b) \quad \text{for} \quad a, b \in SL(n, \mathbf{R}).$$

It is easy to verify that the image of f coincides with the set $\mathfrak{p}(n)$ of all positive-definite matrices with determinant 1 in $\mathfrak{s}(n)$, which is the image $\exp(\mathfrak{s}_0(n))$ where $\mathfrak{s}_0(n) = \{X \in \mathfrak{s}(n) : \operatorname{tr} X = 0\}$.

It is known that $SL(n, \mathbf{R})/SO(n)$ is a symmetric homogeneous space where the involutive automorphism of $SL(n, \mathbf{R})$ is $a \mapsto {}^t a^{-1}$. See [KN2, Chapter XI].

Now we have

Lemma 4.6. $\det \phi(a) = 1$ *for each* $a \in SL(n, \mathbf{R})$.

Proof. We use an inner product in $\mathfrak{s}(n)$ defined by

$$\langle X, Y \rangle = \operatorname{tr}(XY),$$

which is positive-definite. For each $a \in SO(n)$ we see that $\phi(a)$ is an orthogonal transformation of $\mathfrak{s}(n)$ with determinant 1. Now each $a \in SL(n, \mathbf{R})$ can be expressed in the form

$$a = bc, \quad \text{where} \quad b \in SO(n), \ c \in \mathfrak{p}(n).$$

Since c is of the form $c = kdk^{-1}$, where $k \in SO(n)$ and d is a diagonal matrix of determinant 1. We can check that $\det \phi(d) = 1$. It follows that $\det \phi(c) = 1$ and hence $\det \phi(a) = 1$.

Remark 4.3. Lemma 4.6 is immediate if one uses the fact that $SL(n, \mathbf{R})$ coincides with its commutator subgroup.

In view of Lemma 4.6, we may find a volume element in the affine space $\mathbf{R}^m$ that is invariant under $\phi(SL(n, \mathbf{R}))$. (We may show that such a volume element is unique up to a scalar.) In this way we have $\phi : SL(n, \mathbf{R}) \to \mathbf{SA}(m)$. We now consider $f : SL(n, \mathbf{R})/SO(n) \to \mathbf{R}^m = \mathfrak{s}(n)$ with $\xi_x = f(x)$ for each $x \in SL(n, \mathbf{R})/SO(n)$. This transversal vector field is certainly equiaffine and equivariant. We denote by (∇, h, θ, S) the induced structure. It is easy to see that ∇, h, S, and θ are all invariant under $SL(n, \mathbf{R})$ acting on $SL(n, \mathbf{R})/SO(n)$. Thus $\det_\theta h$ is constant and the volume element for h coincides with a constant multiple of θ. In this sense, we may say that f is essentially a Blaschke imbedding as centro-affine hypersurface.

Remark 4.4. If ξ_0 is a transversal vector at $x_0 = f(e) = I_n$ that is invariant under $\phi(SO(n))$ (see Lemma 4.4, (1)), then it is a scalar multiple of I. Thus a choice of transversal vector field that makes f a Blaschke imbedding is essentially unique.

We consider the decomposition of the Lie algebra $\mathfrak{sl}(n, \mathbf{R})$,

(4.14) $$\mathfrak{sl}(n, \mathbf{R}) = \mathfrak{s}_0 \oplus \mathfrak{o}(n),$$

where $\mathfrak{o}(n)$ is the Lie algebra of all skew-symmetric matrices of degree n, and $\mathfrak{s}_0 = \{X \in \mathfrak{s}(n) : \operatorname{tr} X = 0\}$ is invariant under $\operatorname{Ad}(SO(n))$ acting on $\mathfrak{sl}(n, \mathbf{R})$ by adjoint representation. In fact, (4.14) gives the standard decomposition of the Lie algebra according to the involutive automorphism $\sigma_*(X) = -{}^t X$ for the symmetric homogeneous space $M = SL(n, \mathbf{R})/SO(n)$. We can use (4.14) to represent any invariant structure on the space M (see [KN2, Chapter XI]). In particular, $\mathfrak{s}_0$ represents $T_{x_0}(M)$. Suppose $X \in \mathfrak{s}_0$. To find $f_*(X)$, let $a_s = \exp sX$ and $x_s = \pi(a_s)$. Then $f(x_s) = a_s{}^t a_s$ and hence

(4.15) $$f_*(X) = \left(\frac{d}{ds}\right)_{s=0} [a_s{}^t a_s] = 2X.$$

We see that $f_*(T_{x_0}(M))$ consists of symmetric matrices with trace 0.

We have

Proposition 4.7. *The Blaschke structure of the imbedding* $f : M = SL(n, \mathbf{R})/SO(n) \to \mathbf{R}^m = \mathfrak{s}(n)$ *with* $\xi = 4f$ *can be expressed algebraically in terms of the Lie algebra, as follows:*

(4.16) $$\nabla_X Y = XY + YX - \frac{2}{n}\operatorname{tr}(XY)I_n$$

(4.17) $$h(X, Y) = \frac{4}{n}\operatorname{tr} XY, \qquad S = -I.$$

Proof. In order to derive (4.16) and (4.17) we proceed as follows. Let $X \in \mathfrak{s}_0$, $a_s = \exp sX$, and $x_s = \pi(a_s)$. Then for the tangent vector $\dot{x}_s$ to the curve x_s we have

$$f_*(\dot{x}_s) = f_*(a_s X) = \phi(a_s) f_*(X) = 2a_s X a_s$$

by using (4.15). Thus we get

$$\left(\frac{d}{ds}\right)_{s=0} f_*(\dot{x}_s) = 2\left(\frac{d}{ds}\right)_{s=0} a_s X a_s = 4X^2.$$

Since

$$\left(\frac{d}{ds}\right)_{s=0} f_*(\dot{x}_s) = f_*(\nabla_X X) + h(X, X)e,$$

we get again using (4.15)

$$4X^2 = 2\nabla_X X + h(X, X)e.$$

Recalling that $e = I_n$, we get $4 \operatorname{tr} X^2 = nh(X, X)$, from which we have

$$(4.18) \qquad \nabla_X X = 2X^2 - \frac{2}{n} \operatorname{tr} X^2 I_n$$

and

$$(4.19) \qquad h(X, X) = \frac{4}{n} \operatorname{tr} X^2.$$

Polarization gives (4.16) and (4.17) by virtue of $\nabla_X Y = \nabla_Y X$, which follows from the fact that ∇ has torsion 0.

Remark 4.5. (4.19) shows that h is positive-definite. This proper affine hypersphere $SL(n, \mathbf{R})/SO(n)$ is one of the examples of homogeneous hyperbolic affine hyperspheres mentioned in [S1]. See also Note 7.

Remark 4.6. The Levi-Civita connection for h coincides with the canonical invariant connection ∇^0 on the symmetric homogeneous space given by $\nabla^0_X Y = 0$ for $X, Y \in \mathfrak{s}_0$. The Blaschke connection ∇ does not coincide with ∇^0 except for $n = 2$. For $n = 2$, the image $f(SL(2, \mathbf{R})/SO(2))$ is the hyperbolic space of dimension 2 imbedded in $\mathbf{R}^3$ as one component of the two-sheeted hyperboloid. See [IT, vol. II, p. 179].

5. Affine spheres with affine metric of constant curvature

In the preceding sections we have encountered several classes of affine surfaces such as affine spheres, ruled surfaces, and equiaffinely homogeneous surfaces. Here we shall treat the local classification of affine spheres whose affine metrics have constant curvature. As a by-product, we determine all affine spheres, proper or improper, that are ruled by proving the converses of Theorems 1.1 and 1.2. Of course, we consider only Blaschke surfaces in this section.

We begin with the case where the curvature of the affine metric is 0. As affine spheres with flat affine metric, we know already elliptic paraboloids (Example 3.1 of Chapter II), two homogeneous surfaces (I) and (II) of Theorem 3.1, and ruled improper affine spheres in Section 1. The next theorem states that this list exhausts such surfaces:

Theorem 5.1. *Let $f : M^2 \to \mathbf{R}^3$ be an affine sphere with Blaschke structure. If the affine metric h is flat, then the surface is locally affinely congruent to one of the following surfaces:*

(1) $\qquad z = x^2 + y^2,$

(2) $\qquad xyz = 1,$

(3) $\qquad (x^2 + y^2)z = 1,$

(4) $\qquad z = xy + \Phi(x),$

where Φ *is an arbitrary function of* x.

Because of the formula (9.4) of Chapter II, $\hat{\rho} = H + J$, and because H is constant, the Pick invariant J is also constant. Hence, we can prove the theorem by considering separately the case $J \neq 0$ and the case $J = 0$.

The case $J \neq 0$ proceeds as follows. Without loss of generality we may assume $J = 2$ by multiplying the affine normal by a suitable constant. First, choose a local basis of tangent vector fields $\{X_1, X_2\}$ satisfying

$$h(X_i, X_j) = \epsilon_i \delta_{ij}, \qquad \epsilon_1 = 1, \epsilon_2 = \epsilon = \pm 1,$$

and put $C_{ijk} = C(X_i, X_j, X_k)$. By virtue of Lemma 3.2 we may assume that the affine connection ∇ has an expression given in (3.2); the relation (3.3) also remains true. Furthermore, since $\hat{\rho} = 0$ implies $H = -J = -2$, the affine shape operator S is equal to $-2I$. This gives

$$R(X_1, X_2)X_1 = 2X_2 \quad \text{and} \quad R(X_1, X_2)X_2 = -2\epsilon X_1.$$

On the other hand, we have by definition

$$
\begin{aligned}
R(X_1, X_2)X_1 &= \nabla_{X_1}\nabla_{X_2}X_1 - \nabla_{X_2}\nabla_{X_1}X_1 - \nabla_{[X_1, X_2]}X_1 \\
&= \nabla_{X_1}\{-(b+1)X_2\} - \nabla_{X_2}\{X_1 + aX_2\} - \nabla_{\{-\epsilon aX_1 + bX_2\}}X_1 \\
&= 3\epsilon aX_1 + \{\epsilon a^2 + (b+1)(b+2) - X_1(b) - X_2(a)\}X_2.
\end{aligned}
$$

Hence we get

(5.1) $a = 0$ and $(b+1)(b+2) - X_1(b) = 2.$

Taking the first condition into consideration, we have

$$
\begin{aligned}
R(X_1, X_2)X_2 &= \nabla_{X_1}(\epsilon(b-1)X_1) - \nabla_{X_2}(-X_2) - \nabla_{bX_2}X_2 \\
&= \epsilon(X_1(b) - (b-1)(b-2))X_1;
\end{aligned}
$$

hence

(5.2) $(b-1)(b-2) - X_1(b) = 2.$

From (5.1) and (5.2), we easily see $b = 0$. Therefore, the connection ∇ and the shape operator reduce to Case A considered in Section 3. Then the surface is (2) or (3) depending on whether the affine metric is definite or indefinite.

We next treat the case $J = 0$. In this case, $H = 0$. When the affine metric is definite, the cubic form must vanish and the surface is quadratic. The property $H = 0$ implies that the surface is affinely congruent to the surface (1).

It remains to consider the case when $J = 0$ and the affine metric is indefinite. With later applications in mind, we choose an asymptotic coordinate system, namely, a coordinate system $\{x^1, x^2\}$ such that the associated vector fields $X_i = \partial/\partial x^i$, $i = 1, 2$, satisfy

$$h(X_1, X_1) = h(X_2, X_2) = 0, \qquad h(X_1, X_2) = F.$$

The coefficients $C_{ijk} = C(X_i, X_j, X_k)$ satisfy the apolarity condition

$$C_{112} = C_{122} = 0$$

and the Pick invariant is

$$J = \frac{1}{4F^3} C_{111} C_{222}.$$

We need the following lemma.

Lemma 5.2. *Let $f : M^2 \to \mathbf{R}^3$ be an improper affine sphere with indefinite affine metric and $J = 0$. Then any point of M has a neighborhood on which, relative to a coordinate system chosen as above, C_{111} vanishes identically or C_{222} vanishes identically.*

We postpone the proof for a moment and continue the argument. Since the affine metric is flat, we can put $F = 1$ and $C_{222} = 0$ without loss of generality. Let $2\phi = -C_{111}$ for simplicity. Then we have

$$h(\nabla_{X_1} X_1, X_1) = \phi, \qquad h(\nabla_{X_2} X_2, X_2) = 0.$$

From $C_{211} = 0$, we get

$$h(\nabla_{X_2} X_1, X_1) = 0 = h(\nabla_{X_1} X_1, X_2) + h(X_1, \nabla_{X_1} X_2).$$

Similarly, from $C_{122} = 0$ we get

$$h(\nabla_{X_1} X_2, X_2) = 0 = h(\nabla_{X_2} X_1, X_2) + h(X_1, \nabla_{X_2} X_2).$$

In view of $\nabla_{X_1} X_2 = \nabla_{X_2} X_1$, we get

$$\nabla_{X_1} X_2 = \nabla_{X_2} X_1 = \nabla_{X_2} X_2 = 0, \quad \nabla_{X_1} X_1 = \phi X_2.$$

This implies that the immersion f satisfies the system of differential equations

$$f_{uu} = \phi f_v, \quad f_{uv} = \xi, \quad f_{vv} = 0,$$

where we have put $x^1 = u$ and $x^2 = v$. Note that the affine normal ξ is a constant vector. The last two equations imply that f has the form

$$f = uv\xi + Bv + A(u)$$

for a constant vector B. Then the first equation implies that $A'' = \phi(u\xi + B)$. Integration gives

$$f = (uv + \psi)\xi + (v + \varphi)B + Cu + D$$

for certain constant vectors C and D and a function ψ. Since f is an immersion, the vectors ξ, B, and C are linearly independent. Then, choosing C, B, and ξ as the standard affine basis vectors, we get the expression

$$z = xy + \Phi(x)$$

where $\Phi = \psi - u\varphi$, that is, the surface (4) in the theorem. This completes the proof of Theorem 5.1.

Proof of Lemma 5.2. Put $V_1 = \{p \in M : C_{111}(p) = 0\}$ and $V_2 = \{p \in M : C_{222}(p) = 0\}$. It is enough to show that any boundary point of V_1 is an interior point of V_2. Denote the components of the differential $\hat{\nabla}C$ by $C_{ijk,\ell} = (\hat{\nabla}_{X_\ell}C)(X_i, X_j, X_k)$. Since M is an affine sphere, we have $C_{ijk,\ell} = C_{ij\ell,k}$ by Proposition 9.7 of Chapter II. Because the metric connection $\hat{\nabla}$ is given by

$$\hat{\nabla}_{X_1}X_1 = \frac{\partial \log F}{\partial x^1}X_1, \quad \hat{\nabla}_{X_1}X_2 = \hat{\nabla}_{X_2}X_1 = 0, \quad \hat{\nabla}_{X_2}X_2 = \frac{\partial \log F}{\partial x^2}X_2,$$

we see

$$\begin{aligned} C_{111,2} &= C_{112,1} \\ &= X_1(C(X_1, X_1, X_2)) - 2C(\hat{\nabla}_{X_1}X_1, X_1, X_2) - C(X_1, X_1, \hat{\nabla}_{X_1}X_2) \\ &= 0 \end{aligned}$$

and, similarly, $C_{222,1} = 0$. It follows that C_{111} (resp. C_{222}) remains constant on any x^2-curve (resp. x^1-curve) within the asymptotic coordinate system, say for $|x^i| < \delta$. Now let $p = (p^1, p^2)$ be a boundary point of V_1. Then there is a point $q = (q^1, q^2)$ sufficiently near p where $C_{111} \neq 0$. By the property $C_{111,2} = 0$, we see that $C_{111} \neq 0$ at every point with coordinates (q^1, x^2). Hence $C_{222} = 0$ at such points. By the property $C_{222,1} = 0$ we get $C_{222} = 0$ at all points with coordinates (x^1, x^2), namely, in a neighborhood of p; that is, p is an interior point of V_2.

We have

Theorem 5.3. *If $f : M^2 \to \mathbf{R}^3$ is a ruled improper affine sphere, then it is locally of the form* (4) *in Theorem 5.1.*

Proof. By Theorem 11.3 of Chapter II we see that h is indefinite and $J = 0$. From (9.4) of Chapter II, we get $\hat{\rho} = 0$. From Theorem 5.1 (or from its proof) we conclude that M^2 is of the form (4).

We next turn to the classification of affine spheres for which the curvature $\hat{\rho}$ is a nonzero constant. Because H is assumed to be constant, the Pick invariant J is also constant. If $J \neq 0$, then by (10.7) of Chapter II we get $\hat{\rho} = 0$. Since $\hat{\rho} \neq 0$ by assumption, we must have $J = 0$. In the case where the metric is definite, the cubic form vanishes and we see that the surface is affinely congruent to an ellipsoid or to a two-sheeted hyperboloid. Assume from now that the metric is indefinite. Then, with the notation preceding Lemma 5.2, we can assume $C_{222} = 0$ (on a neighborhood of a fixed point). Put $\alpha = \log F$. The metric curvature is $\hat{\rho} = -X_1 X_2 \alpha / F$. It is not hard to see the affine connection is given by

$$\nabla_{X_1} X_1 = X_1(\alpha) X_1 + \frac{\phi}{F} X_2,$$
$$\nabla_{X_1} X_2 = \nabla_{X_2} X_1 = 0,$$
$$\nabla_{X_2} X_2 = X_2(\alpha) X_2,$$

by checking the vanishing of C_{ijk}. Then

$$R(X_1, X_2)X_1 = -X_2 X_1(\alpha) X_1 - \frac{X_2(\phi)}{F} X_2$$

and the Gauss equation implies $R(X_1, X_2)X_1 = F\hat{\rho} X_1$. It then follows that

$$X_2(\phi) = 0.$$

The Gauss equation for $R(X_1, X_2)X_2$ yields nothing new. Hence the immersion f satisfies

$$f_{uu} = \alpha_u f_u + \frac{\phi}{F} f_v,$$
$$f_{uv} = F\xi,$$
$$f_{vv} = \alpha_v f_v,$$

where $u = x^1$ and $v = x^2$. The last equation implies $(f_v/F)_v = 0$ and we can put

$$f_v = FA(u).$$

Then, by the second equation, we get

$$\alpha_u A(u) + A'(u) = \xi;$$

on the other hand, $\xi = -Hf = -\hat{\rho}f$ by properly choosing the origin of $\mathbf{R}^3$ as the center of the proper affine sphere. Thus we may assume

$$f = \alpha_u A(u) + A'(u).$$

We introduce a new coordinate system by $(x, y) = (u, \alpha_u)$; since the Jacobian of this change is $\alpha_{uv} = -F\hat{\rho} \neq 0$, this is well-defined. With respect to this coordinate system, we have

(5.3) $$f = yA(x) + A'(x).$$

This expression inserted into the first equation yields

$$A''' = (y\alpha_{uu} - \phi\hat{\rho} - \alpha_{uuu})A + (y^2 - 2\alpha_{uu})A'.$$

Hence we have

(5.4) $\det [A \ A' \ A''] = \text{const} \neq 0.$

Conversely, the surface given by (5.3) with the condition (5.4) defines an affine sphere of constant curvature, because $J = 0$ and H is constant as we have seen in Theorem 1.2.

Summarizing the arguments above, we have the following theorem.

Theorem 5.4. *Let* $f : M \to \mathbf{R}^3$ *be an affine sphere. Assume the curvature of the affine metric* $\hat{\rho}$ *is a nonzero constant. Then the surface is locally affinely congruent to one of the surfaces listed below.*

(5) $x^2 + y^2 + z^2 = 1,$

(6) $x^2 - y^2 - z^2 = 1,$

(7) $(x, y) \mapsto yA(x) + A'(x)$

       with the condition $\det [A \ A' \ A''] = \text{const} \neq 0.$

As a by-product we obtain the following result due to Radon (see [Bl, p.221]).

Theorem 5.5. *Let* $f : M^2 \to \mathbf{R}^3$ *be a ruled proper affine sphere. Then it is locally affinely congruent to the surface* (7) *in Theorem 5.4.*

Proof. As in the proof of Theorem 5.3, h is indefinite and $J = 0$. Since $H \neq 0$ by assumption, we get $\hat{\rho} \neq 0$.

Remark 5.1. Any surface in class (7) is determined by one space curve defined by the function A. When it describes a plane curve, we can coordinatize the space so that $A = (f_1, f_2, 1)$. Such a curve that defines a surface with the condition (5.4) is given when f_1 and f_2 satisfy an ordinary differential equation of type $y'' + py = 0$. For example, $A = (\sin x, -\cos x, 1)$ (when $p = 1$) defines a one-sheeted hyperboloid; see Example 3.1.

Remark 5.2. A partial list of the classification of surfaces in Theorem 5.1 with additional conditions was given in [L2], [LP], [Ku1]. Theorem 5.1 was first shown by [MR1]. Theorem 5.4 is shown in [Si6]. Refer also to [DMMSV].

Remark 5.3. The surfaces (III) and (IV), $z^2(y - x^2)^3 = \pm 1$, in the list of equiaffinely homogeneous surfaces (Theorem 3.1) are not included above because they are not affine spheres; but, because of homogeneity, the metric curvature $\hat{\rho}$, the affine mean curvature H, and the affine Gauss–Kronecker curvature $K = \det S$ are constant. This raises the problem of classifying such surfaces. We remark that [Vr3] has classified affine nondegenerate surfaces with constant H and constant K and that [DMMSV] has classified affine nondegenerate surfaces with flat affine metric and constant affine mean curvature; the latter, of course, includes Theorem 5.1. Refer also to [Mi], [MMi1]. The first reference here contains the following generalization of Theorem 5.5:

A nondegenerate surface with indefinite affine metric, zero Pick invariant and constant affine mean curvature H is locally of the form

$$f(u,v) = uA(v) + B(v),$$

where $A(v)$ and $B(v)$ are two space curves such that

$$\det [A\ A'\ A''] = H \quad \text{and} \quad \det [A\ B'\ A'] = 1.$$

Remark 5.4. For higher-dimensional cases, one interesting problem is to classify hypersurfaces with constant sectional curvature; in other words, to study Blaschke immersions of Riemannian manifolds of constant sectional curvature as hypersurfaces. A result known in this direction is the following. Refer to [VLS] for the proof.

Theorem 5.6. *Let* $f : M^n \to \mathbf{R}^{n+1}$ *be a locally strictly convex affine hypersphere. If the affine metric is of constant sectional curvature, then the immersion is locally affinely congruent to a quadric or to the surface* $\{(x^1, \ldots, x^{n+1}) \in \mathbf{R}^{n+1} : x^1 x^2 \cdots x^{n+1} = 1\}$.

6. Cayley surfaces

In the preceding sections 1, 3, and 5, we have encountered the Cayley surface $z = xy - \frac{1}{3}x^3$ as a surface belonging to one class or another of nondegenerate surfaces (see Theorems 1.1, 3.1, and 5.1). There is a simple characterization of the Cayley surface, namely,

Theorem 6.1. *Let* M^2 *be a nondegenerate surface with Blaschke structure in* $\mathbf{R}^3$. *If the cubic form C is not 0 and is parallel relative to $\nabla : \nabla C = 0$, then* M^2 *is affinely congruent to the Cayley surface.*

Proof. We make use of results in Section 9 of Chapter II and prove

Proposition 6.2. *Let* M^n, $n \geq 2$, *be a Blaschke hypersurface in* $\mathbf{R}^{n+1}$ *such that* $\nabla C = 0$ *and* $C \neq 0$. *Then* $S = 0$, $\widehat{\text{Ric}} = 0$, $\hat{\rho} = 0$, $J = 0$, *and h is indefinite.*

Proof. By Theorem 9.8 of Chapter II, it follows that $S = 0$. By Proposition 9.4 in the same section, we find $\hat{L} = 0$, where

$$\hat{L}(X,Z) = \text{trace}\{Y \mapsto (\hat{\nabla}_Y K)(X,Z)\}$$

as in (9.7).

Next, from $h(K(X,U),Z) = -\frac{1}{2}C(X,U,Z)$, covariant differentiation ∇_Y gives

$$(\nabla_Y h)(K(X,U),Z) + h((\nabla_Y K)(X,U),Z) = 0.$$

Since the first term is equal to

$$C(Y, K(X, U), Z) = -2h(K(K(X, U), Y), Z),$$

we get

$$2K(K(X, U), Y) = (\nabla_Y K)(X, U)$$

and hence

$$(\nabla_Y K)_X = 2K_Y K_X,$$

from which we get

$$\text{trace}\,\{Y \mapsto (\nabla_Y K)_X(Z)\} = 2\text{tr}\,K_{K_X Z} = 0$$

due to apolarity. Thus $L(X, Z) = 0$ in (9.16).

Finally, from (9.17), we get $\text{trace}\,(K_X K_Z) = 0$. Then from (9.3) the Ricci tensor $\widehat{\text{Ric}}$ and the scalar curvature $\hat{\rho}$ are 0. Consequently, from Proposition 9.3 we get $J = 0$. Since $C \neq 0$, h is an indefinite metric.

Proof of Theorem 6.1. By Proposition 6.2, the surface M^2 has an indefinite metric and zero Pick invariant. Thus by Proposition 11.2 of Chapter II, we have a uniquely determined pair of vector fields $\{X, Y\}$ such that

(6.1) $h(X, X) = 0,$

(6.2) $h(X, Y) = 1,$

(6.3) $h(Y, Y) = 0,$

(6.4) $C(X, U, V) = 0$ for all vector fields $U, V,$

and

(6.5) $C(Y, Y, Y) = 1.$

We are now going to show that X and Y have the following property:

(6.6) $\nabla_X X = \nabla_X Y = \nabla_Y X = 0, \quad \nabla_Y Y = -\dfrac{1}{2}X.$

Now, first, from (6.1) we get $h(\nabla_X X, X) = 0$, which implies $\nabla_X X = \lambda X$ for some function λ. From (6.5) we obtain $C(\nabla_X Y, Y, Y) = 0$, which implies that $\nabla_X Y = \nu X$ for some function ν.

Next, (6.2) leads to $h(\nabla_X X, Y) + h(X, \nabla_X Y) = 0$. Since $h(\nabla_X X, Y) = \lambda$ and $h(X, \nabla_X Y) = h(X, \nu X) = 0$, we get $\lambda = 0$. From (6.3) we get $h(\nabla_X Y, Y) = 0$, which implies $\nu = 0$. We have thus shown that $\nabla_X X = \nabla_X Y = 0$.

From (6.1) we obtain $h(\nabla_Y X, X) = 0$, which implies $\nabla_Y X = \mu X$ for some function μ. From (6.5) we obtain $C(\nabla_Y Y, Y, Y) = 0$, which implies $\nabla_Y Y = \tau X$ for some function τ. Finally, (6.2) gives $h(\nabla_Y X, Y) + h(X, \nabla_Y Y) = 0$, which implies $\mu = 0$, that is, $\nabla_Y X = 0$. From (6.3) we get $(\nabla_Y h)(Y, Y) +$

$2h(\nabla_Y Y, Y) = 0$. By (6.5) and $h(\nabla_Y Y, Y) = h(\tau X, Y) = \tau$, we see that $\tau = -\frac{1}{2}$, that is, $\nabla_Y Y = -\frac{1}{2}X$, completing our arguments.

From (6.6) we see $[X, Y] = 0$. We can also compute $R(X, Y)Y = R(X, Y)X = 0$, which readily follows from $S = 0$. We also know that the affine normal field ξ is parallel in $\mathbf{R}^3$. We may assume that, locally, the surface M^2 is given by

$$(u, v) \mapsto f(u, v) \in \mathbf{R}^3,$$

for which we may assume that $\xi = (0, 0, 1)$. Moreover, we may assume, because $[X, Y] = 0$, that $\{u, v\}$ are local coordinates for M^2 such that $\partial/\partial u = X$ and $\partial/\partial v = Y$. The Gauss equation now reads

$$f_{uu} = 0, \quad f_{uv} = \xi, \quad f_{vv} = -\frac{1}{2}f_u,$$

by virtue of (6.1), (6.2), (6.3), and (6.5). The rest of the proof is very much like the last part of the proof of Theorem 5.1. We get $f_u = v\xi + B$, where B is a constant vector, and hence

$$f = uv\xi + Bu + A(v),$$

where $A(v)$ is a vector function of v. From $f_{vv} = -\frac{1}{2}v\xi - \frac{1}{2}B$, we find

$$A'(v) = -\frac{1}{4}v^2\xi - \frac{1}{2}Bv + C,$$

$$A(v) = -\frac{1}{12}v^3\xi - \frac{1}{4}Bv^2 + Cv + D,$$

where C and D are constant vectors. We may assume that $D = 0$ and ξ, B, C are linearly independent. We get

$$f = (uv - \frac{1}{12}v^3)\xi + (u - \frac{1}{4}v^2)B + Cv.$$

By setting $z = 2(uv - \frac{1}{12}v^3)$, $x = -v$, and $y = -2(u - \frac{1}{4}v^2)$, we get $z = xy - \frac{1}{3}x^3$.

Remark 6.1. To prove Theorem 6.1 in a global form, we make the following observation. Suppose M^2 has the property that each point $p \in M^2$ has a neighborhood U such that $f(U)$ is defined by, say, a cubic equation relative to an affine coordinate system (which can depend on p), then M^2 can be defined by a single cubic equation.

Remark 6.2. Theorem 6.1 was originally proved in [NP3] by using the holonomy group, without appealing to the results in Section 9 of Chapter II.

For a higher dimension $n \geq 3$, let us mention the following example. We consider the graph M^n of

$$x_{n+1} = x_1 x_n + \frac{1}{2}\sum_{i=2}^{n-1}(x_i)^2 + \phi(x_n).$$

This is an improper affine hypersphere ($S = 0$ thus $R = 0$) with Pick invariant $J = 0$; $\nabla C, \nabla^2 C, \ldots$ are all totally symmetric and the affine metric h is Lorentzian and flat (so $\hat{R} = 0$). These assertions follow from the information:

$$h_{ij} = 0 \quad \text{except} \quad h_{1n} = h_{n1} = h_{22} = \cdots = h_{n-1\,n-1} = 1, \; h_{nn} = \phi''(x_n);$$

$$C_{ijk} = 0 \quad \text{except possibly} \quad C_{nnn} = \phi^{(3)};$$

$$C_{ijk;m} = 0 \quad \text{except possibly} \quad C_{nnn;n} = \phi^{(4)};$$

and so on.

If we take ϕ to be a polynomial of degree 3 in x_n, then $C \neq 0$, $\nabla C = 0$. For $n = 3$, Vrancken [Vr1] has shown that a nondegenerate hypersurface M^3 with Blaschke structure in $\mathbf{R}^4$ such that $\nabla C = 0$, $C \neq 0$ is affinely congruent to one of the following graphs:

(i) $x_4 = x_1 x_2 + (x_3)^2 + (x_1)^3$;

(ii) $x_4 = x_1 x_2 + (x_3)^2 + (x_1)^2 x_3$.

For more related results, see [Vr1].

We also add that in Theorem 6.1 we may consider the condition $\hat{\nabla} C = 0$, $C \neq 0$, using the Levi-Civita connection $\hat{\nabla}$ for h. Magid and Nomizu [MN] proved that then the surface M^2 is affinely congruent to one of the following:

(1) the graph of $z = 1/xy$ (h is definite);

(2) the graph of $z = 1/(x^2 + y^2)$ (h is indefinite and $J \neq 0$);

(3) the Cayley surface   (h is indefinite and $J = 0$).

This result was extended to the case where we assume that $\hat{\nabla}^k C = 0$ for some positive integer k. See [Vr2].

7. Convexity, ovaloids, ellipsoids

In this section and the next, we are primarily interested in connected, compact nondegenerate hypersurfaces imbedded in an affine space $\mathbf{R}^{n+1}$. However, we begin with general preliminary discussions.

Let $f : M^n \to \mathbf{R}^{n+1}$ be a hypersurface immersed in $\mathbf{R}^{n+1}$. For each point $x_0 \in M^n$ we take a transversal vector ξ at x_0 and define the affine fundamental form h at x_0 as in Proposition 1.1 of Chapter II. Whether h is definite at x_0 or not is independent of the choice of ξ. If it is, we say that the hypersurface is *definite at* x_0. We shall say that the hypersurface is *definite* if it is definite at each point. This is a property stronger than nondegeneracy that we defined in Definition 2.1 of Chapter II. We have, however, the following:

Lemma 7.1. *Assume that M^n is connected and nondegenerate. If it is definite at one point, so it is at every point.*

Proof. Assume M^n is definite at x_0. Let x_1 be an arbitrary point of M^n and join it to x_0 by a curve x_t. We also fix a Euclidean metric in $\mathbf{R}^{n+1}$. By taking a differentiable family of unit normal vector fields along the curve, we obtain the family of second fundamental forms $h(t)$ and that of Euclidean shape operators $A(t)$. The eigenvalues of $A(t)$ are given by continuous functions $\lambda_1(t) \geq \cdots \geq \lambda_n(t)$. If $h(0)$ is positive-definite, $\lambda_n(0)$ is positive. Since $\lambda_n(t)$ is never 0 (by nondegeneracy), it follows that $\lambda_n(1)$ is positive and thus $h(1)$ is positive-definite. The case where $h(0)$ is negative-definite is similar.

Now assume that the hypersurface is definite. For each point x_0, we may find a neighborhood U of x_0 such that $f(U - x_0)$ lies in one of the open half-spaces that we get by deleting from $\mathbf{R}^{n+1}$ the tangent hyperplane to M^n at x_0. For example, if h is positive-definite for a field of unit normal vectors N around x_0, then $f(U - x_0)$ lies on the open half-space to which N_{x_0} points (see Proposition 5.5 in [KN2, p. 40]). It follows that if the hypersurface is definite, then we may choose a differentiable unit normal field N on M^n such that the second fundamental form h is positive-definite. From the affine point of view, we can choose a differentiable affine normal field ξ on the whole of M^n such that the affine metric h is positive-definite. In any case, it follows that M^n is orientable.

To sum up, we have

Proposition 7.2. *If the hypersurface $f : M^n \to \mathbf{R}^{n+1}$ is definite, then M^n is orientable and we may choose an affine normal field ξ uniquely in such a way that the affine metric h is positive-definite. Each point x_0 has a neighborhood U such that $f(U - x_0)$ lies in the open half-space determined by the tangent hyperplane at x_0 toward which ξ points.*

Definition 7.1. By saying that a hypersurface M^n is *locally strictly convex*, we shall mean that it is definite and is provided with the global affine normal field ξ such that the affine metric h is positive-definite.

Now suppose that $f : M^n \to \mathbf{R}^{n+1}$ is a connected compact hypersurface immersed in $\mathbf{R}^{n+1}$. Again introduce any Euclidean metric in $\mathbf{R}^{n+1}$. By Sard's theorem, there is a height function whose critical points are isolated and nondegenerate. Since M^n is compact, there is a point x_0 where the function has a maximum. Thus M^n must be definite at x_0. (For example, see [CR, pp. 10–12] or [Sp, vol.III, p. 90].) By Lemma 7.1 we conclude that M^n is definite. We may also check that the proof for Theorem 5.6 in Chapter VII of [KN2] is valid for immersions. We can state (see also [Sp, vol.III, p. 94, p. 121])

Proposition 7.3. *Assume that $f : M^n \to \mathbf{R}^{n+1}$ is a connected compact nondegenerate hypersurface immersed in $\mathbf{R}^{n+1}$. Then:*

(1) M^n is locally strictly convex;

(2) M^n is orientable and the spherical map of Gauss is (relative to a Euclidean metric chosen in $\mathbf{R}^{n+1}$) a diffeomorphism of M^n onto the unit sphere S^n;

(3) M^n *is globally strictly convex, that is, for each point* x_0, $f(M^n - x_0)$ *lies in one open half-space determined by the tangent hyperplane at* $f(x_0)$;

(4) f *is an imbedding.*

Definition 7.2. By an *ovaloid* M^n we mean a connected, compact nondegenerate hypersurface imbedded in $\mathbf{R}^{n+1}$.

It is further known that an ovaloid M^n is the boundary of a convex body. We may take an interior point, say o. Since the affine normal field ξ with positive-definite affine metric h is inward, we see that the affine distance function $\rho(x) = v(x)$ with $x \in M^n$ as position vector from o is negative-valued. At a point x_0 where ρ has the smallest value, we know that S is positive-definite as we see from the proof of Proposition 6.1 of Chapter II. Thus both the affine Gauss–Kronecker curvature $K = \det S$ and the affine mean curvature H are positive at x_0.

We have

Proposition 7.4. *For an ovaloid (with positive-definite affine metric), the affine mean curvature is positive somewhere. In particular, there is no ovaloid with vanishing affine mean curvature.*

We are now in a position to prove the following classical result due to Blaschke ($n = 2$) [Bl] and Deicke ($n \geq 3$) [De].

Theorem 7.5. *If an ovaloid* M^n *is an affine hypersphere, then it must be an ellipsoid.*

Proof. We take the affine normal field so that the affine metric h is positive-definite. It is sufficient to show that the Pick invariant J is 0. By Proposition 6.1, Chapter II, S is positive-definite and hence $H > 0$ at some point x_0. Since M^n is an affine hypersphere by assumption, H is a positive constant. For $u = 4n(n-1)J$ we have the estimate in Proposition 10.2 of Chapter II

$$\Delta u \geq 2(n+1)Hu + \frac{(n+1)}{2n(n-1)}u^2.$$

Since M^n is compact, the function u attains a maximum u_0 at some point, where $\Delta u \leq 0$. By the estimate above it follows that $u_0 = 0$ because $H > 0$. Hence C is identically 0.

Remark 7.1. Blaschke indicated several proofs for the theorem ($n = 2$). One of them depends on the formula (10.7) for $\log|J|$ in Chapter II:

$$\Delta \log J = 6(H + J).$$

Here is his proof. We may assume $H > 0$. If $J > 0$ somewhere, then assume that J has a maximum $J_0 > 0$ at x_0. Then $\log J$ defined in a neighborhood of x_0 satisfies $\Delta \log J \leq 0$ at x_0. But $6(H+J) > 0$ at x_0, which is a contradiction. We may extend Theorem 7.5 as follows; see [Cal2].

Theorem 7.6. *Let* M^n *be a locally strictly convex affine hypersphere imbedded in* $\mathbf{R}^{n+1}$. *Assume*

(1) *H is positive for the choice of the affine normal such that h is positive-definite;*

(2) *h is complete (as a Riemannian metric).*

Then M^n is an ellipsoid.

Proof. It suffices to show that M^n is compact. By Proposition 9.2 of Chapter II we have

$$\widehat{\mathrm{Ric}}(X,X) \geq (n-1)Hh(X,X).$$

Since H is a positive constant, we can use Myers's theorem and conclude that M^n is compact.

For Myers's theorem (the case where $n = 2$ is due to Bonnet), see Theorem 3.4 in [KN2, Chapter VIII].

Remark 7.2. A locally strictly convex affine hypersurface is said to be *elliptic* if condition (1) in Theorem 7.6 is satisfied. It is said to be *hyperbolic* if H is negative instead. For complete, hyperbolic locally convex affine hyperspheres, see Note 7 and the work of Calabi [Cal2], Cheng and Yau [CY1], and Sasaki [S1].

We can also prove the following theorem due independently to Calabi [Cal1] and Pogorelov [Pog], which is a generalization of Jörgens's theorem (see Remark 3.3 of Chapter II).

Theorem 7.7. *If the affine metric of an improper affine hypersphere M^n is positive-definite and complete, then M^n is affinely congruent to a paraboloid $x^{n+1} = \sum_{i=1}^{n}(x^i)^2$ in $\mathbf{R}^{n+1}$.*

Proof. In the estimates for the Ricci curvature and for Δu we have

$$\widehat{\mathrm{Ric}} \geq 0 \quad \text{and} \quad \Delta u \geq \frac{(n+1)}{2n(n-1)}u^2,$$

where $u = 4n(n-1)J$. Now we appeal to the following maximum principle due to Cheng and Yau [CY2, p. 353].

Lemma 7.8. *Let M be a complete Riemannian manifold with Ricci curvature bounded from below. Then a non-negative function u on M such that $\Delta u \geq cu^2$, where c is a positive constant, is 0.*

8. Other characterizations of ellipsoids

We give another characterization of an ellipsoid among ovaloids of dimension 2.

Lemma 8.1. *Let (M^2, ∇) be a 2-manifold with an equiaffine connection. Then ∇ is projectively flat if and only if*

(8.1) $$(\nabla_Z R)(X,Y)W = (\nabla_W R)(X,Y)Z$$

for all vectors X, Y, Z and W.

Proof. We know from Theorem 3.3, Chapter I, that ∇ is projectively flat if and only if the Ricci tensor γ satisfies $(\nabla_Z \gamma)(Y, W) = (\nabla_W \gamma)(Y, Z)$. Now for dimension 2 we have

$$(8.2) \qquad R(X, Y)W = \gamma(Y, W)X - \gamma(X, W)Y$$

for all vector fields X, Y, and W. We get

$$(\nabla_Z R)(X, Y)W = (\nabla_Z \gamma)(Y, W)X - (\nabla_Z \gamma)(X, W)Y.$$

We see that (8.1) holds if and only if

$$(\nabla_Z \gamma)(Y, W)X - (\nabla_Z \gamma)(X, W)Y = (\nabla_W \gamma)(Y, Z)X - (\nabla_W \gamma)(X, Z)Y,$$

which holds if and only if $(\nabla_Z \gamma)(Y, W) = (\nabla_W \gamma)(Y, Z)$ for all vectors X, Z, and W. This proves Lemma 8.1.

Corollary 8.2. *If an equiaffine connection ∇ on M^2 is locally symmetric (that is, $\nabla R = 0$), then it is projectively flat.*

We now establish

Proposition 8.3. *Let $f : M^2 \to \mathbf{R}^3$ be a nondegenerate surface with an equiaffine transversal vector field ξ. Let (∇, h, S) be the induced structure. Then ∇ is projectively flat if and only if*

$$(8.3) \qquad \operatorname{tr}_h \nabla S = 0.$$

Proof. From the Gauss equation

$$R(X, Y)Z = h(Y, Z)SX - h(X, Z)SY$$

we get

$$(8.4) \qquad (\nabla_W R)(X, Y)Z = (\nabla_W h)(Y, Z)SX - (\nabla_W h)(X, Z)SY$$
$$+ h(Y, Z)(\nabla_W S)X - h(X, Z)(\nabla_W S)Y.$$

Both $(\nabla_W h)(Y, Z)$ and $(\nabla_W h)(X, Z)$ are symmetric in W and Z. Hence the left-hand side of (8.4) is symmetric in Z and W (that is, (8.1) holds) if and only if

$$(8.5) \qquad h(Y, Z)(\nabla_W S)X - h(X, Z)(\nabla_W S)Y$$
$$= h(Y, W)(\nabla_Z S)X - h(X, W)(\nabla_Z S)Y$$

for all vectors.

We take the trace of the bilinear map which sends (X, W) to each term of the equation above. For this computation we use an h-orthonormal basis $\{e_1, e_2\}$ with $h(e_i, e_i) = \epsilon_i = \pm 1$ and $h(e_1, e_2) = 0$. Then we get

$$\text{trace}_h\{(X, W) \mapsto h(X, Z)(\nabla_W S)Y\} = \sum_i \epsilon_i h(e_i, Z)(\nabla_{e_i} S)Y = (\nabla_Z S)Y$$

and, similarly,

$$\text{trace}_h\{(X, W) \mapsto h(Y, W)(\nabla_Z S)X\} = (\nabla_Z S)Y.$$

Therefore the trace computation leads to

$$h(Y, Z)\text{tr}_h(\nabla S) = 0,$$

and hence (8.3). We have proved that if ∇ is projectively flat, then $\text{tr}_h(\nabla S) = 0$.

To prove the converse, we must show that $\text{tr}_h(\nabla S) = 0$ implies (8.5). We may assume that $W = e_1$, $Z = e_2$. It is also sufficient to check the equation in the following cases:

$$X = e_1, Y = e_2; \quad X = e_2, Y = e_1; \quad X = Y = e_1; \quad X = Y = e_2.$$

In the first and second cases, the equation reduces to $\text{tr}_h(\nabla S) = 0$. In the third and fourth cases, the equation is trivially satisfied. This completes the proof of Proposition 8.3.

Another expression of $\text{tr}_h(\nabla S)$ is given for any dimension by the following.

Proposition 8.4. *For a Blaschke hypersurface M^n, $n \geq 2$, we have*

(8.6) $$h(\text{tr}_h(\nabla S), Y) = Y(\text{tr } S) + 2\,\text{tr}\,(K_Y S).$$

Proof. From Ricci's identity $h(SY, Z) = h(Y, SZ)$ we get

$$(\nabla_X h)(SY, Z) + h((\nabla_X S)Y, Z) = (\nabla_X h)(Y, SZ) + h(Y, (\nabla_X S)Z).$$

We take the trace of the bilinear mapping sending (X, Z) to each term of this equation. By using Codazzi's equation for S, apolarity, and formula (4.3) in Chapter II, we obtain

$$\text{trace}_h\{(X, Z) \mapsto (\nabla_X h)(SY, Z) = C(X, SY, Z)\} = 0,$$

$$\text{trace}_h\{(X, Z) \mapsto h((\nabla_X S)Y, Z) = h((\nabla_Y S)X, Z)\} = \text{tr}\,(\nabla_Y S),$$

$$\text{trace}_h\{(X, Z) \mapsto (\nabla_X h)(Y, SZ) = -2h(X, K_Y SZ)\} = -2\,\text{tr}\,(K_Y S).$$

From all this, we get (8.6) in view of $\text{tr}\,(\nabla_Y S) = Y\,\text{tr}\,S$.

For a Blaschke surface M^2, Propositions 8.3 and 8.4 imply that M^2 is projectively flat if and only if

$$(8.7) \qquad\qquad Y(\operatorname{tr} S) + 2 \operatorname{tr} K_Y S = 0,$$

as observed in [Po2].

Still another interpretation of the condition $\operatorname{tr}_h(\nabla S) = 0$ can be given as follows in the case where the affine metric h for a Blaschke surface M^2 is positive-definite. Around each point in M^2 we take an isothermal coordinate system $\{x, y\}$ for h. We write $X = \partial/\partial x$ and $Y = \partial/\partial y$ so that $h(X, X) = h(Y, Y) = E$ and $h(X, Y) = 0$. If we define a quadratic form

$$(8.8) \qquad\qquad \Psi = \{h(SX, X) - h(SY, Y) - 2ih(SX, Y)\}dz^2,$$

where $z = x + iy$, then we can show by computation that Ψ is independent of the choice of isothermal coordinate system. This means that by (8.8) we can define a global quadratic form Ψ. Then we have

Lemma 8.5. *The form Ψ defined above is holomorphic if and only if* $\operatorname{tr}_h(\nabla S)$ *is identically zero.*

Proof. Let $u = h(SX, X) - h(SY, Y)$ and $v = -2h(SX, Y)$. We show that the Cauchy–Riemann equations are valid. Since $\{x, y\}$ is an isothermal coordinate system, we have $\hat{\nabla}_X X = -\hat{\nabla}_Y Y$. Using $K_X X + K_Y Y = 0$ (apolarity), we obtain

$$(8.9) \qquad \nabla_X X = \hat{\nabla}_X X + K_X X = -\hat{\nabla}_Y Y - K_Y Y = -\nabla_Y Y.$$

Of course, we have also $\nabla_X Y = \nabla_Y X$. Now we have

$$\begin{aligned}
\frac{\partial u}{\partial x} &= X(h(SX, X) - h(SY, Y)) \\
&= (\nabla_X h)(SX, X) - (\nabla_X h)(SY, Y) \\
&\quad + h((\nabla_X S)X, X) + 2h(SX, \nabla_X X) \\
&\quad - h((\nabla_X S)Y, Y) - 2h(SY, \nabla_X Y).
\end{aligned}$$

Similarly, we get

$$\begin{aligned}
-\frac{\partial v}{\partial y} &= Y(h(X, SY) + h(SX, Y)) \\
&= (\nabla_Y h)(X, SY) + (\nabla_Y h)(SX, Y) \\
&\quad + h((\nabla_Y S)Y, X) + 2h(SX, \nabla_Y Y) \\
&\quad + h((\nabla_Y S)X, Y) + 2h(SY, \nabla_Y X).
\end{aligned}$$

Using the Codazzi equations for S and h, the apolarity and (8.9), we get

$$\frac{\partial u}{\partial x} - \frac{\partial v}{\partial y} = h(X, (\nabla_X S)(X)) + h(X, (\nabla_Y S)(Y)) = Eh(\operatorname{tr}_h(\nabla S), X).$$

By similar computation we have

$$\frac{\partial u}{\partial y} + \frac{\partial v}{\partial x} = h(Y, (\nabla_X S)(X) + (\nabla_Y S)(Y)) = Eh(Y, \text{tr}_h(\nabla S)).$$

Thus the Cauchy–Riemann equations are satisfied (and Ψ is holomorphic) if and only if $\text{tr}_h(\nabla S) = 0$.

We now have

Theorem 8.6. *An ovaloid M^2 is an ellipsoid if and only if the induced connection ∇ is projectively flat.*

Proof. We consider the holomorphic form Ψ in Lemma 8.5. Since M^2 is diffeomorphic to the 2-sphere, it follows that Ψ must be identically 0; see [Ho, p. 140]. From its definition, it follows that $S = \lambda I$, that is, M^2 is an affine sphere. Now the theorem of Blaschke, Theorem 7.5, says that M^2 is an ellipsoid.

Corollary 8.7. *An ovaloid M^2 is an ellipsoid if and only if the induced connection ∇ is locally symmetric.*

Proof. This follows from Theorem 8.6 and Corollary 8.2.

Remark 8.1. The results in Theorem 8.6 and Corollary 8.7 appeared first in [OVe], then in [NO2] and [Po2]. For noncompact nondegenerate surfaces with locally symmetric or projective flat connections, see Note 8.

Remark 8.2. For nondegenerate hypersurfaces M^n, $n \geq 3$, with $\nabla R = 0$ see Theorem 4.1 in Section 4, Chapter IV.

9. Minkowski integral formulas and applications

We shall first deal with the case of compact surfaces. Let f and $\bar{f} : M^2 \to \mathbf{R}^3$ be two nondegenerate immersions of a connected compact 2-manifold into $\mathbf{R}^3$ with equiaffine transversal vector fields ξ and $\bar{\xi}$, respectively. We denote the induced structures for (f, ξ) and $(\bar{f}, \bar{\xi})$ by (∇, h, S, θ) and $(\bar{\nabla}, \bar{h}, \bar{S}, \bar{\theta})$. By assumption, we have $\nabla\theta = 0$ and $\bar{\nabla}\bar{\theta} = 0$. As in Section 7, we may assume that h and $\bar{h}$ are positive-definite and ξ and $\bar{\xi}$ are inward. We furthermore fix points o and $\bar{o}$ in the interiors of $f(M^2)$ and of $\bar{f}(M^2)$, respectively. Note that f and $\bar{f}$ are not necessarily Blaschke immersions.

We shall assume that $\nabla = \bar{\nabla}$ and derive some integral formulas. For this purpose, we define a 1-form α on M^2 by

$$(9.1) \qquad\qquad \alpha(X) = \theta(Z, \bar{S}X),$$

where Z denotes the tangential component of the position vector $f(x)$ from o:

$$(9.2) \qquad\qquad f(x) = \rho\xi_x + f_* Z_x,$$

where ρ is the affine distance function; it is defined by $\rho(x) = v(f(x))$, $x \in M^2$, where $v : M^2 \to \mathbf{R}_3$ is the conormal map. We compute $d\alpha$ as follows. From $\alpha(Y) = \theta(Z, \bar{S}Y)$ we obtain

(9.3) $$X\alpha(Y) = (\nabla_X \theta)(Z, \bar{S}Y) + \theta(\nabla_X Z, \bar{S}Y)$$
$$+ \theta(Z, (\nabla_X \bar{S})Y) + \theta(Z, \bar{S}(\nabla_X Y)).$$

Here we have $\nabla_X \theta = 0$. As for $\nabla_X Z$, differentiating (9.2) relative to X, we get

(9.4) $$\nabla_X Z = \rho SX + X \quad \text{and} \quad h(X, Z) = -X\rho.$$

Actually, (9.4) already appeared as (5.10) in Chapter II. We note that the formula is valid without requiring that M^2 is a Blaschke surface. We can rewrite (9.3) as follows:

$$X\alpha(Y) = \rho\,\theta(SX, \bar{S}Y) + \theta(X, \bar{S}Y) + \theta(Z, (\nabla_X \bar{S})Y) + \theta(Z, \bar{S}(\nabla_X Y)).$$

Taking $X\alpha(Y) - Y\alpha(X) - \alpha([X, Y])$, which is equal to $2d\alpha(X, Y)$ by definition, and using the Codazzi equation $(\nabla_Y \bar{S})(X) = (\nabla_X \bar{S})(Y)$ we obtain

(9.5) $$2(d\alpha)(X, Y) = \rho\,[\theta(SX, \bar{S}Y) - \theta(\bar{S}X, SY)] + \theta(X, \bar{S}Y) - \theta(\bar{S}X, Y).$$

Let $\{X_1, X_2\}$ be a unimodular basis in $T_x(M^2)$, that is, $\theta(X_1, X_2) = 1$, and write

$$SX_j = \sum_{i=1}^{2} S_j^i X_i \quad \text{and} \quad \bar{S}X_j = \sum_{i=1}^{2} \bar{S}_j^i X_i.$$

From (9.5) we compute

$$2(d\alpha)(X_1, X_2) = \rho\,[S_1^1 \bar{S}_2^2 - S_1^2 \bar{S}_2^1 + \bar{S}_1^1 S_2^2 - \bar{S}_1^2 S_2^1] + \operatorname{tr} \bar{S}$$
$$= -2\rho\langle S, \bar{S}\rangle + \operatorname{tr} \bar{S},$$

where $\langle, \rangle$ denotes the inner product of signature $(-, -, +, +)$ in the space of all endomorphisms of $T_x(M^2)$ defined by

$$\langle A, B\rangle = \frac{1}{2}[\operatorname{tr}(AB) - \operatorname{tr} A \cdot \operatorname{tr} B].$$

We have also

$$-2\langle A, B\rangle = \det A + \det B - \det(A - B)$$

and finally

(9.6) $$2d\alpha = [\rho\,(\det S + \det \bar{S} - \det(\bar{S} - S)) + \operatorname{tr} \bar{S}]\theta.$$

Thus we have

$$(9.7) \qquad \int_{M^2} [\rho\,(\det S + \det \bar{S}) + \operatorname{tr} \bar{S}]\theta = \int_{M^2} \rho \det (\bar{S} - S)\theta.$$

In particular, setting $\bar{f} = f$ and $\bar{\xi} = \xi$, we get

$$(I) \qquad \int_{M^2} (2\rho \det S + \operatorname{tr} S)\theta = 0.$$

Subtracting (I) from (9.7) we find

$$(II) \qquad \int_{M^2} [\operatorname{tr} \bar{S} - \operatorname{tr} S + \rho\,(\det \bar{S} - \det S)]\theta = \int_{M^2} \rho \det (\bar{S} - S)\theta.$$

If we interchange the role of (f, ξ) and $(\bar{f}, \bar{\xi})$, we get a similar formula

$$(\bar{II}) \qquad \int_{M^2} [\operatorname{tr} S - \operatorname{tr} \bar{S} + \rho\,(\det S - \det \bar{S})]\bar{\theta} = \int_{M^2} \bar{\rho} \det (\bar{S} - S)\bar{\theta}.$$

Finally, we define a 1-form β by $\beta(X) = \theta(Z, X)$, where X is any tangent vector and Z is as in (9.2). We easily get

$$2d\beta = (2 + \rho \operatorname{tr} S)\theta,$$

and hence

$$(III) \qquad \int_{M^2} (2 + \rho \operatorname{tr} S)\theta = 0.$$

We can rewrite (I) and (III) as follows.

$$(Ia) \qquad \int_{M^2} (\rho K + H)\theta = 0,$$

$$(IIIa) \qquad \int_{M^2} (1 + \rho H)\theta = 0,$$

where K is the affine Gaussian curvature and H the affine mean curvature. The formulas (I), (II), (III), (Ia), and (IIIa) are usually called the *Minkowski formulas* for affine surfaces. The formulas for higher dimensions are later given in Propositions 9.4 and 9.5.

As applications of these integral formulas, we now prove two characterization theorems for ellipsoids among the ovaloids with Blaschke structures.

Theorem 9.1. *If the affine Gaussian curvature K of an ovaloid $f : M^2 \to \mathbf{R}^3$ with Blaschke structure is constant, then K is positive and $f(M^2)$ is an ellipsoid.*

Theorem 9.2. *If the affine mean curvature H of an ovaloid $f : M^2 \to \mathbf{R}^3$ with Blaschke structure is constant, then $f(M^2)$ is an ellipsoid.*

Remark 9.1. The original result by Blaschke for Theorem 9.1 assumes that K is a positive constant, see [Bl, p. 248]. Theorem 9.2 is stated in [Bl, p. 201] and proved in the reference cited there.

To prove these theorems for compact Blaschke surfaces, it is sufficient to show that, for a nondegenerate immersion $f : M^2 \to \mathbf{R}^3$ of a compact 2-manifold with an equiaffine transversal vector field, the assumption that $\det S = $ constant or $\operatorname{tr} S = $ constant implies $S = \lambda I$, where λ is a nonzero constant. Then if f is further a Blaschke immersion, it follows that $f(M^2)$ is an ellipsoid by Theorem 7.5.

Proof of Theorem 9.1. We may assume that h is positive-definite. Since S is positive-definite at some point by Proposition 6.1 of Chapter II, it follows that $K = \det S$ is a positive constant. We may assume $K = 1$, without loss of generality; rechoose the affine normal vector ξ by multiplying it with a constant factor, if necessary. Let λ_1 and λ_2 be two eigenvalues of S so that $\lambda_1 \lambda_2 = 1$. Hence $H = \frac{1}{2}(\lambda_1 + \lambda_2) = \frac{1}{2}(\lambda_1 + 1/\lambda_1) \geq 1$, and the equality holds if and only if $\lambda_1 = \lambda_2$.

We know that the affine distance ρ from o interior to $f(M^2)$ is negative-valued on M^2. From (III) and (I) we obtain

$$\int_{M^2} -\rho\theta = \int_{M^2} H\theta \geq \int_{M^2} \theta = \int_{M^2} -\rho H\theta$$

and hence

$$\int_{M^2} \rho(1 - H)\theta \leq 0.$$

But $\rho < 0$ and $1 - H \leq 0$. Hence the integral above is non-negative. It follows that the integral must be 0 and hence $H = 1$. We get $\lambda_1 = \lambda_2$, completing the proof of Theorem 9.1.

Proof of Theorem 9.2. We may assume $1 = H = \frac{1}{2}(\lambda_1 + \lambda_2)$. Then

$$K = \lambda_1 \lambda_2 = \lambda_1(2 - \lambda_1) = 2\lambda_1 - (\lambda_1)^2 \leq 1,$$

and the equality holds if and only if $\lambda_1 = 1$. By (III) and (II) we get

$$\int_{M^2} -\rho\theta = \int_{M^2} \theta = \int_{M^2} -\rho K\theta,$$

which implies $\int_{M^2} \rho(1 - K)\theta = 0$. Since $\rho < 0$ as before and $1 - K \geq 0$, it follows that $K = 1$ and hence $\lambda_1 = \lambda_2 = 1$, as desired. This concludes the proof of Theorem 9.2.

Now we shall deal with higher-dimensional cases. Let $f : M^n \to \mathbf{R}^{n+1}$ be a nondegenerate immersion with an equiaffine transversal vector field ξ. Let (∇, S, h, θ) denote the induced structure as before. Let ρ be the affine distance relative to a point o. When M^n is compact, we take o in the interior of $f(M^n)$ so that ρ is negative-valued. We have $\rho(x) = v(x - o)$, where v is the affine conormal vector. We also take the vector field Z defined by (9.2); we have $Z = -\mathrm{grad}\,\rho$ from (9.4).

We denote by P the complete polarization of the determinant function. Given endomorphisms $A^1, \ldots, A^n$, it is defined by the property

$$(9.8) \qquad \frac{1}{n!} \sum_{\tau \in \mathscr{S}_n} \theta(A^{\tau(1)}X_1, \ldots, A^{\tau(n)}X_n) = P(A^1, \ldots, A^n)\theta(X_1, \ldots, X_n),$$

where $\mathscr{S}_n$ is the symmetric group of degree n. P is symmetric in its arguments and $P(A, \ldots, A) = \det A$. In particular, when I denotes the identity transformation, we set

$$(9.9) \qquad \sigma_p(A) = P(I, \ldots, I, \overbrace{A, \ldots, A}^{p}) = \binom{n}{p}^{-1} \sum a_{i_1} \cdots a_{i_p},$$

where $\{a_1, \ldots, a_n\}$ is the set of eigenvalues of A.

Now, fixing a field of endomorphism of TM denoted by T, we define an $(n-1)$-form γ by

$$(9.10) \qquad \gamma(X_1, \ldots, X_{n-1}) = \theta(Z, TX_1, \ldots, TX_{n-1}),$$

which is a formal generalization of the 1-forms α and β defined previously.
Lemma 9.3. *Assume T satisfies the condition $(\nabla_X T)Y = (\nabla_Y T)X$. Then*

$$(9.11) \qquad d\gamma = [\rho P(S, T, \ldots, T) + P(I, T, \ldots, T)]\theta.$$

Proof. First, recall that

$$n\, d\gamma(X_0, \ldots, X_{n-1})$$
$$= \sum (-1)^j X_j(\gamma(X_0, \ldots, \widehat{X}_j, \ldots, X_{n-1}))$$
$$+ \sum_{j<k} (-1)^{j+k} \gamma([X_j, X_k], X_0, \ldots, \widehat{X}_j, \ldots, \widehat{X}_k, \ldots, X_{n-1})$$

and that covariant differentiation of (9.10) gives

$$X(\gamma(X_1, \ldots, X_{n-1}))$$
$$= \theta(\nabla_X Z, TX_1, \ldots, TX_{n-1})$$
$$+ \sum_i \theta(Z, TX_1, \ldots, (\nabla_X T)X_i + T(\nabla_X X_i), \ldots, TX_{n-1})$$

because of $\nabla\theta = 0$. Hence, we see

$$
\begin{aligned}
nd\gamma(X_0,\ldots,X_{n-1}) \\
= \sum_j [\theta(\nabla_{X_j}Z, TX_0,\ldots,\widehat{TX_j},\ldots,TX_{n-1}) \\
+ \sum_{j\neq k} \theta(Z, TX_0,\ldots,(\nabla_{X_j}T)X_k + T(\nabla_{X_j}X_k),\ldots,\widehat{TX_j},\ldots,TX_{n-1})] \\
+ \sum_{j<k} (-1)^{j+k}\theta(Z, T[X_j,X_k], TX_0,\ldots,\widehat{TX_j},\ldots,\widehat{TX_k},\ldots,TX_{n-1}) \\
= \sum_j \theta(TX_0,\ldots,\nabla_{X_j}Z,\ldots,TX_{n-1}) \\
+ \sum_{j>k} (-1)^{j+k}\theta(Z, (\nabla_{X_j}T)X_k + T(\nabla_{X_j}X_k), \\
TX_0,\ldots,\widehat{TX_k},\ldots,\widehat{TX_j},\ldots,TX_{n-1}) \\
- \sum_{j<k} (-1)^{j+k}\theta(Z, (\nabla_{X_j}T)X_k + T(\nabla_{X_j}X_k), \\
TX_0,\ldots,\widehat{TX_j},\ldots,\widehat{TX_k},\ldots,TX_{n-1}) \\
+ \sum_{j<k} (-1)^{j+k}\theta(Z, T(\nabla_{X_j}X_k - \nabla_{X_k}X_j), \\
TX_0,\ldots,\widehat{TX_j},\ldots,\widehat{TX_k},\ldots,TX_{n-1}) \\
= \sum_j \theta(TX_0,\ldots,\nabla_{X_j}Z,\ldots,TX_{n-1}) \\
- \sum_{j<k} (-1)^{j+k}\theta(Z, (\nabla_{X_j}T)X_k - (\nabla_{X_k}T)X_j, \\
TX_0,\ldots,\widehat{TX_j},\ldots,\widehat{TX_k},\ldots,TX_{n-1}).
\end{aligned}
$$

Now, the assumption implies that the second term of the last equation disappears, and (9.40) implies

$$
\begin{aligned}
d\gamma(X_0,\ldots,X_{n-1}) = \frac{\rho}{n}\sum_j \theta(TX_0,\ldots,SX_j,\ldots,TX_{n-1}) \\
+ \frac{1}{n}\sum_j \theta(TX_0,\ldots,X_j,\ldots,TX_{n-1}).
\end{aligned}
$$

The definition of the polar form P shows the result.

Definition 9.1. For the affine shape operator S, we call

$$
(9.12) \qquad\qquad H_p(S) = \sigma_p(S) = P(I,\ldots,I,\overbrace{S,\ldots,S}^{p})
$$

the *p-th affine mean curvature*. $H_1(S)$ is the affine mean curvature H and $H_n(S)$ is the affine Gauss–Kronecker curvature K.

We apply Lemma 9.3 to the case when $T = I + tS$ where t is a parameter; the condition of Lemma 9.3 is satisfied because of Codazzi's equation for S. Note that we have the following expansions:

$$P(S, T, \ldots, T) = \sum_i \binom{n-1}{i} P(I, \ldots, I, \overbrace{S, \ldots, S}^{i+1}) t^i,$$

$$P(I, T, \ldots, T) = \sum_i \binom{n-1}{i} P(I, \ldots, I, \overbrace{S, \ldots, S}^{i}) t^i,$$

$$\gamma(X_1, \ldots, X_n) = \sum_i \binom{n-1}{i} \beta_i(S) t^i,$$

where

(9.13) $$\beta_i(S) = \binom{n-1}{i}^{-1} \sum \theta(Z, S^{a_1} X_1, \ldots, S^{a_{n-1}} X_{n-1});$$

the last summation runs for all $\{a_p\}$ such that $a_p = 0$ or 1 and $\sum_p a_p = i$. The $(n-1)$-form β_1 is the same as the form β which we defined in Remark 6.1 of Chapter II. Comparing the coefficients of t^i in the expansion of (9.11), we have

(9.14) $$d\beta_i(S) = (\rho H_{i+1} + H_i)\theta.$$

Proposition 9.4. *Let $f : M^n \to \mathbf{R}^{n+1}$ be an equiaffine nondegenerate immersion of a compact manifold M. Then*

$$\int_M (\rho H_{i+1}(S) + H_i(S))\theta = 0 \quad \text{for each } i, \ 1 \le i \le n-1.$$

We next consider two equiaffine nondegenerate immersions of a manifold M^n, $f : M^n \to \mathbf{R}^{n+1}$ and $\bar{f} : M^n \to \mathbf{R}^{n+1}$, and assume the induced affine connection under f is equal to that under $\bar{f}$. Then the field of endomorphisms $T = I + tS + s\bar{S}$ also satisfies the condition of Lemma 9.3. Now we have the following expansions:

$$P(S, T, \ldots, T) = \sum_{i,j} \binom{n-1}{i} \binom{n-1-i}{j} H_{i+1,j}(S, \bar{S}) t^i s^j,$$

$$P(I, T, \ldots, T) = \sum_{i,j} \binom{n-1}{i} \binom{n-1-i}{j} H_{i,j}(S, \bar{S}) t^i s^j,$$

where

$$(9.15) \qquad H_{i,j}(S, \bar{S}) = P(\overbrace{I, \ldots, I}^{i}, \overbrace{S, \ldots, S}^{j}, \bar{S}, \ldots, \bar{S}),$$

and the expansion

$$\gamma(X_1, \ldots, X_n) = \sum_{i,j} \binom{n-1}{i} \binom{n-1-i}{j} \beta_{i,j}(S, \bar{S}) t^i s^j.$$

The form $\beta_{i,j}(S, \bar{S})$ has a similar expression to that of $\beta_i(S)$, which we do not write down here. Then we have

Proposition 9.5. *Let $f, \bar{f} : M^n \to \mathbf{R}^{n+1}$ be two equiaffine nondegenerate immersions of a compact manifold M and assume that the induced connections coincide with each other. Then*

$$\int_M [\rho H_{i+1,j}(S, \bar{S}) + H_{i,j}(S, \bar{S})]\theta = 0$$

for all (i, j) with $i + j + 1 \le n$.

In the special case $n = 2$, we have $H_{0,1}(S, \bar{S}) = \frac{1}{2} \mathrm{tr}\, \bar{S}$. By setting $i = 0$ and $j = 1$ in Proposition 9.5, we get formula (9.7). If we let $i = 1$ in Proposition 9.4 and note that $H_1(S) = \frac{1}{2} \mathrm{tr}\, S$, we get formula (I).

The next algebraic lemma is well-known.

Lemma 9.6. *Let A be an $n \times n$ matrix. Then*

$$\sigma_{p-1}(A)\sigma_{p+1}(A) - (\sigma_p(A))^2 \le 0,$$

and the equality holds if and only if all eigenvalues are equal.

We can now prove a generalization of Theorem 9.2.

Theorem 9.7. *Let $f : M \to \mathbf{R}^{n+1}$ be an ovaloid with Blaschke structure. Assume that the affine mean curvature H is constant. Then $f(M)$ is an ellipsoid.*

Proof. We may assume h is positive-definite. Since M is convex, we may choose a point o in the interior of the convex body bounded by $f(M)$ so that the affine distance function ρ from o is negative-valued on M. Since $H_2 \le H^2$ from Lemma 9.6, we get

$$H \int \theta = \int H\theta = \int -\rho H_2 \theta$$

$$\le \int -\rho H^2 \theta = H \int (-\rho H)\theta = H \int \theta.$$

Hence we have the equality and thus $H_2 = H^2$. Lemma 9.6 implies that $S = \lambda I$ for some constant λ, that is, $f(M)$ is an affine hypersphere. Theorem 7.5 proves the theorem.

Moreover, we can prove

Theorem 9.8. *If the p-th affine mean curvature of an ovaloid* $f : M \to \mathbf{R}^{n+1}$ *with Blaschke structure is constant, then* $f(M)$ *is an ellipsoid.*

We remark that we get Theorem 9.1 for $n = p = 2$ and Theorem 9.2 for $n = 2$, $p = 1$.

The procedure is the same as that for the Euclidean case. We need the following algebraic fact due to Gårding ([Ga]).

Let $\sigma_p : \mathbf{R}^n \to \mathbf{R}$ be the p-th elementary symmetric function and let $\mathbf{e}$ denote the vector $(1, \ldots, 1)$ in $\mathbf{R}^n$. We put

$$C_p = \{x \in \mathbf{R}^n : \sigma_p(t\mathbf{e} + x) \neq 0 \quad \text{for} \quad t \geq 0\}.$$

This set is a convex cone and equal to the connected component of the set $\{x \in \mathbf{R}^n : \sigma_p(x) \neq 0\}$ including the vector $\mathbf{e}$. We have the identity

$$\frac{d}{dt}\sigma_p(t\mathbf{e} + x) = p\sigma_{p-1}(t\mathbf{e} + x)$$

and the roots of the algebraic equation $\sigma_p(t\mathbf{e}+x) = 0$ relative to t are all real. Hence, for $x \in C_p$, the roots of $\sigma_{p-1}(t\mathbf{e} + x) = 0$ are also real and negative. This means $x \in C_{p-1}$.

Proposition 9.9. *Let* $f : M \to \mathbf{R}^{n+1}$ *be an ovaloid with Blaschke structure. Assume* $H_p > 0$ *for some p. Then*

(1) $H_q > 0$ *for* $q < p$.

(2) $(H_q)^{q-1} \leq (H_{q-1})^q$ *for* $2 \leq q \leq p$ *and the equality holds for at least one q if and only if all eigenvalues of S are equal.*

Consider the mapping $\lambda : M \to \mathbf{R}^n$ sending a point x to the vector $(\lambda_1(x), \ldots, \lambda_n(x))$ where $\lambda_i(x)$ are eigenvalues of the shape operator S. This mapping is continuous and the image contains a point of C_p because of Proposition 6.1, Chapter II. Hence the image itself is included in C_p. Then the reasoning above shows (1). Once the positivity is known, the property (2) follows from Lemma 9.6, whose proof we leave to the reader. Refer to [HLP, p. 52].

Proof of Theorem 9.8. Assume H_p is constant. In view of Proposition 6.1, Chapter II, $H_p > 0$. Put $H_p = c^p$, $c > 0$. Then Proposition 9.9 implies $H_q > 0$ for $q < p$ and $H_q \geq c^q$; in particular, $H_{p-1} \geq c^{p-1}$ and $H = H_1 \geq c$. Therefore,

$$\int \rho H_p \theta = -\int H_{p-1}\theta \leq -c^{p-1}\int \theta \quad \text{(Proposition 9.4 when } i = p - 1)$$

and

$$\int \rho H_p \theta = c^{p-1}\int \rho c\theta \geq c^{p-1}\int \rho H\theta = -c^{p-1}\int \theta$$

(Proposition 9.4 when $i = 0$).

Hence the equalities must hold in every place and $f(M)$ is an affine hypersphere. This completes the proof in view of Theorem 7.5.

10. The Blaschke–Schneider theorem

Another proof of Theorem 7.5 relies on a result, due to Blaschke ($n = 2$) [Bl] and Schneider ($n \geq 3$) [Schn1], concerning the differential equation

$$(10.1) \qquad \triangle u + nHu = 0,$$

for a function u on a nondegenerate hypersurface $f : M^n \to \mathbf{R}^{n+1}$ with affine mean curvature H. The Laplacian $\triangle$ is relative to the affine metric h.

Let us first observe that if $v : M^n \to \mathbf{R}_{n+1}$ is the conormal map, then for any constant vector $\mathbf{a} \in \mathbf{R}^{n+1}$ the function $u(x) = v_x(\mathbf{a})$ on M^n satisfies (10.1). Indeed, using the notation in Section 5, Chapter II, we have for any $X, Y \in \mathfrak{X}(M)$

$$Yu = (v_*(Y))(\mathbf{a}),$$
$$XYu = [D_X(v_*(Y))](\mathbf{a}) = [v_*(\bar{\nabla}_X Y)](\mathbf{a}) + [\bar{h}(X, Y)(-v)](\mathbf{a})$$

and

$$(\hat{\nabla}_X Y)(u) = [v_*(\bar{\nabla}_X Y)](\mathbf{a}) + 2v_*(K_X Y)(\mathbf{a})$$

so that

$$\text{Hess}_u(X, Y) = -h(SX, Y)v(\mathbf{a}) - 2v_*(K_X Y)(\mathbf{a}).$$

Threfore by apolarity we get

$$\triangle u = -(\text{tr } S)u,$$

i.e., (10.1).

Now we want to prove

Theorem 10.1. *Let* $f : M^n \to \mathbf{R}^{n+1}$ *be an ovaloid with Blaschke structure. Then, for any solution u of the equation*

$$\triangle u + nHu = 0,$$

there exists a constant vector $\mathbf{a}$ such that $u(x) = v_x(\mathbf{a})$.

Proof. Set

$$\mathbf{a} = \text{grad}_h u - u\xi.$$

Since $u(x) = -v_x(\mathbf{a})$, it is enough to see that $\mathbf{a}$ is constant. We have

$$D_X \mathbf{a} = \{f_*(\nabla_X \text{grad}_h u) + h(\text{grad}_h u, X)\xi\} - \{(Xu)\xi + u(-f_*(SX))\}$$
$$= f_*(\nabla_X \text{grad}_h u + uSX).$$

We define an endomorphism T of $T_x M$ by

$$T(X) = \nabla_X \operatorname{grad}_h u + u S X$$

so that $f_* T X = D_X \mathbf{a}$. Since

$$\operatorname{trace}\{X \mapsto D_X \mathbf{a}\}$$
$$= \operatorname{trace}\{X \mapsto \hat{\nabla}_X \operatorname{grad}_h u\} + \operatorname{trace}\{X \mapsto K_X \operatorname{grad}_h u\} + u \operatorname{tr} S$$
$$= \Delta u + n H u$$
$$= 0,$$

we have $\operatorname{tr} T = 0$. We take a fixed point o lying in the interior of $f(M)$ and consider the affine distance ρ from o. We define an $(n-1)$-form α by

$$\alpha(X_2, \ldots, X_n)$$
$$= \frac{1}{(n-1)!} \sum_i \operatorname{sign}(i_2, \ldots, i_n) \omega(f - o, \mathbf{a}, X_{i_2}\mathbf{a}, X_{i_3}f, \ldots, X_{i_n}f)$$

where ω is the parallel volume form of $\mathbf{R}^{n+1}$ and $i = (i_2, \ldots, i_n)$ is a permutation of $(2, \ldots, n)$. The derivation rule

$$n d\alpha(X_1, \ldots, X_n) = \sum_j (-1)^{j-1} (D_{X_j}\alpha)(X_1, \ldots, \widehat{X_j}, \ldots, X_n)$$

implies that

$$n d\alpha(X_1, \ldots, X_n)$$
$$= \sum_{j_1} (-1)^{j_1-1} \frac{1}{(n-1)!} \sum_{j'} \operatorname{sign}(j') \cdot \{\omega(X_{j_1}f, \mathbf{a}, X_{j_2}\mathbf{a}, X_{j_3}f, \ldots, X_{j_n}f)$$
$$+ \omega(f - o, X_{j_1}\mathbf{a}, X_{j_2}\mathbf{a}, X_{j_3}f, \ldots, X_{j_n}f)\},$$

where j' denotes permutation of $(1, \ldots, \hat{j}, \ldots, n)$ to $(j_2, \ldots, j_n)$. In view of $(-1)^{j_1-1}\operatorname{sign}(j') = \operatorname{sign}(j)$ where $j = (j_1, j_2, \ldots, j_n)$, we have

$$d\alpha(X_1, \ldots, X_n)$$
$$= \frac{1}{n!} \sum_j \operatorname{sign}(j) \cdot \{\omega(X_{j_1}f, \mathbf{a}, X_{j_2}\mathbf{a}, X_{j_3}f, \ldots, X_{j_n}f)$$
$$+ \omega(f - o, X_{j_1}\mathbf{a}, X_{j_2}\mathbf{a}, X_{j_3}f, \ldots, X_{j_n}f)\}.$$

Let $\{X_1, \ldots, X_n\}$ be a unimodular basis. Then, since $X_j f = f_* X_j$ and $X_j \mathbf{a}$ are tangent to the immersion, we get

$$\omega(X_{j_1}f, \mathbf{a}, X_{j_2}\mathbf{a}, X_{j_3}f, \ldots, X_{j_n}f) = u\omega(\xi, f_* X_{j_1}, f_* T X_{j_2}, \ldots, f_* X_{j_n})$$
$$= (-1)^{n+1} u T_{j_2}^{j_2} \theta(X_{j_1}, \ldots, X_{j_n})$$
$$= (-1)^{n+1} u T_{j_2}^{j_2} \operatorname{sign}(j).$$

Similarly, by the definition $f - o = Z + \rho\xi$,

$$\omega(f - o, X_{j_1}\mathbf{a}, X_{j_2}\mathbf{a}, X_{j_3}f, \ldots, X_{j_n}f) = (-1)^n \rho(T_{j_1}^{j_1} T_{j_2}^{j_2} - T_{j_1}^{j_2} T_{j_2}^{j_1}) \operatorname{sign}(j).$$

Therefore,

(10.2)
$$d\alpha = (-1)^n \left[\frac{\rho}{n(n-1)} H_2(T) - \frac{u}{n} \operatorname{tr} T \right] \theta.$$

Now the condition $\operatorname{tr} T = 0$ implies, on the one hand,

$$\int \rho H_2(T)\theta = 0$$

and, on the other hand,

$$0 = (\operatorname{tr} T)^2 = H_1(T)^2 \geq H_2(T).$$

Hence, $H_2(T) = 0$ and the equality $H_1(T)^2 = H_2(T)$ holds; this means that $T = 0$ by Lemma 9.6. Thus $\mathbf{a}$ is constant.

Now we shall give another proof of Theorem 7.5.

Proof of Theorem 7.5. Let $f : M^n \to \mathbf{R}^{n+1}$ be a nondegenerate immersion of a connected compact manifold M^n, and assume that it is an affine hypersphere. From Section 7 we may choose the affine normal field ξ that points inward so that the affine metric h is positive-definite and the center o of the affine hypersphere is in the interior of the convex body bounded by $f(M^n)$. For convenience's sake, we assume that o is the origin of $\mathbf{R}^{n+1}$. This means $f(x) + \xi_x/H = 0$, that is, $\xi_x = -Hf(x)$. By Theorem 6.5, Chapter II, we get $\Delta f = n\xi$ and hence

$$\Delta f + nHf = 0.$$

Now, by Theorem 10.1, for each element $e \in \mathbf{R}_{n+1}$, there exists a vector $A(e) \in \mathbf{R}^{n+1}$ such that $e(f(x)) = v_x(A(e))$ for all x; in fact,

$$A(e) = -\operatorname{grad} e(f(x)) + e(f(x))\xi$$

and $A : \mathbf{R}_{n+1} \to \mathbf{R}^{n+1}$ is a linear mapping. If $A(e) = 0$, then $e(f) = 0$ and $e(f_*X) = 0$ for every $X \in T_xM$. Since $\xi = -Hf$ is transversal to f, we have $e = 0$. This means that A is nonsingular. Therefore we have $A(v(x)) = f(x)$: in fact,

$$A(v(x)) = -\operatorname{grad} [v(x)(f(x))] + v(x)(f(x))\xi$$
$$= -\operatorname{grad}(-1/H) - \xi_x/H = f(x),$$

because H is constant. By setting $B = A^{-1}$, we have

$$1 = v(\xi) = \langle v, \xi \rangle = -H \langle Bf, f \rangle.$$

Hence f lies in a nondegenerate ellipsoid. The compactness implies that $f(M)$ coincides with an ellipsoid.

11. Affine minimal hypersurfaces and paraboloids

Let $f : M \to \mathbf{R}^{n+1}$ be a nondegenerate immersion. The affine volume of the immersion is the integral

$$A(f) = \int_M \omega_h,$$

where ω_h is the volume form of the affine metric h. If the triple (∇, h, S) is the Blaschke structure relative to the affine normal field ξ, then the volume form

$$\theta(X_1, \ldots, X_n) = \tilde{\omega}(X_1, \ldots, X_n, \xi)$$

plays the role of ω_h. The aim of this section is to provide a variational formula for the affine volume regarded as a functional and to define the notion of affine minimal immersions as a critical point of this functional. Then we prove a theorem characterizing a paraboloid among the global affine minimal hypersurfaces.

In order to state the variational formula, let us fix the notation: Let M be an orientable manifold, possibly with smooth boundary ∂M. Let $f : M \to \mathbf{R}^{n+1}$ be a nondegenerate immersion. A smooth variation of f is a differentiable mapping

$$F : I \times M \longrightarrow \mathbf{R}^{n+1}, \qquad I = (-\epsilon, \epsilon),$$

with the property: for $f_t(x) = F(t, x)$

(1) f_t is a nondegenerate immersion,

(2) $f_0 = f$,

(3) $f_t = f$ outside a compact set.

Let ξ_t be the affine normal field for f_t, which depends smoothly on t. Set

$$(11.1) \qquad \theta_t(X_1, \ldots, X_n) = \tilde{\omega}(f_{t*}X_1, \ldots, f_{t*}X_n, \xi_t).$$

Here we have $\xi_0 = \xi$ and $\theta_0 = \theta$. Assume that the immersion f_t is Blaschke relative to (ξ_t, θ_t). The affine volume of the immersion f_t is then

$$A(t) = \int_M \theta_t.$$

Denote by $V = F_*(\partial/\partial t)|_{t=0}$ the variation vector field, which is decomposed as

$$(11.2) \qquad V = V_0 + v\xi, \text{ where } V_0 \text{ is tangential.}$$

With the notation above we have the following formula.

Proposition 11.1.

$$\left.\frac{\partial A}{\partial t}\right|_{t=0} = -\frac{n(n+1)}{n+2}\int_M vH\theta.$$

The proof will be given at the end of this section. This formula lets us define affine minimality as follows.

Definition 11.1. A Blaschke immersion is said to be *affine minimal* when the affine mean curvature H vanishes everywhere, that is, when the functional A above is critical at this immersion.

With reference to the formula, it is better to say that the immersion is critical; however, we follow the traditional term affine minimal.

Example 11.1. Obviously, every improper affine hypersphere is affine minimal. In particular, a paraboloid is affine minimal.

Example 11.2. Let us recall the graph of a function treated as Blaschke immersion in Example 3.3 of Chapter II. It is defined by

$$f : G \ni (x^1,\ldots,x^n) \mapsto (x^1,\ldots,x^n,F(x^1,\ldots,x^n)) \in \mathbf{R}^{n+1},$$

where G is a domain in $\mathbf{R}^n$; we recall that the immersion is nondegenerate if and only if $\phi = |\det [F_{ij}]|^{1/(n+2)}$ does not vanish. In this case, the affine metric $\bar{h}$ is given by $\bar{h}_{ij} = \bar{h}(\partial_i,\partial_j) = \phi^{-1}F_{ij}$ where $F_{ij} = \partial^2 F/\partial x^i \partial x^j$ and the affine normal field $\bar{\xi}$ is given in the form

$$\bar{\xi} = -\sum_{k,j=1}^n (F^{kj}\phi_j)f_*(\partial_k) + \phi\xi,$$

where $\xi = (0,\ldots,0,1)$, $\partial_k = \partial/\partial x^k$, and $\phi_j = \partial\phi/\partial x^j$. Then we have

$$D_{\partial_i}\bar{\xi} = -\sum_{j,k} \partial_i(F^{kj}\phi_j)f_*(\partial_k)$$

and

(11.3)
$$S(\partial_i) = \sum_{j,k} \partial_i(F^{kj}\phi_j)\partial_k.$$

Hence we see

$$H = \frac{1}{n}\operatorname{tr} S = \frac{1}{n}\sum_{i,j} \partial_i(F^{ij}\phi_j).$$

Set $u = 1/\phi$ and let $\triangle$ be the Laplace operator relative to the affine metric. Then

$$\triangle u = \sum \frac{1}{\sqrt{\mathbf{h}}} \partial_i \left(\sqrt{\mathbf{h}} \bar{h}^{ij} \partial_j u \right) \qquad \text{where} \qquad \mathbf{h} = \det [\bar{h}_{ij}] = \phi^2$$

$$= -\frac{1}{\phi} \sum \partial_i (F^{ij} \phi_j)$$

$$= -nHu.$$

This follows also from (10.1); when v denotes the conormal map, we get $1 = v(\bar{\xi}) = \phi v(\xi)$ and $u = 1/\phi = v(\xi)$ for a constant vector ξ. From this equation we have

Proposition 11.2. *The graph of f is affine minimal if and only if the function u defined above is harmonic relative to the affine metric.*

As examples, we have affine minimal immersions when the function F is a (nondegenerate) quadratic polynomial, because u turns out to be constant. A less trivial example of this class is given when

$$F(x^1, \ldots, x^n) = x^1 \tanh x^2 + \frac{1}{2} \sum_{i=3}^{n} (x^i)^2.$$

In fact, we see $F_{ij} = 0$ except $F_{12} = F_{21} = 1/(\cosh x^2)^2$, $F_{ii} = 1$ $(i \geq 3)$, and $\phi = (\cosh x^2)^{4/(n+2)}$. Hence the affine metric is indefinite, the shape operator is of rank 1, and $H = 0$.

Remark 11.1. The last example is due to [Ko]. We know numerous other examples; refer, say, to [VeVr], [Bl], and Notes 2 and 3.

Referring to these examples, one can pose the following problem, which is an analogue of Bernstein's theorem in Euclidean geometry.

Problem 11.1. Consider the surface given as a graph over $\mathbf{R}^2$,

$$\mathbf{R}^2 \ni (x^1, x^2) \mapsto (x^1, x^2, F(x^1, x^2)) \in \mathbf{R}^3.$$

If the affine metric is definite and the surface is affine minimal, that is, the function F satisfies a differential equation $\triangle u = 0$ as above, then is the surface affinely equivalent to an elliptic paraboloid?

In addition to the assumption of the problem, assume that the affine metric is complete. The formula in Proposition 9.2 of Chapter II then implies $\widehat{\mathrm{Ric}}(X, Y) = \mathrm{tr}(K_X K_Y)$ is positive-definite and, by the theorem of Blanc and Fiala (refer to [BF]), any positive harmonic function is constant. This means that the function u is constant and therefore the shape operator is trivial; hence Theorem 7.7 shows the following.

Theorem 11.3. *Any affine minimal surface given as a graph over $\mathbf{R}^2$,*

$$\mathbf{R}^2 \ni (x^1, x^2) \mapsto (x^1, x^2, F(x^1, x^2)) \in \mathbf{R}^3,$$

is an elliptic paraboloid provided that the affine metric is definite and complete.

The following result has already been proved as Proposition 7.4. We provide a different proof.

Proposition 11.4. *Let $f : M^n \to \mathbf{R}^{n+1}$ be a Blaschke immersion of a compact manifold M^n. Then f cannot be affine minimal.*

Proof. Let ρ be the affine distance function on M^n relative to an arbitrary point in $\mathbf{R}^{n+1}$. Proposition 6.2 of Chapter II says that

$$\triangle \rho + n(1 + H\rho) = 0.$$

If f is affine minimal, then Green's formula implies that $0 = \int \triangle \rho \theta = -n \int \theta$, which is a contradiction.

We shall prove

Theorem 11.5. *Any locally strictly convex, topologically closed, affinely complete, affine minimal surface in $\mathbf{R}^3$ is an elliptic paraboloid.*

Proof. Proposition 11.4 says that the surface is noncompact. On the other hand, the surface being locally strictly convex and topologically closed, it can be written as the graph of a strictly convex function defined on the whole of $\mathbf{R}^2$ – due to a generalization of Hadamard's theorem (see [Sac]). Then Theorem 11.3 implies the result.

Remark 11.2. Theorem 11.5 is due to [Cal4]. Problem 11.1 is yet to be completely solved. Refer to papers [Cal5], [L3], [MMi2]. Analogously to the Euclidean case, the affine minimal surface has an integral representation, the so-called affine Weierstrass formula. See Note 4.

Proof of Proposition 11.1. Let us recall the notation fixed in the beginning. We write the fundamental equations for each t as follows:

(11.4)
$$D_X f_{t*} Y = f_{t*}(\nabla^t_X Y) + h_t(X, Y)\xi_t,$$
$$D_X \xi_t = -f_{t*}(S^t X).$$

Let $\{X_1, \ldots, X_n\}$ be a unimodular basis of f: $\theta(X_1, \ldots, X_n) = 1$. It has the property $|\det [h(X_i, X_j)]| = 1$.

We extend each vector field X_i to $I \times M$ depending trivially on t. Then

(11.5)
$$\left[X_i, \frac{\partial}{\partial t} \right] = 0.$$

Set

$$\varphi(t) = |\det [h_t(X_i, X_j)]|.$$

Determine the function λ on $I \times M$ so that

$$\theta_t(\lambda X_1, \ldots, \lambda X_n) = 1.$$

Then, by the assumption that ξ_t is the affine normal field for each t, we have $|\det [h_t(\lambda X_i, \lambda X_j)]| = 1$. On the other hand, since both θ_t and θ are volume forms on M, there exists a function μ_t such that

$$\theta_t = \mu_t \theta.$$

It is easy to see $\mu_t = \varphi^{\frac{1}{2}}$. Now we have

$$\frac{\partial}{\partial t} \theta_t = \frac{1}{2} \varphi^{-\frac{1}{2}} \frac{\partial \varphi}{\partial t} \theta;$$

hence at $t = 0$,

$$\frac{\partial}{\partial t} \theta_t \Big|_{t=0} = \frac{1}{2} \frac{\partial \varphi}{\partial t} \Big|_{t=0} \theta.$$

Here note that

(11.6)
$$\frac{\partial \varphi}{\partial t} \Big|_{t=0} = \sum h^{ij} \frac{\partial}{\partial t} h_t(X_i, X_j) \Big|_{t=0}.$$

Differentiation of (11.1) gives another representation: set

$$\frac{\partial}{\partial t} \xi_t \Big|_{t=0} = f_*(Z) + a\xi$$

where Z is the tangent vector to M. Then

$$\frac{\partial}{\partial t} \theta_t(X_1, \ldots, X_n) \Big|_{t=0}$$

$$= \sum \tilde{\omega}(X_1, \ldots, D_V f_{t*} X_i, \ldots, \xi) + \tilde{\omega}(X_1, \ldots, X_n, \frac{\partial}{\partial t} \xi_t)$$

$$= \sum \tilde{\omega}(X_1, \ldots, D_{X_i} V, \ldots, \xi) + \tilde{\omega}(X_1, \ldots, X_n, a\xi) \quad \text{by (11.5)}$$

$$= \sum \tilde{\omega}(X_1, \ldots, \nabla_{X_i} V_0 - vSX_i, \ldots, \xi) + \tilde{\omega}(X_1, \ldots, X_n, a\xi)$$

$$= (\operatorname{div} V_0 - v \operatorname{tr} S + a)\theta(X_1, \ldots, X_n),$$

where we use the identity

$$\sum \tilde{\omega}(X_1, \ldots, \nabla_{X_i} V_0, \ldots, X_n, \xi)$$

$$= \operatorname{div} V_0 + \sum \tilde{\omega}(X_1, \ldots, K_{X_i} V_0, \ldots, X_n, \xi)$$

$$= \operatorname{div} V_0 \quad \text{(apolarity)}.$$

This implies

(11.7)
$$a = \frac{1}{2} \frac{\partial}{\partial t} \varphi \Big|_{t=0} - \operatorname{div} V_0 + v \operatorname{tr} S.$$

Now we differentiate the first equation of (11.4) relative to t. The left-hand side is

$$D_{\frac{\partial}{\partial t}} D_X f_{t*} Y = D_X D_{\frac{\partial}{\partial t}} f_{t*} Y \quad \text{because of } R^D = 0 \text{ and } (11.5)$$
$$= D_X D_Y V \quad \text{by } (11.5)$$
$$= D_X (\nabla_Y V_0 + h(V_0, Y)\xi - vSY + Y(v)\xi)$$
$$\equiv [X(h(V_0, Y)) + h(X, \nabla_Y V_0) + XY(v) - vh(X, SY)]\xi$$

modulo tangent space. The derivative at $t = 0$ of the right-hand side of (11.4) is equal to

$$D_{\frac{\partial}{\partial t}}(\nabla_X^t Y)\Big|_{t=0} + \frac{\partial}{\partial t} h_t(X, Y)\Big|_{t=0} \xi + h(X, Y)(Z + a\xi)$$
$$= D_{\nabla_X Y}(V_0 + v\xi)|_{t=0} + \frac{\partial}{\partial t} h_t(X, Y)\Big|_{t=0} \xi + h(X, Y)(Z + a\xi)$$
$$\equiv \left[\frac{\partial}{\partial t} h_t(X, Y)\Big|_{t=0} + h(\nabla_X Y, V_0) + (\nabla_X Y)(v) + ah(X, Y) \right]\xi$$

modulo tangent space. Therefore by equating the both sides, we get

$$\frac{\partial}{\partial t} h_t(X, Y)\Big|_{t=0} + ah(X, Y)$$
$$= X(h(V_0, Y)) + h(X, \nabla_Y V_0) + XY(v)$$
$$\quad - vh(X, SY) - (\nabla_X Y)v - h(\nabla_X Y, V_0)$$
$$= C(X, Y, V_0) + h(\nabla_X V_0, Y) + h(X, \nabla_Y V_0) + \text{Hess}_v(X, Y) - vh(X, SY).$$

Then taking trace$_h$ of the bilinear function represented by each term shows the identity

$$\frac{\partial}{\partial t}\varphi\Big|_{t=0} + na = 2\sum h^{ij}h(\nabla_{X_i} V_0, X_j) + \triangle v - v\,\text{tr}\,S$$
$$= 2\,\text{div}\,V_0 + \triangle v - v\,\text{tr}\,S.$$

Here we have used

$$\sum 2h^{ij}h(\nabla_{X_i} V_0, X_j) = \sum h^{ij}h(\hat{\nabla}_{X_i} V_0 + K_{X_i} V_0, X_j)$$
$$= \sum h^{ij}h(\hat{\nabla}_{X_i} V_0, X_j) \quad \text{(apolarity)}$$
$$= \text{div}\,V_0.$$

Combining this identity with equation (11.7), we get

(11.8) $$\frac{1}{2}\frac{\partial}{\partial t}\varphi\Big|_{t=0} = \text{div}\,V_0 + \frac{1}{n+2}\triangle v - \frac{n(n+1)}{n+2}v\,\text{tr}\,S.$$

The boundary condition (3) in the definition of variation implies the proposition in view of Green's theorem.

IV

Affine-geometric structures

In this chapter we emphasize the geometry of affine immersions that has been developed in the last ten years. In Section 1 we are concerned with the general properties of possibly degenerate hypersurfaces. We study, in Section 2, affine immersions of $\mathbf{R}^n$ into $\mathbf{R}^{n+1}$, which are analogous to the classification problems for isometric immersions of Euclidean n-space $\mathbf{E}^n$ into $\mathbf{E}^{n+1}$ and of Lorentz–Minkowski n-space $\mathbf{L}^n$ into $\mathbf{L}^{n+1}$. As a matter of fact, we apply the affine result to these metric cases to obtain a unified proof. Section 3 proves the Cartan–Norden theorem saying, roughly, that if there is an affine immersion $(M^n, \nabla) \to \mathbf{R}^{n+1}$, where ∇ is the Levi-Civita connection of a pseudo-Riemannian metric g on M^n, then g can be extended to a pseudo-Euclidean metric on $\mathbf{R}^{n+1}$. In Section 4, we give a generalization of the theorem on locally symmetric Blaschke hypersurfaces to the case of an arbitrary transversal connection. Section 5 extends the classical Cohn-Vossen rigidity theorem for an ovaloid in Euclidean 3-space to equiaffine immersion. In Section 6, the classical Pick–Berwald theorem on vanishing cubic form is extended in such a way that it will later lead to a generalization in a projective setting. Sections 7 and 8 together provide an introduction to projective differential geometry by using the method of affine geometry as follows. Section 7 studies the notion of projective immersion, and Section 8 introduces several projective invariants and projectively invariant properties for a nondegenerate hypersurface in $\mathbf{P}^{n+1}$ (including a projective version of the Pick–Berwald theorem). In Section 9, we first clarify various notions on connections on a complex manifold and then discuss holomorphic immersions of a complex manifold into complex affine space $\mathbf{C}^{n+1}$. Depending on whether we choose a holomorphic or antiholomorphic transversal $(1,0)$-vector field, the induced connection on M^n is holomorphic or affine Kähler.

1. Hypersurfaces with parallel nullity

Given an immersion $f : M \to \mathbf{R}^{n+1}$ and a transversal vector field ξ, we consider the condition that the kernel of h be parallel relative to the connection

∇ induced by ξ. It turns out that this condition is independent of the choice of ξ and, then, the kernel defines an integrable totally geodesic distribution. The condition that the induced connection on each leaf of the distribution is complete is also independent of the choice of ξ. The aim of this section is to give a fundamental structural theorem for such an immersion given in [NO4].

Let $f : M \to \mathbf{R}^{n+1}$ be an affine immersion of a connected manifold M and ξ an arbitrarily chosen transversal vector field. Recall the formulas of Gauss and Weingarten

$$D_X f_*(Y) = f_*(\nabla_X Y) + h(X, Y)\xi$$
$$D_X \xi = -f_*(SX) + \tau(X)\xi;$$

see Section 1 of Chapter II. For each x, define the *null space*

$$(1.1) \qquad N(x) = \{X \in T_x(M) : h(X, Y) = 0 \quad \text{for all} \quad Y \in T_x(M)\}.$$

Because of Proposition 2.5, Chapter II, the space $N(x)$ is defined independently of ξ. We define the distribution $\mathcal{N} : x \mapsto N(x)$ along f and call it the *relative nullity foliation*; indeed, we prove

Lemma 1.1. *The distribution $\mathcal{N}$ is integrable and totally geodesic.*

Proof. It is sufficient to show that $\mathcal{N}$ is totally geodesic, that is, for vector fields Y and Z belonging to $\mathcal{N}$, $\nabla_Y Z$ has values in $\mathcal{N}$. In the equation of Codazzi for h,

$$(\nabla_X h)(Y, Z) + \tau(X)h(Y, Z) = (\nabla_Y h)(X, Z) + \tau(Y)h(X, Z),$$

take $Y, Z \in \mathcal{N}$. Then we get $h(X, \nabla_Y Z) = 0$. This being valid for all X, we have $\nabla_Y Z \in \mathcal{N}$. It then follows that $[Y, Z] = \nabla_Y Z - \nabla_Z Y \in \mathcal{N}$, hence $\mathcal{N}$ is integrable.

We say that $\mathcal{N}$ is *complete* if each leaf L is complete relative to ∇, that is, every ∇-geodesic in L extends infinitely for its affine parameter. The equation (2.9) of Chapter II implies that the induced connection on each leaf is defined independently of the choice of ξ and, in particular, the property that $\mathcal{N}$ is complete is independent of the choice of ξ. The formula of Gauss restricted to a leaf L is

$$D_X f_* Y = f_*(\nabla_X Y).$$

This implies that $f(L)$ is a totally geodesic submanifold in $\mathbf{R}^{n+1}$ and, if $\mathcal{N}$ is complete, then $f(L)$ is an entire affine subspace.

We say that $\mathcal{N}$ is *parallel* if, for every curve from x to y, parallel translation along the curve relative to ∇ maps $N(x)$ to $N(y)$. This is the case if and only if $\nabla_X Y \in \mathcal{N}$ for every vector X and for every vector field $Y \in \mathcal{N}$. The equation (2.9) of Chapter II shows that this condition is also independent of the choice of ξ. So we simply say that $\mathcal{N}$ is parallel.

Definition 1.1. We say that the immersion satisfies the condition of *parallel nullity* if $\mathcal{N}$ is parallel.

Under the assumption of parallel nullity, we show that for two distinct leaves L_1 and L_2 of $\mathcal{N}$, $f(L_1)$ and $f(L_2)$ are affine subspaces that are D-parallel in $\mathbf{R}^{n+1}$. To prove this, let x_t be an arbitrary curve from $x \in L_1$ to $y \in L_2$, and let $Y_t \in \mathcal{N}$ be a family of ∇-parallel vectors along the curve. Then

$$D_t f_*(Y_t) = f_*(\nabla_t Y_t) + h(\dot{x}_t, Y_t) = 0,$$

which shows that $f_*(Y_t)$ is D-parallel in $\mathbf{R}^{n+1}$. Hence $f(L_1)$ and $f(L_2)$ are parallel affine subspaces in $\mathbf{R}^{n+1}$.

Lemma 1.2. *Assume $\bar{\xi} = \xi$ mod $\mathcal{N}$, that is, $\bar{\xi} = \xi + f_*(Z)$ for $Z \in \mathcal{N}$. Then we have*

$$(1.2) \qquad \bar{h} = h, \qquad \bar{\tau} = \tau,$$
$$(1.3) \qquad \bar{\nabla}_X Y \equiv \nabla_X Y \text{ mod} \mathcal{N} \quad \text{for all} \quad X \text{ and } Y, \quad \bar{\nabla}\bar{h} = \nabla h.$$

Assume, further, $\mathcal{N}$ is parallel. Then

$$(1.4) \qquad \bar{S}X \equiv SX \text{ mod} \mathcal{N}, \quad \bar{R}(X,Y)Z \equiv R(X,Y)Z \text{ mod} \mathcal{N}$$

for all X, Y, and Z. Moreover, if ξ satisfies $S\mathcal{N} \subset \mathcal{N}$, then

$$(1.5) \qquad (\bar{\nabla}_X \bar{S})(Y) \equiv (\nabla_X S)(Y) \text{ mod} \mathcal{N},$$
$$(1.6) \qquad (\bar{\nabla}_X \bar{R})(Y,Z)W \equiv (\nabla_X R)(Y,Z)W \text{ mod} \mathcal{N}$$

for all X, Y, Z, and W.

Proof. We leave this to the reader; refer to Proposition 2.5 of Chapter II.

With the preparation above, we can state the main theorem of this section.

Theorem 1.3. *Let $f : M^n \to \mathbf{R}^{n+1}$ be a connected hypersurface satisfying the condition of parallel nullity. Set $s = \dim \mathcal{N}$ and $r = n - s$. Assume $\mathcal{N}$ is complete. Then we can express $f : M^n \to \mathbf{R}^{n+1}$ as follows: $M^n = M^r \times L$, $f = f_1 \times f_0$, where $f_1 : M^r \to \mathbf{R}^{r+1}$ is a connected nondegenerate hypersurface and f_0 is a connection-preserving map of a leaf L of $\mathcal{N}$ onto $\mathbf{R}^s$, and $\mathbf{R}^{n+1} = \mathbf{R}^{r+1} \times \mathbf{R}^s$. Such a representation is unique up to equiaffine transformation of $\mathbf{R}^{n+1}$ so that a nondegenerate hypersurface M^r is determined uniquely up to a equiaffine transformation of $\mathbf{R}^{r+1}$.*

Proof. Let x_0 be an arbitrary but fixed point of M^n. For the leaf L_0 through x_0 of $\mathcal{N}$, $f(L_0)$ is an entire affine subspace of dimension $s = n - r$ through $o = f(x_0)$ in $\mathbf{R}^{n+1}$. Call it $\mathbf{R}^s$. For any point $p \in \mathbf{R}^{n+1}$ we denote by $\mathbf{R}^s(p)$ the s-dimensional affine subspace through p that is parallel to $\mathbf{R}^s$. We know that if $x \in M^n$, then the image under f of the leaf L through x coincides with $\mathbf{R}^s(f(x))$. Let us choose an affine subspace of dimension $r + 1$, say $\mathbf{R}^{r+1}$, through $f(x_0)$ that is transversal to $\mathbf{R}^s$. For any p in $\mathbf{R}^{n+1}$, the $(r + 1)$-dimensional affine subspace through p and parallel to

$\mathbf{R}^{r+1}$ will be denoted by $\mathbf{R}^{r+1}(p)$. The mapping $f : M^n \to \mathbf{R}^{n+1}$ is then transversal to $\mathbf{R}^{r+1}$. In fact, for any $x \in M^n$ such that $p = f(x) \in \mathbf{R}^{r+1}$ we have $T_p(\mathbf{R}^{n+1}) = T_p(\mathbf{R}^{r+1}) + f_*(T_x(M^n))$, because $f_*(T_x(M^n))$ contains $\mathbf{R}^s(p) = f(L)$, where L is the leaf of $\mathcal{N}$ through x. By a well-known theorem concerning transversal mappings (for example, see [Hi, p.22]), it follows that $M^r = \{x \in M^n : f(x) \in \mathbf{R}^{r+1}\}$ is an r-dimensional submanifold of M^n. We see that the restriction of f to M^r gives rise to a hypersurface $f_1 : M^r \to \mathbf{R}^{r+1}$; we shall show in a moment that M^r is connected . In the case where the original immersion is an imbedding, we may think of M^r as the intersection of M^n with $\mathbf{R}^{r+1}$.

Now we define a one-to-one map $\Phi : M^n \to M^r \times L_0$ as follows. We consider $o = f(x_0)$ as the origin of $\mathbf{R}^{n+1}$, $\mathbf{R}^s$, and $\mathbf{R}^{r+1}$, whenever we need a reference point in each of these affine spaces. Now for any $x \in M^n$ we define

$$\Phi(x) = (y, z) \in M^r \times L_0,$$

where y and z are determined as follows. Consider $p = f(x)$. For the leaf $L(x)$ of $\mathcal{N}$ through x, $f(L(x))$ is the affine subspace $\mathbf{R}^s(p)$, which meets $\mathbf{R}^{r+1}$ at a certain unique point, say q. Since f maps one-to-one on $L(x)$, there is a unique point $y \in L(x) \subset M^n$ such that $f(y) = q$. This means $y \in M^r$. On the other hand, the vector from q to p is parallel to the vector from o to z, where z is a certain uniquely determined point of $\mathbf{R}^s$. It is now easy to find the inverse map $M^r \times L_0 \to M^n$ of Φ. Since Φ is differentiable, the existence of the projection $M^n \to M^r$ shows that M^r is connected. So we get a representation $\Phi : M^n \to M^r \times L_0$, where $\mathbf{R}^{n+1} = \mathbf{R}^{r+1} \times \mathbf{R}^s$, $f_1 : M^r \to \mathbf{R}^{r+1}$ is a hypersurface, id : $L_0 \to \mathbf{R}^s$ is a diffeomorphism preserving the flat connections, and thus $f = (f_1 \times \mathrm{id}) \circ \Phi$. We shall call Φ a *cylinder representation* with a *profile hypersurface* M^r.

We have yet to prove the uniqueness of such a representation. For this purpose we use the following lemma in analytic geometry, which is easy to prove.

Lemma 1.4. *Let $\mathbf{R}^s$ be a fixed affine subspace of the affine space $\mathbf{R}^{n+1}$. Suppose $\mathbf{R}^{r+1}$ and $\bar{\mathbf{R}}^{r+1}$ are two affine subspaces that are transversal to $\mathbf{R}^s$. We define a map F of $\mathbf{R}^{r+1}$ onto $\bar{\mathbf{R}}^{r+1}$ as follows: for each $x \in \mathbf{R}^{r+1}$, let $\mathbf{R}^s(x)$ denote the affine subspace through x that is parallel to $\mathbf{R}^s$. We let $\bar{x}$ be a uniquely determined point of intersection with $\bar{\mathbf{R}}^{r+1}$ and set $F(x) = \bar{x}$. Then F is an affine transformation of $\mathbf{R}^{r+1}$ onto $\bar{\mathbf{R}}^{r+1}$. Moreover, F is volume-preserving if we fix a parallel volume element ω_{n+1} on $\mathbf{R}^{n+1}$ and a parallel volume element ω_s on $\mathbf{R}^s$ and define parallel volume elements ω_{r+1} and $\bar{\omega}_{r+1}$ on $\mathbf{R}^{r+1}$ and $\bar{\mathbf{R}}^{r+1}$, respectively, such that ω_{n+1} is the direct product of ω_{r+1} and ω_s as well as of $\bar{\omega}_{r+1}$ and ω_s.*

Now suppose $\bar{\Phi} : M^n \to \bar{M}^r \times \bar{L}$ is another cylinder representation, where $\bar{f}_1 : \bar{M}^r \to \bar{\mathbf{R}}^{r+1}$ is a nondegenerate hypersurface of $\bar{\mathbf{R}}^{r+1}$ and $\bar{f}_0 : \bar{L} \to \bar{\mathbf{R}}^s$ is a connection-preserving map of a leaf $\bar{L}$ of $\mathcal{N}$ onto an affine subspace $\bar{\mathbf{R}}^s$ transversal to $\bar{\mathbf{R}}^{r+1}$. We may assume, without loss of generality, that $L = \bar{L}$, $\mathbf{R}^s = \bar{\mathbf{R}}^s$, and $f_0 = \bar{f}_0$. Now we get an equiaffine transformation

$F_1 : \mathbf{R}^{r+1} \to \bar{\mathbf{R}}^{r+1}$ in the manner of Lemma 1.4. Combining this with the identity map, we get an equiaffine transformation, denoted by F, of $\mathbf{R}^{n+1}$ onto itself. It is now clear that $F_1(M^r) = \bar{M}^r$ and $\bar{\Phi} = F \circ \Phi$. This completes the proof of the theorem.

Corollary 1.5. *Under the assumptions of the theorem, we can find a transversal vector field ξ for M^n with the following properties:*

(1) *The affine shape operator vanishes on $\mathcal{N}$, that is, ξ is D-parallel in the direction of $\mathcal{N}$.*

(2) *The restriction of ξ to a profile hypersurface M^r coincides with the affine normal field of the nondegenerate hypersurface M^r.*

Such ξ is unique once a profile hypersurface is chosen.

Remark 1.1. If we do not assume the completeness for $\mathcal{N}$, then for any point x_0 of M^n, we can get a local cylinder decomposition of a neighborhood of x_0 in the form $V \times W$, where V is a nondegenerate hypersurface in $\mathbf{R}^{r+1}$ and W is an open subset of $\mathbf{R}^s$.

A cylinder representation as in the theorem will help us better understand the relationship between the geometry of M^n and that of M^r. For this purpose, continuing the notation in the proof of the theorem, we define a distribution $\mathcal{T}$ by

$$\mathcal{T}(x) = (f_{*x})^{-1}(\mathbf{R}^{r+1}) \quad \text{for each} \quad x \in M^n,$$

where $\mathbf{R}^{r+1}$ is now considered as the vector subspace instead of the affine space through $f(x_0)$. This distribution is integrable. We denote by π the projection of the vector space $\mathbf{R}^{n+1}$ onto $\mathbf{R}^{r+1}$ parallel to the subspace $\mathbf{R}^s$. We also denote by the same symbol the projection of $f_*^{-1}(\mathbf{R}^{r+1})$ onto $\mathcal{T}$ parallel to $\mathcal{N}$ so that $f_* \circ \pi = \pi \circ f_*$. Let ξ be a transversal vector field of f. We define $\bar{\xi} = \pi \circ \xi$. Then $\bar{\xi}$ is also transversal to f and equal to ξ mod $\mathcal{N}$. By the formulas in Lemma 1.2 we have the following proposition.

Proposition 1.6. *With the notation as above, the following relations hold:*

$$\bar{h} = h, \quad \bar{\tau} = \tau, \quad \bar{S} = \pi \circ S,$$
$$\bar{\nabla}_X Y = \pi(\nabla_X Y), \quad \bar{R}(X,Y)Z = \pi(R(X,Y)Z),$$

and

$$(\bar{\nabla}_X \bar{S})(Y) = \pi(\nabla_X S)(Y), \quad (\bar{\nabla}_W \bar{R})(X,Y)Z = \pi(\nabla_W R)(X,Y)Z;$$

for the last two identities we need to assume that ξ satisfies the condition $S\mathcal{N} \subset \mathcal{N}$. Moreover, the same relations hold if $\bar{\nabla}$ is considered as the connection on M^r.

Now we give an application. Recalling the Gauss equation, we easily see that

(1.7)
$$N(x) \subset \bigcap_{X,Y \in T_x(M)} \ker R(X,Y).$$

Lemma 1.7. *If* rank $S > 1$, *then the subspaces on both sides are equal.*

Proof. Choose $Z \in \bigcap \ker R(X, Y)$, that is, $R(X, Y)Z = 0$ for all X and Y. The Gauss equation implies $h(Y, Z)SX = h(X, Z)SY$. If $h(X, Z) \neq 0$ for some X, then this means that the image of S is 1-dimensional or $S = 0$. This contradicts the assumption rank $S > 1$. So $Z \in N(x)$.

Corollary 1.8. *Assume ξ is an equiaffine transversal vector field to a hypersurface* $f : M^n \to \mathbf{R}^{n+1}$ *such that* $\nabla R = 0$. *If* rank $S > 1$ *everywhere, then M^n is locally a cylinder $M^r \times \mathbf{R}^s$ and $\bar{\nabla}$ restricted to M^r is locally symmetric.*

Proof. Because of Lemma 1.7 and the condition $\nabla R = 0$, the distribution $\mathcal{N}$ is parallel. The last property of Proposition 1.6 implies that the restricted connection is also locally symmetric.

2. Affine immersions $\mathbf{R}^n \to \mathbf{R}^{n+1}$; applications

We start this section with a model of affine immersions $\mathbf{R}^n \to \mathbf{R}^{n+1}$.

Example 2.1. *Affine cylinder.* An affine cylinder in $\mathbf{R}^{n+1}$ is a hypersurface generated by a parallel family of affine $(n-1)$-spaces $\mathbf{R}^{n-1}(t)$ each passing through the point $\gamma(t)$ of a curve $\gamma(t)$. We describe this more precisely.

Let $\gamma(t)$ be a smooth curve in $\mathbf{R}^{n+1}$ and $\xi(t)$ a vector field along $\gamma(t)$. Let $\mathbf{R}^{n-1}$ be a fixed affine $(n-1)$-space in $\mathbf{R}^{n+1}$ and denote by $\mathbf{R}^{n-1}(p), p \in \mathbf{R}^{n+1}$, the affine $(n-1)$-space through p that is parallel to $\mathbf{R}^{n-1}$. We assume that

(a) $\gamma'(t)$, $\xi(t)$ and $\mathbf{R}^{n-1}(\gamma(t))$ are linearly independent;

(b) $\gamma''(t) = \rho(t)\xi(t)$, where $\rho = \rho(t)$ is a certain differentiable function.

We define a mapping $f : \mathbf{R}^n \to \mathbf{R}^{n+1}$ as follows. Write $\mathbf{R}^n = \mathbf{R} \times \mathbf{R}^{n-1}$ so that every point of $\mathbf{R}^n$ is written as (t, y), $t \in \mathbf{R}$, $y \in \mathbf{R}^{n-1}$. We set

$$f(t, y) = \gamma(t) + y.$$

This defines an immersion of $\mathbf{R}^n$. We choose a transversal field ξ by translating the vector $\xi(t)$ to the point $f(t, y)$; the transversality follows from (a). We easily see that f is an affine immersion of $\mathbf{R}^n \to \mathbf{R}^{n+1}$. Along the curve $x(t) = (t, 0)$ in $\mathbf{R}^n$, we have

$$D_t f(\dot{x}_t) = \gamma'' = \rho(t)\xi(t) \quad \text{and} \quad h\left(\frac{\partial}{\partial t}, \frac{\partial}{\partial t}\right) = \rho(t).$$

In the special case where we take $\xi(t) = \gamma''$ and furthermore γ'' and γ''' are linearly independent, we call it a *proper affine cylinder*. In this case, since

$$D_t \xi = \gamma''' = -f_*(S\left(\frac{\partial}{\partial t}\right)) + \tau\left(\frac{\partial}{\partial t}\right)\gamma'',$$

the shape operator S does not vanish. We also see that $h(\partial/\partial t, \partial/\partial t) = 1$ for every t.

Now we are going to characterize this immersion; in fact, we are interested in classifying all affine immersions $f : M = \mathbf{R}^n \to \mathbf{R}^{n+1}$, both manifolds being equipped with the standard flat affine connections. For such f, we may always choose an equiaffine transversal field ξ (see Example 2.3 of Chapter II).

We observe: if h is identically 0, then f is totally geodesic and $f(\mathbf{R}^n)$ is an affine hyperplane; if S is identically 0, then f is a graph immersion by Proposition 2.8 of Chapter II. Hence in the following we consider the case where the set $\Omega = \{x \in M : S_x \neq 0, h_x \neq 0\}$ is not empty.

Lemma 2.1. *For each $x \in \Omega$, the null space $N(x)$ of h coincides with $\ker S_x$ and its dimension is $n - 1$. The distribution $\mathcal{N} : x \mapsto N(x)$ is a totally geodesic foliation in M.*

Proof. Since $R = 0$, the Gauss equation is $h(Y, Z)SX = h(X, Z)SY$. The equality $N(x) = \ker S_x$ follows directly. Assume $\operatorname{rank} S_x \geq 2$ for some $x \in \Omega$. Then we can find two vectors X and Y such that SX and SY are linearly independent. By the Gauss equation, we have $h(Y, Z) = 0 = h(X, Z)$ for every Z. This means that $X, Y \in N(x) = \ker S_x$, which is a contradiction; therefore, $\operatorname{rank} S = 1$ at every point of Ω. The second assertion has been proved in Lemma 1.1 of this chapter.

We show that the relative nullity foliation $\mathcal{N}$ on Ω is complete, that is,

Proposition 2.2. *Each leaf L of the foliation $\mathcal{N}$ on Ω in Lemma 2.1 is complete.*

Proof. Let $x(t)$ be a geodesic starting at x_0 in L. To show that $x(t)$ extends for all values of t in L, first extend it as a geodesic in M. It is sufficient to show that $x(t)$ lies in Ω.

Suppose there is $b > 0$ such that $x(b) \notin \Omega$ but $x(t) \in \Omega$ for all $t < b$. Fix an open subset W of Ω containing the geodesic $x(t)$, $0 \leq t < b$, and let X be a vector field such that $\nabla_X X = 0$, $X \in \mathcal{N}$, and X at $x(t)$ equals the tangent vector $\dot{x}(t)$ for $0 \leq t < b$. We choose a parallel vector field U on $M = \mathbf{R}^n$ that is transversal to the hyperplane $H = \mathbf{R}^{n-1}$ containing L in $M = \mathbf{R}^n$.

In the situation above, we have the following:

(i) Write $\nabla_U X = \mu U + Z$ at each point $p \in W \cap H$, where $Z(p) \in N(p)$. Then the function μ satisfies $X\mu = -\mu^2$ along the geodesic $x(t)$ for $0 \leq t < b$.

(ii) Write $SU = \lambda U + V$ at each point $p \in W \cap H$, where $V(p) \in \mathcal{N}(p)$. Then the function λ satisfies $X\lambda = -\mu\lambda$ along $x(t)$ for $0 \leq t < b$.

(iii) Let $\rho = h(U, U)$ on $W \cap H$. Then $X\rho = -\mu\rho$ along $x(t)$ for $0 \leq t < b$.

We prove these assertions.

(i) Since U is parallel and $\mathcal{N}$ is totally geodesic (Lemma 2.1), we get

$$\nabla_X(\nabla_U X) = \nabla_X(\mu U + Z) = (X\mu)U + \nabla_X Z \equiv (X\mu)U \mod \mathcal{N}.$$

Since $R = 0$, we have along $x(t)$

$$\nabla_X(\nabla_U X) = \nabla_{[X,U]} X = -\nabla_{\nabla_U X} X = -\nabla_{\mu U + Z} X$$
$$= -\mu\nabla_U X - \nabla_Z X \equiv -\mu^2 U \mod \mathcal{N}.$$

Hence $(X\mu)U \equiv -\mu^2 U \mod \mathcal{N}$ and $X\mu = -\mu^2$.

(ii) From the Codazzi equation for S

$$\nabla_X(SU) - S(\nabla_X U) = \nabla_U(SX) - S(\nabla_U X),$$

we get along $x(t)$

$$(X\lambda)U + \lambda(\nabla_X U) + \nabla_X V = -\mu SU = -\mu(\lambda U + V)$$

and $(X\lambda)U \equiv -\mu\lambda U \bmod \mathscr{N}$. Thus $X\lambda = -\mu\lambda$ along $x(t)$.

(iii) We have along $x(t)$

$$\begin{aligned}
X\rho &= (\nabla_X h)(U, U) - 2h(\nabla_X U, U) = (\nabla_X h)(U, U) = (\nabla_U h)(X, U) \\
&= U(h(X, U)) - h(\nabla_U X, U) - h(X, \nabla_U U) = -\mu h(U, U) = -\mu\rho.
\end{aligned}$$

Now we can prove that $x(b) \in \Omega$. Indeed, the equations in (i), (ii), and (iii) are

$$\frac{d\mu}{dt} = -\mu^2, \quad \frac{d\lambda}{dt} = -\lambda\mu, \quad \frac{d\rho}{dt} = -\rho\mu \quad \text{for } 0 \leq t < b.$$

Thus μ is identically 0 or $\mu = 1/(t + a)$ for some constant a. It follows that $\lambda =$ constant (nonzero, since $\lambda_{x_0} \neq 0$) or $\lambda = 1/c(t + a)$ and the same for ρ. In all cases, neither λ nor ρ approaches 0 as $t \to b$. This means, at the point $p = x(b)$, $SU \neq 0$ as well as $h(U, U) \neq 0$. Thus $x(b) \in \Omega$.

It now follows that $x(t) \in L$ for all t. As we complete the proof of Proposition 2.2, we make the following remark. The possibility of $\mu = 1/(t + a)$ above is now excluded. Hence $\mu \equiv 0$ and thus λ and ρ are non-zero constants on the leaf L.

We now have

Theorem 2.3. *Let* $f : \mathbf{R}^n \to \mathbf{R}^{n+1}$ *be an affine immersion such that S and h vanish nowhere. Then f is affinely congruent to a proper affine cylinder immersion.*

Proof. We have $\Omega = \mathbf{R}^n$ in this case. By Proposition 2.2, $\mathscr{N}$ is complete. Thus each leaf is a hyperplane in $M = \mathbf{R}^n$ and all leaves are parallel. Let U be a vector transversal to all these hyperplanes and let us consider a line $x(t)$ in the direction of U. We denote by $\mathbf{R}^{n-1}(t)$ the leaf through the point $x(t)$. Since each leaf is mapped totally geodesically, $f(\mathbf{R}^{n-1}(t))$ is an affine $(n - 1)$-space in $\mathbf{R}^{n+1}$. If Y_t is a parallel vector field along $x(t)$ such that $Y_t \in T_{x(t)}(\mathbf{R}^{n-1}(t))$, then $f_*(Y_t)$ is also parallel in $\mathbf{R}^{n+1}$, because

$$D_t f_*(Y_t) = f_*(\nabla_t Y_t) + h(U, Y_t) = 0.$$

This shows that all subspaces $f(\mathbf{R}^{n-1}(t))$ are parallel to each other.

Now it is easy to verify that f is affinely congruent to a proper affine cylinder immersion based on the parallel family $f(\mathbf{R}^{n-1}(t))$ and the curve $\gamma(t) = f(x(t))$.

We shall also state

Theorem 2.4. *Let $f : \mathbf{R}^n \to \mathbf{R}^{n+1}$ be an affine immersion. Then $\Omega = \{x \in \mathbf{R}^n : S_x \neq 0, h_x \neq 0\}$, is the union of parallel hyperplanes. Each connected component Ω_α of Ω is a strip consisting of parallel hyperplanes and $f : \Omega_\alpha \to \mathbf{R}^{n+1}$ is affinely congruent to a proper affine cylinder immersion.*

The results above were obtained in [NP2].

We shall now apply our results on affine immersions $\mathbf{R}^n \to \mathbf{R}^{n+1}$ to obtain the classical results in the metric cases, namely, the theorem of Hartman and Nirenberg [HN] on isometric immersions between Euclidean spaces $E^n \to E^{n+1}$ and that of Graves [Gr1, 2] on isometric immersions of Lorentz–Minkowski spaces $L^n \to L^{n+1}$.

First, we prove the theorem of Hartman and Nirenberg that the image of an isometric immersion $E^n \to E^{n+1}$ is a cylinder over a plane curve. More precisely, we prove

Theorem 2.5. *Let $f : E^n \to E^{n+1}$ be an isometric immersion between Euclidean spaces. Then there exist isometric immersions*

$$f_1 : E^1 \to E^2 \quad and \quad f_0 = \mathrm{id} : E^{n-1} \to E^{n-1}$$

such that

$$f : E^n = E^1 \times E^{n-1} \to E^{n+1} = E^2 \times E^{n-1}$$

can be expressed in the form $f = f_1 \times f_0$.

Proof. Choosing a unit normal vector field as ξ, we see that f is an affine immersion $\mathbf{R}^n \to \mathbf{R}^{n+1}$ (see Example 2.1 of Chapter II). If g denotes the Euclidean metric on E^{n+1} as well as that on E^n, then the usual second fundamental form h is given by $h = g(SX, Y)$, where S is the shape operator in the sense of isometric immersion that coincides with the affine shape operator for ξ. If $S \equiv 0$, we are done. Otherwise, as in Lemma 2.1 we have the relative nullity foliation $\mathcal{N}$ on $\Omega = \{x : S_x \neq 0\}$, which is complete as in Proposition 2.2. The leaves $L(x)$ of $\mathcal{N}$ in each connected component Ω_α are disjoint, hence parallel, hyperplanes in $M = E^n$. Now looking back at the proof of Proposition 2.2, we pick a point o and choose U to be a unit vector field orthogonal to the hyperplane $H(o)$ that contains the leaf $L(o)$ through o. In the present case, S is symmetric and 0 on $\mathcal{N}$. Hence $SU = \lambda U$, where $\lambda = \mathrm{tr}\, S/n$ is a nonzero eigenvalue of S. As remarked at the end of the proof of Proposition 2.2, we see that λ is constant on the entire line x_t (knowing, as we do, that L is complete). Now let $\ell = (u_t)$ be the line in E^n through o and in the direction of U. We have the following situation. (From this point on, we follow the arguments in the appendix of [N1].)

(i) For each point $p \in \Omega_\alpha$ we have a hyperplane $H(p) \subset \Omega_\alpha$ and the nonzero eigenvalue λ of S is constant on $H(p)$.

(ii) All the hyperplanes $H(p), p \in \Omega$, are parallel. In fact, if $p, q \in \Omega_\alpha$, then $H(p)$ and $H(q)$ are parallel as we already know. If p and q are in two distinct components Ω_α and Ω_β, then $H(p)$ and $H(q)$ are disjoint. If they have a point r in common, then $r \in \Omega_\alpha \cap \Omega_\beta$ and hence $\Omega_\alpha = \Omega_\beta$, contrary to the assumption.

(iii) The line ℓ is orthogonal to $H(p)$ at every point $p \in \ell \cap \Omega$. Indeed, since $\lambda(p) \neq 0$, p belongs to Ω_α for some α and the hyperplane $H(p)$ is parallel to $H(o)$. Since ℓ is orthogonal to $H(o)$, it follows that ℓ is orthogonal to $H(p)$.

(iv) For each point $p \in \ell$ that does not belong to Ω, we define $H(p)$ to be the hyperplane through p and parallel to $H(o)$. Then $\lambda(q) = 0$ for every point $q \in H(p)$. In fact, suppose there is a point $q \in H(p)$ with $\lambda(q) \neq 0$. Then $q \in \Omega$ and $H(q)$, being parallel to $H(o)$, must coincide with $H(p)$. Since λ is constant on $H(q)$, we must have $\lambda(p) \neq 0$, which is a contradiction.

After these preparations, it is fairly easy to see how f maps all $H(p)$ into E^{n+1}. Let Y_t be a vector field along $\ell = (u_t)$ that is parallel to $Y \in N(o)$. We have

$$D_t f_*(Y_t) = f_*(\nabla_t Y_t) + h(\dot{u}_t, Y_t)\xi = h(\dot{u}_t, Y_t)\xi,$$

since $\nabla_t Y_t = 0$. If $\lambda(\dot{u}_t) \neq 0$, then, in a neighborhood of the point, Y_t belongs to $N(u_t)$ and $\dot{u}_t$ belongs to $N(u_t)^\perp$. Thus $h(\dot{u}_t, Y_t) = 0$. If $\lambda(\dot{u}_t) = 0$, then h is 0 at the point u_t. Hence $h(\dot{u}_t, Y_t) = 0$. Therefore, for each point of ℓ, we have $D_t f_*(Y_t) = 0$. Hence $f_*(Y_t)$ is parallel in E^{n+1}. It follows that $f(H(p))$, $p \in \ell$, are all parallel to the fixed subspace $E^{n-1} = f(H(o))$.

Since ℓ is orthogonal to all $H(p)$ and since f is isometric, we see that $\gamma = f(\ell)$ is a curve on the plane, say E^2, through $f(o)$ and perpendicular to E^{n-1}. From the fact that $f_*(Y_t)$ is parallel whenever Y_t is parallel along ℓ, it follows that $f(u_t, Y) = (f_1(u_t), f_0(y))$ for all $(u_t, y) \in M = \ell \times H(o)$, where f_1 is the restriction of f to ℓ and f_0 is the identity. This completes the proof of Theorem 2.5.

We shall now proceed to the result of Graves.

Theorem 2.6. *Let* $f : L^n \rightarrow L^{n+1}$ *be an isometric immersion. Then we have the following three cases:*

(1) $f : L^n = E^1 \times L^{n-1} \rightarrow L^{n+1} = E^2 \times L^{n-1}$ *is the product of isometric immersions*

$$f_1 : E^1 \rightarrow E^2 \quad and \quad f_0 = \mathrm{id} : L^{n-1} \rightarrow L^{n-1}.$$

Thus the image of f *is a cylinder over a plane curve in* E^2 *in* L^{n+1} *with* $(n-1)$-*dimensional generators that are perpendicular to* E^2 *through each point of the curve.*

(2) $f : L^n = L^1 \times E^{n-1} \rightarrow L^{n+1} = L^2 \times E^{n-1}$ *is the product of isometric immersions*

$$f_1 : L^1 \rightarrow L^2 \quad and \quad f_0 = \mathrm{id} : E^{n-1} \rightarrow E^{n-1}.$$

Thus the image of f *is a cylinder over a timelike curve in* L^2 *in* L^{n+1} *with* $(n-1)$-*dimensional generators that are perpendicular to* L^2.

(3) $f : L^n = L^2 \times E^{n-2} \rightarrow L^{n+1} = L^3 \times E^{n-2}$ *is the product of isometric immersions*

$$f_1 : L^2 \rightarrow L^3 \quad and \quad f_0 = \mathrm{id} : E^{n-2} \rightarrow E^{n-2}.$$

Here f_1 is described, relative to a null coordinate system $\{u, v\}$ in L^2 and a null coordinate system $\{u^1, z, v^1\}$ with spacelike axis z, by

$$f_1(u, v) = (u, \psi(u), \frac{1}{2}\phi(u) + v),$$

where $(\psi')^2 - \phi' = 0$. (We shall later describe f_1 as the B-scroll of a certain null curve in L^3.)

Proof. For any isometric immersion $f : L^n \to L^{n+1}$ we choose a spacelike unit normal vector field ξ. Then f is an affine immersion $\mathbf{R}^n \to \mathbf{R}^{n+1}$ relative to ξ. The hypersurface theory in L^{n+1} can be developed in analogy to Euclidean hypersurface theory. In particular, the second fundamental form h and the shape operator S can be defined. Excluding the case where $S \equiv 0$, we have the relative nullity foliation $\mathcal{N}$ on $\Omega = \{x : S_x \neq 0\}$, which is complete, as in Lemma 2.1 and Proposition 2.2. It follows that the leaves of $\mathcal{N}$ are affine $(n-1)$-subspaces parallel to a fixed affine $(n-1)$-subspace H. Thus the signature of the induced metric tensor g on each of the leaves remains constant. If g on H is nondegenerate, it is either positive-definite or Lorentzian. If g on H is degenerate, then (H, g) is isometric to $E^{n-2} \oplus \mathrm{Span}\{V\}$, where g on E^{n-2} is positive-definite, and $g(E^{n-2}, V) = 0$ and $g(V, V) = 0$.

In the first two cases, we can take, in the proof of Proposition 2.2, a timelike or spacelike unit normal vector field as U and can obtain $SU = \lambda U$, where λ is a nonzero eigenvalue of S. Together with the completeness of $\mathcal{N}$ we know that λ is constant on each leaf. We can now follow the same kind of arguments as in the proof of Theorem 2.5, including the steps (i)–(iv).

When the foliation $\mathcal{N}$ is spacelike, the leaves are spacelike $(n-1)$-subspaces that are mapped by f onto parallel spacelike $(n-1)$-subspaces in L^{n+1}. The line ℓ is timelike and is mapped by f into the 2-plane L^2 perpendicular to the family of parallel spacelike $(n-1)$-subspaces in L^{n+1} above.

When the foliation is Lorentzian, the leaves are parallel Lorentzian $(n-1)$-subspaces in L^n; they are mapped by f onto parallel Lorentzian $(n-1)$-subspaces in L^{n+1}. The line ℓ is spacelike and is mapped by f into the Euclidean plane E^2 perpendicular to the family of parallel Lorentzian $(n-1)$-subspaces in L^{n+1} above.

It remains to check the case where the foliation has the form $E^{n-2} \oplus \mathrm{Span}\{V\}$ in the manner indicated before. The leaves are parallel affine $(n-1)$-subspaces with the signature of $E^{n-2} \oplus \mathrm{Span}\{V\}$. We take the line ℓ through o determined by a vector U such that $U \perp E^{n-2}$, $g(U, V) = -1$, and $g(U, U) = 0$. We can show that f maps the parallel E^{n-2}-spaces onto parallel Euclidean $(n-2)$-subspaces in L^{n+1}. We also see that f maps $\mathrm{Span}\{U\}$ and $\mathrm{Span}\{V\}$, and thus the subspace $L^2 = \mathrm{Span}\{V, U\}$, into the subspace L^3 perpendicular to the family of Euclidean $(n-2)$-subspaces in L^{n+1} above. We may then express the map $f_1 : L^2 \to L^3$ by using the null coordinate system $\{u, v\}$ in L^2 and a null coordinate system $\{u^1, z, v^1\}$ by

$$(u, v) \mapsto (u, \psi(u), \frac{1}{2}\phi(u) + v),$$

where $u \mapsto (u, \psi(u), \frac{1}{2}\phi(u))$ is a null curve so that

(2.1) $$(\psi')^2 - \phi' = 0.$$

One can easily verify that f_1 subject to (2.1) is an isometric immersion of L^2 into L^3.

Remark 2.1. Let us denote by $x(u)$ the curve

(2.2) $$x(u) = (u, \psi(u), \frac{1}{2}\phi(u))$$

subject to (2.1). Let

$$A(u) = \frac{dx}{du} = (1, \psi', \frac{1}{2}\phi'),$$
$$B(u) = (0, 0, 1),$$
$$C(u) = (0, 1, \psi').$$

We can verify that $\{A, B, C\}$ forms a null frame field along the curve

(2.3) $$\langle A, A \rangle = \langle B, B \rangle = 0, \langle A, B \rangle = -1, \langle A, C \rangle = \langle B, C \rangle = 0, \langle C, C \rangle = 1$$

that satisfies the following differential equations:

(2.4) $$\frac{dx}{du} = A, \quad \frac{dA}{du} = kC, \quad \frac{dB}{du} = 0, \quad \frac{dC}{du} = kB.$$

In fact, $k = \psi''$. Graves called a null curve with frame field satisfying (2.3) and (2.4) a *generalized null cubic*. If k is a constant, then $x(u)$ turns out to be a cubic. For any generalized null cubic $x(u)$, the *B-scroll* is the surface

(2.5) $$(u, v) \mapsto f(u, v) = x(u) + vB,$$

which can be shown to be an isometric immersion $L^2 \to L^3$ with degenerate relative nullity foliation.

3. The Cartan–Norden theorem

This section presents a theorem concerning affine immersions of a metric connection that gives a precise statement of the results hinted at by E. Cartan [Car1] and indicated by Norden [No]. The reader may recall the notions of conjugate connection (Section 4, Chapter I) and conormal immersion (Section 5, Chapter II). First we rephrase Theorem 5.2 of Chapter I as follows.

Lemma 3.1. *Let (M, h) be a pseudo-Riemannian manifold and let ∇ and $\bar{\nabla}$ be two torsion-free affine connections on M that are conjugate relative to h. If g is another pseudo-Riemannian metric on M, let B be a nonsingular $(1,1)$-tensor such that*

$$(3.1) \qquad g(X, Y) = h(BX, Y)$$

and define $\bar{g}$ by

$$(3.2) \qquad \bar{g}(X, Y) = h(X, B^{-1}Y).$$

If $\nabla g = 0$, then $\bar{\nabla}\bar{g} = 0$.

Proof. We recall from Section 5 of Chapter I that $\lambda_h : T(M) \to T^*(M)$ is defined by $\lambda_h(X)(Y) = h(X, Y)$. Likewise, we have $\lambda_g(X)(Y) = g(X, Y)$. The dual metric g^* of g is given by

$$g^*(\lambda_g X, \lambda_g Y) = g(X, Y).$$

From (3.1) we get $\lambda_g(X) = \lambda_h(BX)$ and hence

$$g^*(\lambda_h X, \lambda_h Y) = g^*(\lambda_h B B^{-1} X, \lambda_h B B^{-1} Y) = g(B^{-1}X, B^{-1}Y)$$
$$= h(X, B^{-1}Y)$$
$$= \bar{g}(X, Y),$$

by (3.2). This shows that $\bar{g}$ corresponds to the metric g^* by λ_h. From the proof of Theorem 5.2 of Chapter I it follows that $\bar{\nabla}\bar{g} = 0$.

Theorem 3.2. *Let (M^n, g) be a pseudo-Riemannian manifold, ∇ its Levi-Civita connection and $f : (M, \nabla) \to \mathbf{R}^{n+1}$ an affine immersion with a transversal field ξ. If f is nondegenerate, we have either*

(1) ∇ is flat and f is a graph immersion,

or

(2) ∇ is not flat and $\mathbf{R}^{n+1}$ admits a parallel pseudo-Riemannian metric relative to which f is an isometric immersion and ξ is perpendicular to $f(M)$.

Proof. We may assume that ξ is equiaffine, as in Example 2.3 of Chapter II. We consider the conormal immersion $\nu : (M, \bar{\nabla}) \to \mathbf{R}_{n+1}$, where $\mathbf{R}_{n+1}$ is the dual vector space of $\mathbf{R}^{n+1}$. The affine connection $\bar{\nabla}$ is conjugate to ∇ relative to the affine fundamental form h for the immersion f. Since h is nondegenerate, we may write

$$g(X, Y) = h(BX, Y),$$

where B is a certain nonsingular $(1,1)$-tensor symmetric relative to h. We define a pseudo-Riemannian metric $\bar{g}$ by

$$\bar{g}(X, Y) = h(B^{-1}X, Y).$$

By Lemma 3.1, $\bar{\nabla}$ is the Levi-Civita connection for $\bar{g}$.

Now the conormal immersion being a centro-affine immersion, we know that $\bar{\nabla}$ is projectively flat. Since $\bar{\nabla}$ is the Levi-Civita connection for $\bar{g}$, the metric $\bar{g}$ must be of constant sectional curvature, say c. The form h for the conormal immersion is equal to the normalized Ricci tensor (see the argument preceding Proposition 5.2, Chapter II), which is equal to $c\bar{g}$. Thus $\bar{h} = c\bar{g}$.

Case (1): $c = 0$. Then $\bar{\nabla}$ is flat. Since $\bar{h} = 0$ and $\bar{h}(X, Y) = h(SX, Y)$, the shape operator vanishes and by the Gauss equation ∇ is flat. By Proposition 2.8, Chapter II, we conclude that f is a graph immersion.

Case (2): $c \neq 0$. We have

$$\bar{h}(X, Y) = h(SX, Y) = c\bar{g}(X, Y) = c\,h(B^{-1}X, Y),$$

which implies

(3.3) $$S = c\,B^{-1}.$$

Now we define a symmetric $(0, 2)$-tensor $\langle\ ,\ \rangle$ along the immersion f as follows:

$$\langle f_* Y, f_* Z \rangle = g(Y, Z), \quad \langle f_* Y, \xi \rangle = 0, \quad \langle \xi, \xi \rangle = c.$$

We shall now show that this tensor field is parallel in $\mathbf{R}^{n+1}$ and thus defines a pseudo-Euclidean metric on $\mathbf{R}^{n+1}$; the first of the defining equations says that f is isometric, the second that ξ is perpendicular to $f(M^n)$, and the third that ξ is non-null relative to the pseudo-Euclidean metric. Now in order to prove our assertion, we want to show that

$$X\langle U, V \rangle = \langle D_X U, V \rangle + \langle U, D_X V \rangle$$

for all vector fields U, V along f. It is sufficient to verify this in the following three cases:

(i) $U = f_*(Y), V = f_*(Z)$, (ii) $U = f_*(Y), V = \xi$, (iii) $U = V = \xi$,

where Y and Z are vector fields on M. The only non-trivial case is (ii). In this case, we get $X\langle f_*(Y), \xi \rangle = 0$ and

$$\begin{aligned}
\langle D_X f_* Y, \xi \rangle + &\langle f_* Y, D_X \xi \rangle \\
&= \langle f_*(\nabla_X Y), \xi \rangle + h(Y, Z)\langle \xi, \xi \rangle + \langle f_* Y, -f_*(SX) \rangle \\
&= c\,h(X, Y) - g(Y, SX),
\end{aligned}$$

which is 0 by virtue of (3.3). The proof of the theorem is now complete.

We state the following corollary.

Corollary 3.3. *Let (M^n, g) be a pseudo-Riemannian manifold, ∇ its Levi-Civita connection, and $f : (M, \nabla) \to \mathbf{R}^{n+1}$ an affine immersion. If the Ricci tensor of g is nondegenerate, then $\mathbf{R}^{n+1}$ admits a parallel pseudo-Riemannian metric such that f is an isometric immersion and the transversal field is perpendicular to $f(M)$.*

Proof. From $\mathrm{Ric}(Y, Z) = h(Y, Z) \operatorname{tr} S - h(SY, Z)$, it follows that h is nondegenerate if the Ricci tensor is nondegenerate.

The following two corollaries are immediate applications.

Corollary 3.4. *Let g be the standard Riemannian metric on S^n with constant sectional curvature 1 and ∇ the Levi-Civita connection. For every affine immersion $f : (S^n, \nabla) \to \mathbf{R}^{n+1}$, the image $f(S^n)$ is an ellipsoid.*

Corollary 3.5. *Let (H^n, g) be the hyperbolic space with standard Riemannian metric of constant sectional curvature -1 and ∇ its Levi-Civita connection. Then every affine immersion $f : (H^n, \nabla) \to \mathbf{R}^{n+1}$ is an isometric immersion of (H^n, g) into $\mathbf{R}^{n+1}$ with flat Lorentz metric. If $n \geq 3$, $f(H^n)$ is affinely congruent to one component of the two-sheeted hyperboloid $-(x^0)^2 + (x^1)^2 + \cdots + (x^n)^2 = -1, \, x^0 > 0$.*

The result in this section were originally obtained in [NP2]. For isometric immersions of H^2 into H^3, see [N2] and [Fe].

4. Affine locally symmetric hypersurfaces

Let (M^n, ∇) be an n-manifold with a locally symmetric affine connection ∇, that is, $\nabla R = 0$. The aim of this section is to determine affine immersions $M^n \to \mathbf{R}^{n+1}, n \geq 3$. The case of $n = 2$ will be treated in Note 8.

The first result in this problem was the following in [VeVe].

Theorem 4.1. *Let $f : M \to \mathbf{R}^{n+1}$, $n \geq 3$, be a nondegenerate immersion with Blaschke structure. The Blaschke connection ∇ is locally symmetric if and only if M is an improper affine hypersphere or a nondegenerate quadratic hypersurface.*

Later, the result was extended to the non-Blaschke case. More precisely, we have ([NO3])

Theorem 4.2. *Let $f : M^n \to \mathbf{R}^{n+1}$, $n \geq 3$, be a nondegenerate hypersurface endowed with a transversal vector field ξ. If the connection ∇ induced by ξ is locally symmetric, then ∇ is locally flat, or it is the Blaschke connection and furthermore $f(M^n)$ is an open part of a hyperquadric with center.*

For a further extension of Theorem 4.2 to the case of degenerate immersions of rank ≥ 2, we refer the reader to [NO3]. We shall now give a proof of Theorem 4.2.

In order to prove Theorem 4.2 we recall Propositions 1.1, 1.2 and Theorem 2.1, Chapter II, in particular, the Gauss equation, the two Codazzi equations and the Ricci equation for (∇, h, S) induced by the given transversal vector field ξ. The assumption $\nabla R = 0$ implies $R(X, Y) \cdot R = 0$ for all $X, Y \in T_x(M^n)$, where $R(X, Y)$ acts on R as derivation. For all $X, Y, Z, W \in T_x(M^n)$ we obtain by using the Gauss equation

$$
\begin{aligned}
(4.1) \qquad 0 = {}& (R(X, Y) \cdot R)(Z, V)W \\
= {}& [h(V, W)h(Y, SZ) - h(Z, W)h(Y, SV)]SX \\
& + [h(Z, W)h(X, SV) - h(V, W)h(X, SZ)]SY \\
& + [h(Y, V)h(Z, W) - h(Y, Z)h(V, W)]S^2 X \\
& + [h(X, Z)h(V, W) - h(X, V)h(Z, W)]S^2 Y \\
& + [h(X, V)h(SY, W) - h(Y, V)h(SX, W) \\
& + h(X, W)h(SY, V) - h(Y, W)h(SX, V)]SZ \\
& + [h(Y, Z)h(SX, W) - h(X, Z)h(SY, W) \\
& + h(Y, W)h(SX, Z) - h(X, W)h(SY, Z)]SV.
\end{aligned}
$$

We first prove

Lemma 4.3. *At each point x of M^n the endomorphism S_x is either nonsingular or identically 0 on $T_x(M^n)$.*

Proof. For any $X \in \ker S_x$ the equation (4.1) gives

$$
\begin{aligned}
(4.2) \qquad 0 = {}& [h(Z, W)h(X, SV) - h(V, W)h(X, SZ)]SY \\
& + [h(X, Z)h(V, W) - h(X, V)h(Z, W)]S^2 Y \\
& + [h(X, V)h(SY, W) + h(X, W)h(SY, V)]SZ \\
& - [h(X, Z)h(SY, W) + h(X, W)h(SY, Z)]SV.
\end{aligned}
$$

Assume that $\ker S_x \neq \{0\}$. In the sequel we omit the letter x. Consider the two cases I and II.

Case I: $h|_{\ker S} = 0$. Then rank $S \geq 2$ because $n \geq 3$ and h is nondegenerate. Let $0 \neq X \in \ker S$. Take $W = X$, $Z = Y$. Equation (4.2) yields

$$
(4.3) \qquad h(Y, X)h(X, SV)SY - h(X, Y)h(SY, X)SV = 0
$$

where Y and V are arbitrary. Therefore we get

$$
(4.4) \qquad h(X, Y)h(SY, X) = 0
$$

for every Y. Now there exists a basis $\{e_1, \ldots, e_n\}$ of $T_x(M^n)$ such that $h(X, e_i) \neq 0$ for every $i = 1, \ldots, n$, as we see by the following argument. Let $\langle X \rangle$ be the space spanned by X and let $\langle X \rangle^*$ the space of all vectors that are h-orthogonal to X. Then $\dim \langle X \rangle^* = n - 1$ and $X \in \langle X \rangle^*$. Let $e_1 \notin \langle X \rangle^*$ and let $\langle X \rangle'$ be an algebraic complement to $\langle X \rangle$ in $\langle X \rangle^*$. Then $\langle X \rangle \oplus \langle X \rangle' \oplus \langle e_1 \rangle = T_x(M^n)$. Let $\{\bar{e}_2, \ldots, \bar{e}_{n-1}\}$ be a basis of $\langle X \rangle'$. Set $\bar{e}_n = X$. Then $\{e_1, e_2 = e_1 + \bar{e}_2, \ldots, e_n = e_1 + \bar{e}_n\}$ is a basis satisfying the desired condition.

By substituting vectors of this basis as Y into (4.4) we get

$$
(4.5) \qquad h(SY, X) = 0 \text{ for every } Y \in T_x(M^n) \text{ and for every } X \in \ker S.
$$

Now we go back to (4.2) and set $Z = X$. By also using (4.5) we obtain

$$h(X, V)h(X, W)S^2 Y = 0$$

for all Y, V, and W. Take $V = W$ such that $h(X, V) \neq 0$. Then $S^2 Y = 0$. Hence $S^2 = 0$ on $T_x(M^n)$. Using once again (4.2), (4.5), we get

(4.6) $[h(X, V)h(SY, W) + h(X, W)h(SY, V)]SZ$
$$- [h(X, Z)h(SY, W) + h(X, W)h(SY, Z)]SV = 0$$

for all Y, Z, W, and V. Let $\{e_1, \ldots, e_n\}$ be a basis of $T_x(M^n)$ such that $h(e_i, X) \neq 0$ for every $i = 1, \ldots, n$ and let V be an arbitrary vector of this basis.

If $SV \neq 0$, then there is a vector Z of the basis such that SV and SZ are linearly independent. By substituting these V and Z into (4.6) and setting $W = V$ we get $h(X, V)h(SY, V) = 0$. Since $h(X, V) \neq 0$, we have $h(SY, V) = 0$. If $SV = 0$, then we still have $h(SY, V) = 0$ because of (4.5). Hence for every $V \in T_x(M^n)$ we have

(4.7) $$h(SY, V) = 0.$$

It follows that S vanishes on $T_x(M^n)$, which is impossible in the case under consideration. Thus Case I does not occur.

Case II: $h|_{\ker S}$ is not identically 0. There is an $X \in \ker S$ such that $h(X, X) \neq 0$. By putting such X into (4.2) and setting $Z = X$ we obtain

(4.8) $0 = h(X, W)h(X, SV)SY$
$$+ [h(X, X)h(V, W) - h(X, V)h(X, W)]S^2 Y$$
$$- [h(X, X)h(SY, W) + h(X, W)h(SY, X)]SV,$$

for every Y, V, and W. Take any $Y \notin \ker S$. Since $n \geq 3$ and h is nondegenerate, there exist $W \neq 0$ and V such that $h(X, W) = 0$, $h(SY, W) = 0$ and $h(V, W) \neq 0$. Then by (4.8) we get $S^2 Y = 0$. Thus

(4.9) $$S^2 = 0 \quad \text{on} \quad T_x(M^n).$$

Take now an arbitrary Y and set $V = Y$. By (4.8) and (4.9) we get $h(SY, W)SY = 0$ for every W. Then $SY = 0$. Consequently, S vanishes on $T_x(M^n)$. This completes the proof of Lemma 4.3.

We now prove Theorem 4.2. From the Gauss equation and from rank $h \geq 2$, we see

(4.10) $$\text{im } S_x = \text{im } R_x,$$

where im R_x denotes the subspace of $T_x(M^n)$ spanned by all vectors $R(X, Y)Z$ for all X, Y, $Z \in T_x(M^n)$. Since R is parallel relative to ∇ by assumption, so is im S_x. Thus the dimension of im S_x is constant on M^n.

By Lemma 4.3, it follows that either S vanishes on M^n (and thus ∇ is flat) or S is nonsingular at every point $x \in M^n$. Assume the second alternative. Let X, Y, and Z be mutually h-orthogonal and set $V = X$ and $W = Y$. Using (4.1) we obtain

$$
(4.11) \qquad
\begin{aligned}
0 =& [h(X,X)h(SY,Y) - h(Y,Y)h(SX,X)]SZ \\
& + h(Y,Y)h(SX,Z)SX.
\end{aligned}
$$

Let $\{e_1, \ldots, e_n\}$ be an h-orthonormal basis of $T_x(M^n)$ and let X, Y, and Z be distinct vectors of the basis. Since S is nonsingular, (4.11) implies the equality $h(SX,Z) = 0$. Hence for every vector X of the basis, the vector SX is parallel to X. Since the basis is arbitrary, it follows that S is a multiple of the identity: $S = \lambda I$ for some nowhere-vanishing function λ. Using the Ricci equation we get $d\tau = 0$. Therefore for every $x \in M^n$ there is a neighborhood U of x and a function ψ on U such that $d\psi = \tau|_U$. Let $\bar{\xi} = e^{-\psi}\xi|_U$. It is easily seen that $\bar{\xi}$ is equiaffine and induces the same connection ∇ on U. Let $\bar{h}$ and $\bar{S}$ correspond to $\bar{\xi}$ as in Propositions 1.1 and 1.2, Chapter II. By applying all the arguments above to the objects induced by $\bar{\xi}$ and using the Codazzi equation for $\bar{S}$ we obtain $\bar{S} = \bar{\lambda} I$, where $\bar{\lambda}$ is a constant. The assumption $\nabla R = 0$ and the Gauss equation yield

$$
0 = (\nabla_V R)(X,Y)Z = \bar{\lambda}[\nabla \bar{h}(V,Y,Z)X - \nabla \bar{h}(V,X,Z)Y].
$$

Hence $\nabla \bar{h} = 0$, in particular, the volume element of $\bar{h}$ is parallel relative to ∇. Thus $\bar{\xi}$ differs from the affine normal by a constant scalar. Since ξ and $\bar{\xi}$ have the same direction, ∇ is obviously the Blaschke connection on M^n. Therefore we could take ξ to be the affine normal from the beginning of the proof. Now using the classical theorem of Pick and Berwald (Theorem 4.5, Chapter II), we conclude the proof of Theorem 4.2.

Remark 4.1. When $n = 2$, neither Theorem 4.1 nor Lemma 4.3 holds. Consider the helicoidal surface defined by

$$
f(u,v) = (u \cos v, u \sin v, v),
$$

with $\xi = (\sin v, -\cos v, 0)$. By writing $\partial_u = \partial/\partial u$ and $\partial_v = \partial/\partial v$, we have

$$
\nabla_u(\partial_u) = \nabla_u(\partial_v) = 0, \quad \nabla_v(\partial_v) = -u\partial_u,
$$

$$
h(\partial_u, \partial_u) = 0, \quad h(\partial_u, \partial_v) = -1, \quad h(\partial_v, \partial_v) = 0,
$$

$$
S(\partial_u) = 0, \qquad S(\partial_v) = -\partial_u.
$$

The affine metric h is flat. The only nonzero component of the curvature tensor R of ∇ is given by

$$
R(\partial_u, \partial_v)\partial_v = -\partial_u.
$$

We have also $\nabla R = 0$. It is amusing to compute the geodesics of ∇. Their equations are

$$\frac{d^2u}{dt^2} - u\left(\frac{dv}{dt}\right)^2 = 0 \quad \text{and} \quad \frac{d^2v}{dt^2} = 0.$$

For example, through the origin of the (u, v)-plane, the pregeodesics of ∇ are the u-axis, the v-axis, and the curves of the form $u = a \sinh v$, $a \neq 0$.

5. Rigidity theorem of Cohn-Vossen type

A well-known theorem of Cohn-Vossen in Euclidean differential geometry says the following. Let $f : M^2 \to E^3$ be an ovaloid whose induced metric has positive Gaussian curvature. Then every isometric immersion $\bar{f} : M^2 \to E^3$ is congruent to f. (See, for example, Chern [Ch1].) In this section, we establish an affine analogue of this theorem.

Theorem 5.1. *Let* $f : M^2 \to \mathbf{R}^3$ *be an ovaloid such that, as Blaschke surface, the affine Gaussian curvature* $K = \det S$ *vanishes nowhere. If a Blaschke immersion* $\bar{f} : M \to \mathbf{R}^3$ *has the same induced connection as that of* f, *then* $\bar{f}$ *is affinely congruent to* f.

Actually, this follows from the following more general result for nondegenerate immersions with equiaffine transversal vectors fields (which are not necessarily affine normal fields). See [NO2].

Theorem 5.2. *Let* M^2 *be a connected compact 2-manifold, and let* f *and* $\bar{f} : M^2 \to \mathbf{R}^3$ *be two nondegenerate immersions with equiaffine transversal vector fields* ξ *and* $\bar{\xi}$. *Assume that*

(1) *the induced connections* ∇ *and* $\bar{\nabla}$ *coincide;*

(2) *for the affine shape operators* S *and* $\bar{S}$, *we have* $\det S = \det \bar{S}$ *at every point and* $\det S$ *is never* 0.

Then f *and* $\bar{f}$ *are affinely congruent.*

Results of this type seem to be missing in classical literature of affine differential geometry. The proofs of Theorems 5.1 and 5.2 rely on the integral formulas provided in Section 9, Chapter III, and on certain lemmas that are peculiar to the 2-dimensional case. We first prove

Lemma 5.3. *Let* g *and* $\bar{g}$ *be positive-definite scalar products on a 2-dimensional vector space* V *and* A *and* $\bar{A}$ *endomorphisms, both positive-definite, or both negative-definite, and symmetric relative to* g *and* $\bar{g}$, *respectively. Assume that* g, $\bar{g}$, A, *and* $\bar{A}$ *are related by*

$$(5.1) \qquad g(Y, Z)AX - g(X, Z)AY = \bar{g}(Y, Z)\bar{A}X - \bar{g}(X, Z)\bar{A}Y$$

for every X, Y, $Z \in V$. *If* $\det A = \det \bar{A}$ *or* $\operatorname{tr} A = \operatorname{tr} \bar{A}$, *then* $\det(\bar{A} - A) \leq 0$, *and the equality holds if and only if* $A = \bar{A}$.

Proof. Let $\{X_1, X_2\}$ be an orthonormal basis relative to $\bar{g}$. We put $g_{ij} = g(X_i, X_j)$. From (5.1), we have

$$(5.2) \qquad \bar{A}X_1 = g_{22}AX_1 - g_{12}AX_2, \qquad \bar{A}X_2 = -g_{12}AX_1 + g_{11}AX_2.$$

This means that the endomorphism $G = \bar{A}A^{-1}$ is expressed, relative to the basis $\{AX_1, AX_2\}$, by the matrix $\hat{G} = \begin{bmatrix} g_{22} & -g_{12} \\ -g_{12} & g_{11} \end{bmatrix}$. From $\bar{A} = GA$, we get $\det(\bar{A} - A) = \det A \det(I - G)$. Since $\det A > 0$, it is enough to prove $\det(I - G) \le 0$ under the assumption that $\det A = \det \bar{A}$ or $\operatorname{tr} A = \operatorname{tr} \bar{A}$.

Consider first the case where $\det A = \det \bar{A}$, which implies $\det G = 1$. Hence,

$$\begin{aligned} \det(I - G) &= \det(I - \hat{G}) \\ &= (1 - g_{11})(1 - g_{22}) - g_{12}^2 \\ &= 1 - (g_{11} + g_{22}) + \det G \\ &= 2 - (g_{11} + g_{22}). \end{aligned}$$

Since $g_{ii} > 0$ by the definiteness, we have $g_{11} + g_{22} \ge 2\sqrt{g_{11}g_{22}}$, while $g_{11}g_{22} \ge g_{11}g_{22} - g_{12}^2 = 1$. Thus

$$g_{11} + g_{22} \ge 2\sqrt{g_{11}g_{22}} \ge 2.$$

Hence $\det(I - G) \le 0$, and the equality holds if and only if $G = I$, that is, if and only if $A = \bar{A}$.

In the second case where $\operatorname{tr} A = \operatorname{tr} \bar{A}$, we may assume that $\bar{A}$ is diagonalizable relative to $\{X_1, X_2\}$, that is,

$$\bar{A}X_1 = \lambda X_1, \qquad \bar{A}X_2 = \mu X_2.$$

From (5.2) we have

$$g_{22}AX_1 - g_{12}AX_2 = \lambda X_1, \qquad -g_{12}AX_1 + g_{11}AX_2 = \mu X_2.$$

Writing vectors in V as column vectors, we may rewrite the equation above in the matrix form

$$(5.3) \qquad [AX_1 \ AX_2]\hat{G} = [\lambda X_1 \ \mu X_2].$$

Multiplying this equation on the right by the inverse of the matrix $\hat{G}$,

$$(5.4) \qquad \hat{G}^{-1} = \frac{1}{\det G} \begin{bmatrix} g_{11} & g_{12} \\ g_{12} & g_{22} \end{bmatrix},$$

we obtain

$$[AX_1 \ AX_2] = \frac{1}{\det G}[\lambda g_{11}X_1 + \mu g_{12}X_2 \ \lambda g_{12}X_1 + \mu g_{22}X_2],$$

which means that A can be expressed, relative to $\{X_1, X_2\}$, by the matrix

$$\hat{A} = \frac{1}{\det G} \begin{bmatrix} \lambda g_{11} & \lambda g_{12} \\ \mu g_{12} & \mu g_{22} \end{bmatrix}.$$

Hence we have

$$\frac{1}{\det G}(g_{11}\lambda + g_{22}\mu) = \operatorname{tr} A = \operatorname{tr} \bar{A} = \lambda + \mu,$$

i.e.,

$$(g_{11} - \det G)\lambda = (\det G - g_{22})\mu.$$

Since λ and μ are positive, we see $(\det G - g_{11})(\det G - g_{22}) \le 0$. Then using (5.4) we get

$$(\det G)^2 \det(\hat{G}^{-1} - I) = (\det G - g_{11})(\det G - g_{22}) - g_{12}^2 \le 0.$$

On the other hand, we have

$$(\det G)^2 \det(G^{-1} - I) = \det G \cdot \det(I - G),$$

where $\det G > 0$. Therefore we have $\det(I - G) \le 0$ and the equality holds if and only if $G = I$.

Remark 5.1. The part under the assumption $\operatorname{tr} A = \operatorname{tr} \bar{A}$ is not used here but has applications in [NO1].

Proof of Theorem 5.2. Let h and $\bar{h}$ be the affine fundamental forms for (f, ξ) and $(\bar{f}, \bar{\xi})$. They are definite, since M is compact (see Section 7, Chapter III). We can assume that they are positive-definite; if, for instance, h is negative-definite, we can replace f by $-f$. (See Appendix 3.) We may also suppose that the induced volume forms for f and $\bar{f}$ coincide: $\theta = \bar{\theta}$. In fact, $\nabla \theta = \bar{\nabla} \bar{\theta} = 0$ implies that $\bar{\theta} = c\theta$ with some constant c and it is enough to multiply $\bar{f}$ and $\bar{\xi}$ by the constant $c^{\frac{1}{3}}$ and achieve $\theta = \bar{\theta}$. (See Appendix 3 again.)

We may choose a point o in the interior of the convex body bounded by $f(M)$ so that the affine distance function ρ for (f, ξ) relative to o is negative. (See Section 7, Chapter III.) By Proposition 6.1, Chapter II, S is positive-definite at some point and, since $K = \det S$ vanishes nowhere, it follows that S is positive-definite everywhere. For the same reason, $\bar{S}$ is positive-definite on M. Taking h and $\bar{h}$ as g and $\bar{g}$ and S and $\bar{S}$ as A and $\bar{A}$ in Lemma 5.3, we see that the Gauss equation for $\nabla = \bar{\nabla}$ takes the form (5.1). By Lemma 5.3 we get $\rho \det(\bar{S} - S) \ge 0$. In the integral formula II of Section 9, Chapter III, namely,

$$(5.5) \qquad \int_{M^2} (\operatorname{tr} \bar{S} - \operatorname{tr} S)\theta = \int_{M^2} \rho \det(\bar{S} - S)\theta,$$

the right-hand side is non-negative. Interchanging the role of S and $\bar{S}$, we have in view of $\theta = \bar{\theta}$

$$(5.6) \qquad \int_{M^2} (\operatorname{tr} S - \operatorname{tr} \bar{S})\theta = \int_{M^2} \bar{\rho} \det (\bar{S} - S)\theta,$$

where the right-hand side is again non-negative. Since the left-hand sides of (5.5) and (5.6) have opposite signs, it follows that

$$\int_{M^2} \rho \det (\bar{S} - S)\theta = 0,$$

which implies that $\det (\bar{S} - S) = 0$ on M. Again using Lemma 5.3 we get $\bar{S} = S$. Hence, by the Gauss equation, we get $h = \bar{h}$. Now Proposition 1.3, Chapter II, completes the proof of Theorem 5.2.

Proof of Theorem 5.1. We first prove the following result that is essentially due to Ślebodziński [Sl1].

Lemma 5.4. *Let $f : M^2 \to \mathbf{R}^3$ be a Blaschke surface. Then the affine Gaussian curvature K is equal to $\epsilon \det R(X_1, X_2)$, where $\epsilon = \pm 1$ depending on whether h is definite or not, and $\{X_1, X_2\}$ is a unimodular basis, i.e., $\theta(X_1, X_2) = 1$, $\det R(X_1, X_2)$ being independent of the choice of such $\{X_1, X_2\}$.*

Proof. Let $\{X_1, X_2\}$ be an orthonormal basis in $T_x(M)$ relative to h, that is, $h(X_1, X_1) = 1$, $h(X_1, X_2) = 0$, $h(X_2, X_2) = \epsilon$, where $\epsilon = \pm 1$. From the Gauss equation, we have

$$R(X_1, X_2)X_1 = -SX_2, \qquad R(X_1, X_2)X_2 = \epsilon SX_1.$$

Then, the extension of the endomorphism $R(X_1, X_2)$ on $\overset{2}{\wedge} T_x(M)$ is

$$R(X_1, X_2)(X_1 \wedge X_2) = \epsilon \det S \cdot X_1 \wedge X_2,$$

thus proving the lemma; the independence assertion is obvious.

Remark 5.2. Lemma 5.4 can be generalized to the Blaschke immersion of a manifold of even dimension: $\det S$ depends only on the connection ∇ and the volume form θ; see [O4].

Now to derive Theorem 5.1 from Theorem 5.2, we have only to note the following. For ovaloids $f(M)$ and $\bar{f}(M)$, both ξ and $\bar{\xi}$ are chosen so as to make h and $\bar{h}$ positive-definite. Since f and $\bar{f}$ have the same induced connection by assumption, Lemma 5.4 implies that they have the same affine Gauss–Kronecker curvature: $\det S = \det \bar{S}$. Thus Theorem 5.1 follows from Theorem 5.2.

Remark 5.3. Theorem 5.2 actually includes the classical Cohn-Vossen theorem. In fact, an isometric immersion is an affine immersion with a unit normal field as equiaffine transversal field. Moreover, positive Gauss–Kronecker curvature implies that the immersion is nondegenerate. The assumptions of Theorem 5.2 are thus satisfied. Once we know that (f, ξ) and $(\bar{f}, \bar{\xi})$ are affinely congruent, it follows that this congruence must be Euclidean congruence.

Remark 5.4. Simon [Si7] gives an entirely different proof to the results in this section without assuming that det S vanishes nowhere.

Remark 5.5. For $n \geq 3$, it is known that an isometric immersion of a Riemannian manifold M^n into Euclidean space E^{n+1} is rigid provided its type number (namely, the rank of the second fundamental form) is at least 3. See [KN2, pp. 43–46]. For an affine immersion, local rigidity results are known in various forms. For instance, Opozda [O3] proves that an affine immersion $f : (M^n, \nabla) \to \mathbf{R}^{n+1}$ is rigid if rank $h \geq 2$ and rank $S \geq 3$; recently, Nomizu and Vrancken have shown that f is rigid if rank $h = n \geq 3$ and rank $S \geq 2$.

6. Extensions of the Pick–Berwald theorem

In this section, we discuss various extensions of the Pick–Berwald theorem (Theorem 4.5 of Chapter II). First of all, we recall from Remark 4.3 of Chapter II that this theorem is valid for any nondegenerate immersion $f : M^n \to \mathbf{R}^{n+1}$ with equiaffine transversal vector field ξ, because the proofs we gave in Section 4, Chapter II, do not depend on apolarity. We shall now prove

Theorem 6.1. *Let* $f : M^n \to \mathbf{R}^{n+1}$ *be an immersion with an arbitrary transversal vector field. Assume that*

(1) *f has rank $r \geq 2$ everywhere;*

(2) *$\nabla h = 0$, where ∇ is the induced connection.*

Then $f(M^n)$ lies in a quadratic hypersurface; more precisely, we have an affine coordinate system $\{x^1, \ldots, x^r, x^{r+1}, \ldots, x^{n+1}\}$ such that $f(M^n)$ satisfies a quadratic equation

$$F(x^1, \ldots, x^{r+1}) = 0.$$

Moreover, if ∇ is complete, then we can write

$$M^n = M^r \times L^s,$$

where M^r is an r-dimensional submanifold of M^n, L^s is an s-dimensional totally geodesic submanifold $(r + s = n)$ with flat induced connection of M^n, and

$$f = f_1 \times f_0, \quad \mathbf{R}^{n+1} = \mathbf{R}^{r+1} \times \mathbf{R}^s,$$

where $f_1 : M^r \to \mathbf{R}^{r+1}$ is a nondegenerate immersion of M^r as a hyperquadric in an affine subspace of $\mathbf{R}^{r+1}$ of $\mathbf{R}^{n+1}$ and $f_0 : L^s \to \mathbf{R}^s$ is an affine imbedding of L^s onto $\mathbf{R}^s$.

Proof. We first remark that conditions (1) and (2) imply that ξ is equiaffine, that is, $\tau = 0$. In fact, by Codazzi's equation (2.2) in Theorem 2.1 of Chapter

II, the assumption $\nabla h = 0$ implies that $\tau(X)h(Y,Z)$ is symmetric in all three variables. We get $\tau = 0$ by the following.

Lemma 6.2. *Let α be a 1-form and let h be a symmetric bilinear form of rank ≥ 2 on a real n-dimensional vector space. If*

$$\alpha(X)h(Y,Z) = \alpha(Y)h(X,Z) \quad \text{for all} \quad X, Y, Z,$$

then $\alpha = 0$.

Proof of Lemma 6.2. We take an h-orthonormal basis

$$(6.1) \qquad \{X_1, \ldots, X_r, X_{r+1}, \ldots, X_n\} \quad \text{with} \quad h(X_i, X_j) = \epsilon_i \delta_{ij},$$

where

$$\epsilon_i = \pm 1 \quad \text{for} \quad 1 \leq i \leq r \quad \text{and} \quad \epsilon_k = 0 \quad \text{for} \quad r+1 \leq k \leq n.$$

Now for any i, $1 \leq i \leq r$, take $j \neq i$, $1 \leq j \leq r$. Then

$$\alpha(X_i)h(X_j, X_j) = \alpha(X_j)h(X_i, X_j) = 0$$

implies that $\alpha(X_i) = 0$. This argument works also for $r+1 \leq i \leq n$ and $1 \leq j \leq r$. Hence $\alpha = 0$.

After this point, we may pursue various paths.

First proof. We prove a global version of Theorem 6.1, namely, the case where ∇ is assumed to be complete by appealing to the results in Section 1 of this chapter. The assumption $\nabla h = 0$ implies that the distribution $\mathcal{N}$ given by the null spaces of h_x, $x \in M^n$, is parallel. Since ∇ on M^n is complete, it follows that $\mathcal{N}$ also is complete. Hence we get a decomposition as stated in Theorem 1.3 of Section 1:

$$M^n = M^r \times L^s \quad \text{and} \quad f = f_1 \times f_0,$$

where $n = r + s$, $f_1 : M^r \to \mathbf{R}^{r+1}$ is a nondegenerate immersion into an affine subspace $\mathbf{R}^{r+1}$ of $\mathbf{R}^{n+1}$ and $f_0 : L^s \to \mathbf{R}^s$ is an affine imbedding of L^s onto an affine subspace $\mathbf{R}^s$ such that $\mathbf{R}^{n+1} = \mathbf{R}^{r+1} \times \mathbf{R}^s$.

It remains only to prove that $f_1(M^r)$ is a quadratic hypersurface in $\mathbf{R}^{r+1}$. We may choose a transversal vector field $\bar{\xi}$ to M^r in $\mathbf{R}^{r+1}$ and obtain $\bar{h} = h$, $\bar{\nabla}_X Y = \nabla_X Y$ as in Proposition 1.6. It follows that $\bar{\nabla} \bar{h} = 0$. Thus we know that $\bar{\xi}$ is equiaffine. It is also obvious that $\bar{h}$ is nondegenerate, that is, of rank r. Theorem 4.5 of Chapter II now shows that $f(M^r)$ lies in a hyperquadric in $\mathbf{R}^{r+1}$.

When the completeness for (M^n, ∇) is not assumed, we may still use the local version of the decomposition (see Remark 1.1) and prove that each point of M^n has a neighborhood U such that $f(U)$ is part of a hyperquadric, say, $F(x^1, \ldots, x^{r+1}) = 0$, where F is a quadratic function in part of the coordinates of an affine coordinate system in $\mathbf{R}^{n+1}$. But then it follows that $f(M^n)$ lies in one and the same hyperquadric in $\mathbf{R}^{n+1}$.

Second proof. We first remark that if h is nondegenerate, then the arguments in Lemma 4.6 of Chapter II based on

$$(6.2) \qquad h(Y,Y)h(SX,Y) = h(X,Y)h(SY,Y),$$

are valid when ξ is equiaffine and not necessarily Blaschke. Now if the rank of h is $r \geq 2$, choose a basis as in (6.1). If $1 \leq i, j \leq r$, $i \neq j$, then we get $h(SX_i, X_j) = 0$, from which we get $SX_i = \rho_i X_i$ mod $\ker h$, where $\ker h$ is spanned by $\{X_{r+1}, \ldots, X_n\}$. By modifying Lemma 3.7 of Chapter II, we can show that $\rho_1, \ldots, \rho_r$ are all equal; call this number ρ. Now let $1 \leq j \leq r$, $r+1 \leq i \leq n$. From (6.2) we get $h(SX_i, X_j) = 0$. This shows that $SX_i \in \ker h$. Hence $S(\ker h) \subset \ker h$. We have therefore

$$(6.3) \qquad SX = \rho X \text{ mod } \ker h \quad \text{for all} \quad X \in T_x(M).$$

We want to show that ρ is a constant. Since ξ is equiaffine, we have Codazzi's equation in the form $(\nabla_X S)Y = (\nabla_Y S)X$. We extend a basis as in (6.1) to vector fields in a neighborhood of a point with the property that $\ker h$ is spanned by $\{X_{r+1}, \ldots, X_n\}$ at each point. Then

$$\begin{aligned}
(\nabla_{X_i} S)X_j &= \nabla_{X_i}(SX_j) - S(\nabla_{X_i} X_j) = \nabla_{X_i}(\rho X_j + Z) - S(\nabla_{X_i} X_j) \\
&= (X_i \rho)X_j + \rho \nabla_{X_i} X_j + \nabla_{X_i} Z - S(\nabla_{X_i} X_j) \\
&= (X_i \rho)X_j \text{ mod } \ker h,
\end{aligned}$$

where $Z \in \ker h$ and $\nabla_X Z \in \ker h$, since $\ker h$ is parallel. Thus by Codazzi's equation we get

$$(6.4) \qquad (X_i \rho)X_j = (X_j \rho)X_i \text{ mod } \ker h.$$

This holds for all i and j. Now if $1 \leq i \leq r$, take $j \neq i$, $1 \leq j \leq r$. Then (6.4) implies $X_i \rho = 0$. If $r+1 \leq i \leq n$, take j, $1 \leq j \leq r$. Then (6.4) implies $X_i \rho = 0$. We have hence shown that $X\rho = 0$ for every $X \in T_x(M)$, that is, ρ is a constant.

At this point we can go in either direction, that is, follow the first proof or the second proof of Theorem 4.5 in Chapter II. In the first case, we define the Lie quadric $\mathfrak{F}_x$ in the same way. In the computation we have, instead of (4.9) of Chapter II,

$$\nabla_X U = (\mu\rho - 1)X \text{ mod } \ker h,$$

and thus

$$h(\nabla_X U, U) = (\mu\rho - 1)h(X, U).$$

Since (4.10) of Chapter II is still valid, we can show $X\Phi = 0$. This completes the proof.

Remark 6.1. We can specify the equation of the quadric in the following way.

Case $\rho = 0$: The equation is $h(U, U) = 2\mu$. Let $\{x^1, \ldots, x^n\}$ be an affine coordinate system in the tangent hyperplane such that

$$h = (x^1)^2 \pm \cdots \pm (x^r)^2$$

and write x^{n+1} for μ. Then the equation of the quadric is

$$x^{n+1} = \frac{1}{2}((x^1)^2 \pm \cdots \pm (x^r)^2).$$

Case $\rho \neq 0$: The equation of the quadric is $(\mu-1)^2 + h(U, U) = 1$. Relative to a suitable affine coordinate system $\{x^1, \ldots, x^{n+1}\}$ we have

$$(x^1)^2 \pm \cdots \pm (x^r)^2 + \rho\left(x^{n+1} - \frac{1}{\rho}\right)^2 = \frac{1}{\rho},$$

without $\{x^{r+1}, \ldots, x^n\}$ appearing.

Remark 6.2. As for the argument following the second proof of Theorem 4.5, we refer the reader to [NP3], where Theorem 6.1 was first proved (assuming that ξ is equiaffine).

We shall now extend Theorem 4.5 of Chapter II so as to make the assumption independent of the choice of transversal vector field.

Let $f : M^n \to \mathbf{R}^{n+1}$ be an immersion. If we choose a transversal vector field ξ, we get the induced structure (∇, h, S, τ) and the cubic form C given by $C(X, Y, Z) = (\nabla_X h)(Y, Z) + \tau(X)h(Y, Z)$.

Definition 6.1. We say that C is *divisible* by h, also denoted by $h|C$, if there is a 1-form μ such that for all X, Y, and Z.

$$(6.5) \qquad C(X, Y, Z)$$
$$= \mu(X)h(Y, Z) + \mu(Y)h(Z, X) + \mu(Z)h(X, Y)$$

Lemma 6.3. *The property that C is divisible by h does not depend on the choice of ξ.*

Proof. If we change ξ to

$$(6.6) \qquad \bar{\xi} = \phi\xi + f_*(U),$$

where U is a vector field on M^n and ϕ a nonvanishing function, then computation shows that the structure induced by $\bar{\xi}$ is given as follows:

$$(6.7) \qquad \bar{\nabla}_X Y = \nabla_X Y - \frac{1}{\phi}h(X, Y)U,$$

$$(6.8) \qquad \bar{h} = \frac{1}{\phi}h,$$

$$(6.9) \qquad \bar{\tau} = \tau + \eta + d\log|\phi|,$$

$$(6.10) \qquad \phi\bar{C}(X, Y, Z) = C(X, Y, Z)$$
$$+ \eta(X)h(Y, Z) + \eta(Y)h(Z, X) + \eta(Z)h(X, Y),$$

where η is the 1-form such that $\eta(X) = h(X, U)/\phi$ for all X.

Now if C is divisible by h, then these identities show that $\bar{C}$ is divisible by $\bar{h}$.

We now state

Theorem 6.4. *Let* $f : M^n \to \mathbf{R}^{n+1}$ *be a nondegenerate immersion. If* C *is divisible by* h, *then* $f(M^n)$ *lies in a hyperquadric.*

To prove this theorem, we need the following lemma.

Lemma 6.5. *For a nondegenerate immersion* $f : M^n \to \mathbf{R}^{n+1}$ *with a transversal vector field* ξ, *assume that*

(1) *the volume element* ω_h *for* h *coincides with the induced volume element* θ;

(2) C *vanishes.*

Then $\tau = 0$. *As a consequence,* ξ *is the affine normal vector field and* $f(M^n)$ *lies in a hyperquadric.*

Proof. Recall $C(X, Y, Z) = (\nabla_X h)(Y, Z) + \tau(X)h(Y, Z)$ and $\nabla_X \theta = \tau(X)\theta$. By using the Levi-Civita connection $\hat{\nabla}$ and the difference tensor K, $K_X = \nabla_X - \hat{\nabla}_X$, we get

$$(\nabla_X h)(Y, Z) = -h(K_X Y, Z) - h(Y, K_X Z).$$

From $C = 0$ we have

$$\tau(X)h(Y, Z) = h(K_X Y, Z) + h(Y, K_X Z).$$

Taking an orthonormal basis $\{X_1, \ldots, X_n\}$ with $h(X_i, X_i) = \epsilon_i = \pm 1$, let $Y = X_i$ and $Z = \epsilon_i X_i$ in the preceding equation and sum over i. We obtain $n\tau(X) = 2 \operatorname{tr} K_X$.

On the other hand, we have

$$\tau(X)\theta = \nabla_X \theta = K_X \omega_h = -\operatorname{tr} K_X \omega_h = -\operatorname{tr} K_X \theta,$$

that is, $\tau(X) = -\operatorname{tr} K_X$. Comparing this with the equation above we get $\operatorname{tr} K_X = 0$ and hence $\tau = 0$.

Proof of Theorem 6.4. By assumption, we have (6.5), say. Since h is nondegenerate, we may choose the vector field U such that $h(X, U) = \mu(X)$ for all X. For $\bar{\xi}$ given in (6.6), where ϕ is still arbitrary, we have $C = 0$ from (6.10). Moreover, by choosing ϕ suitably, we may arrange the volume element induced by $\bar{\xi}$ to coincide with the volume element of $\bar{h}$. We can now use Lemma 6.5 to conclude that $f(M^n)$ lies in a hyperquadric.

Finally, we state the following theorem that unifies Theorems 6.1 and 6.4.

Theorem 6.6. *Let* $f : M^n \to \mathbf{R}^{n+1}$ *be an immersion with rank* ≥ 2 *everywhere. If* C *is divisible by* h, *then* $f(M^n)$ *lies in a hyperquadric.*

For the proof, which is somewhat involved, we refer the reader to [NP3]. We shall later see a further extension of this result to the case of hypersurfaces immersed in a real projective space.

7. Projective structures and projective immersions

The real projective space $\mathbf{P}^{n+1}$ is the quotient space of $\mathbf{R}^{n+2} - \{0\}$ by an equivalence relation: $x \sim y$ in $\mathbf{R}^{n+2} - \{0\}$ if and only if $y = \lambda x$ for some $\lambda \neq 0$. We denote by π the natural projection $\mathbf{R}^{n+2} - \{0\} \to \mathbf{P}^{n+1}$. By introducing the homogeneous coordinates $[x^1, \ldots, x^{n+2}]$ in $\mathbf{P}^{n+1}$, we may write

$$(7.1) \qquad \mathbf{P}^{n+1} = \bigcup_k U_k, \quad \text{where} \quad U_k = \{[x^1, \ldots, x^{n+2}] : x^k \neq 0\}.$$

We define $\phi_k : U_k \to \mathbf{R}^{n+2}$, where $\mathbf{R}^{n+2}$ is the affine space with the usual coordinate system $\{y^1, \cdots, y^{n+2}\}$, by

$$(7.2) \qquad \phi_k([x^1, \ldots, x^{n+2}]) = \left(\frac{x^1}{x^k}, \ldots, \frac{x^{k-1}}{x^k}, 1, \frac{x^{k+1}}{x^k}, \ldots, \frac{x^{n+2}}{x^k} \right).$$

Then ϕ_k is a diffeomorphism of U_k onto the affine hyperplane $H_k : y^k = 1$ in $\mathbf{R}^{n+2}$.

By transferring the flat affine connection on H_k induced from the standard flat affine connection on $\mathbf{R}^{n+2}$ to U_k by ϕ_k, we obtain a flat affine connection ∇^k on U_k. Now in the intersection $U_k \cap U_m, k \neq m$, the two affine connections ∇^k and ∇^m are projectively equivalent, that is, we have

$$(7.3) \qquad \nabla^k_X Y = \nabla^m_X Y + \rho(X)Y + \rho(Y)X, \quad \text{where} \quad \rho = -d \log y^m$$

for any vector fields X and Y (see Definition 3.3 of Chapter I). Geometrically, this follows because $\phi_m \circ (\phi_k)^{-1}$ is a projective transformation from the hyperplane H^k into the hyperplane H^m given by

$$(7.4) \qquad (y^1, \ldots, y^{k-1}, 1, y^{k+1}, \ldots, y^{n+2}) \longmapsto$$
$$(z^1, \ldots, z^{m-1}, 1, z^{m+1}, \ldots, z^{n+2}) = \frac{1}{y^m}(y^1, \ldots, y^{k-1}, 1, y^{k+1}, \ldots, y^{n+2}),$$

which takes lines into lines. The centro-affine immersions ϕ_k and ϕ_m of $U_k \cap U_m$ are related just like M and $\bar{M}$ in (3.13) of Chapter I. Thus we can derive (7.3) in the same way as we did for (3.14).

The considerations above lead to the following general notion. We say that a differentiable manifold M^n has a *projective structure* if M^n has an open covering $\{U_\alpha\}$, where each open set U_α has a torsion-free affine connection ∇^α, in such a way that in any non-empty intersection $U_\alpha \cap U_\beta$ the two affine connections ∇^α and ∇^β are projectively equivalent; that is, there is a 1-form ρ (not necessarily closed) on $U_\alpha \cap U_\beta$ such that

(7.5) $$\nabla^\beta_X Y = \nabla^\alpha_X Y + \rho(X)Y + \rho(Y)X$$

for any vector fields X, Y on $U_\alpha \cap U_\beta$. We may call such a family $\{(U_\alpha, \nabla^\alpha)\}$ an *atlas* for a projective structure P on M^n. We also write $\nabla^\alpha \in P$ for simplicity. An atlas can be enlarged to a maximal one by adjoining all pairs (U, ∇_U) such that ∇_U and ∇^α are projectively equivalent in any non-empty intersection $U \cap U_\alpha$. Each pair (U, ∇_U) belonging to a maximal atlas is called an *allowable chart* for the projective structure. We have

Proposition 7.1. *If M^n has a projective structure P, we can find a torsion-free affine connection ∇ on M^n that is projectively equivalent to every allowable chart (U, ∇_U). Moreover, given a volume element ω on M^n, there is a unique affine connection $\bar{\nabla}$ such that*

(1) *$\bar{\nabla}$ is projectively equivalent to ∇;*

(2) *$\bar{\nabla}\omega = 0$.*

Proof. Choose a partition of unity $\{\phi_\alpha\}$ subordinate to an atlas $\{(U_\alpha, \nabla^\alpha)\}$ for P and get a global affine connection $\nabla = \sum_\alpha \phi_\alpha \nabla^\alpha$. Each point x of M^n has a compact neighborhood on which all but a finite number of ϕ_α's vanish so that ∇ is locally a finite convex sum of affine connections that are projectively equivalent to each other. Thus ∇ is projectively equivalent to every chart of the atlas. To prove the additional statement, verify that

$$\rho(X) = \frac{1}{n+1} \frac{(\nabla_X \omega)(X_1, \ldots, X_n)}{\omega(X_1, \ldots, X_n)}$$

is well-defined independently of the choice of basis $\{X_1, \ldots, X_n\}$ in the tangent space at each point. We can get the desired connection $\bar{\nabla}$ by using (7.3) with this 1-form ρ.

We shall say that a projective structure P on a manifold M^n is *flat* if each chart for P is projectively flat, that is, projectively equivalent to a flat affine connection. For a flat structure P we may find an atlas consisting of all charts (U, ∇_U) such that ∇_U is flat. What we saw in the beginning is a natural atlas for $\mathbf{P}^{n+1}$ consisting of a finite number of charts (U_k, ∇^k), where ∇^k is a flat connection and the 1-form ρ in (7.3) is closed.

For an arbitrary projective structure P on M^n, we may also choose an atlas consisting of charts $(U_\alpha, \nabla^\alpha)$ such that ∇^α has symmetric Ricci tensor. Recall that an affine connection has symmetric Ricci tensor if and only if it is locally equiaffine (that is, has a parallel local volume element). When two such connections are projectively equivalent, the 1-form ρ that appears in the relation (7.5) is closed. (See Proposition 3.1 and Definition 3.3 of Chapter I. See also Appendix 1.)

Remark 7.1. In Section 4 of Chapter III, we considered $SL(n, \mathbf{R})$ with a projectively flat bi-invariant affine connection. The flat projective structure determined by the connection is hence bi-invariant. In [NP1] it was shown that a semi-simple Lie group that admits a bi-invariant, projectively flat affine connection is essentially $SL(n, \mathbf{R})$ or $SL(n, \mathbf{H})$, where $\mathbf{H}$ is the quarternion field. It gave the classification of all manifolds that admit a flat projective structure invariant under a Lie group acting on both sides (with left and right actions commuting). It dealt also with the problem, considered by E. Cartan, of determining all Lie groups whose (0)-connections are projectively flat.

We shall discuss the notion of projective immersion that is analogous to that of affine immersion. Let (M, P) and $(\tilde{M}, \tilde{P})$ be differentiable manifolds each with a projective structure defined in terms of an atlas of local affine connections $(U_\alpha, \nabla^\alpha)$ and $(\tilde{U}_\beta, \tilde{\nabla}^\beta)$, respectively. We set $n = \dim M$ and $n + p = \dim \tilde{M}$.

Definition 7.1. We shall say that an immersion $f : M \to \tilde{M}$ is a *projective immersion* if the following condition is satisfied.

(A) for each point $x_0 \in M$, there exist local charts $(U, \nabla) \in P$ and $(\tilde{U}, \tilde{\nabla}) \in \tilde{P}$, where U is a neighborhood of x_0 and $\tilde{U}$ a neighborhood of $f(x_0)$, such that $f : (U, \nabla) \to (\tilde{U}, \tilde{\nabla})$ is an affine immersion.

This means that there is a field of transversal subspaces $x \mapsto N_x$ of dimension p such that

$$(7.6) \qquad T_{f(x)}\tilde{M} = f_*(T_x(M)) + N_x$$

and such that for any vector fields X, Y on M we have

$$(7.7) \qquad \tilde{\nabla}_X f_*(Y) = f_*(\nabla_X Y) + \alpha(X, Y) \quad \text{with } \alpha(X, Y) \in N_x.$$

In the case where $p = 1$, there is a transversal vector field ξ on U such that

$$(7.8) \qquad \tilde{\nabla}_X f_*(Y) = f_*(\nabla_X Y) + h(X, Y)\xi.$$

Obviously, an affine immersion $(M, \nabla) \to (\tilde{M}, \tilde{\nabla})$ is a projective immersion relative to the projective structures P and $\tilde{P}$ determined by ∇ and $\tilde{\nabla}$ on M and $\tilde{M}$, respectively.

Condition (A) may be replaced by a practically convenient condition (B) or (C) as follows. The proof is not difficult; see [NP4].

Proposition 7.2. *Suppose $f : (M, P) \to (\tilde{M}, \tilde{P})$ is a projective immersion. Then*

(B) *for any point $x_0 \in M$ and for any chart $(\tilde{U}, \tilde{\nabla}) \in \tilde{P}$ with $f(x_0) \in \tilde{U}$, there exists a chart $(U, \nabla) \in P$, with $x_0 \in U$ such that $f : (U, \nabla) \to (\tilde{U}, \tilde{\nabla})$ is an affine immersion;*

(C) *for any point $x_0 \in M$ and for any chart $(U, \nabla) \in P$, where U is a sufficiently small neighborhood of x_0, there exists a chart $(\tilde{U}, \tilde{\nabla}) \in \tilde{P}$, where $\tilde{U}$ is a neighborhood of $f(x_0)$, such that $f : (U, \nabla) \to (\tilde{U}, \tilde{\nabla})$ is an affine immersion.*

We say that a projective immersion $f : (M, P) \to (\tilde{M}, \tilde{P})$ is *totally geodesic* at $x_0 \in M$ if there is some $\tilde{\nabla} \in \tilde{P}$ around $f(x_0)$ such that, for any vector fields X and Y on M, $[\tilde{\nabla}_X f_*(Y)]_{x_0}$ is tangent to M, that is, of the form $f_*(Z)$, where $Z \in T_{x_0}(M)$. By condition (B) we see that this condition is equivalent to $\alpha = 0$ in the equation (7.7) for some chart $(U, \nabla) \in P$ around x_0. Obviously, the condition that f is totally geodesic at x_0 is independent of the choice of $(\tilde{U}, \tilde{\nabla})$ around $f(x_0)$. We say that f is *totally geodesic* if it is so at every point $x_0 \in M$.

For a projective structure (M, P) a curve x_t is called a *path* if it is a pregeodesic relative to any $\nabla \in P$. This notion is well-defined because a pregeodesic remains a pregeodesic under a projective change (refer to Section 3 of Chapter I).

Proposition 7.3. *A projective immersion $f : (M, P) \to (\tilde{M}, \tilde{P})$ is totally geodesic if and only if for each path x_t in M the image $f(x_t)$ is a path in $\tilde{M}$.*

The proof is left to the reader.

From now on, we shall assume that the codimension p is equal to 1. We shall prove

Proposition 7.4. *Let $f : (M, P) \to (\tilde{M}, \tilde{P})$ be a projective immersion of codimension 1. Then there is a uniquely determined transversal direction field $[\xi]$ except in the interior V of the set of points where f is totally geodesic.*

Definition 7.2. It follows that on the set $M - V$ we have a uniquely determined *transversal direction field* $[\xi]$. The conformal class $[h]$ of the affine fundamental form h for ∇ is uniquely defined and is 0 where f is totally geodesic. We call it the *fundamental form* of the projective immersion f. The rank of $[h]$ is called the *rank* of f. If it is equal to n at every point, we say that f is *nondegenerate*.

Proof. For any point x_0, choose a chart $(U, \nabla) \in P$ and a chart $(\tilde{U}, \tilde{\nabla}) \in \tilde{P}$ with $x_0 \in U$ and $f(x_0) \in \tilde{U}$ such that $f : (U, \nabla) \to (\tilde{U}, \tilde{\nabla})$ is an affine immersion with a transversal vector field ξ. Suppose we have another choice ∇' in U and $\tilde{\nabla}'$ in $\tilde{U}$ such that f is an affine immersion $(U, \nabla') \to (\tilde{U}, \tilde{\nabla}')$ relative to a transversal vector field ξ'. In particular, we have

$$\tilde{\nabla}'_X f_* Y = f_*(\nabla'_X Y) + h'(X, Y)\xi'.$$

Then there exist a 1-form ρ on U and a 1-form $\tilde{\rho}$ on $\tilde{U}$ such that

$$\tilde{\nabla}'_X(f_*(X)) = \tilde{\nabla}_X(f_*(X)) + 2\tilde{\rho}(f_*X)f_*X$$

and

$$\nabla'_X X = \nabla_X X + 2\rho(X)X.$$

Writing $\xi' = \phi\xi + f_* Z$, where ϕ is a nonvanishing function and Z a vector field, we obtain

$$2\tilde{\rho}(f_*X)f_*(X) = 2\rho(X)f_*X + h'(X, X)f_* Z.$$

In order to prove that $Z = 0$, assume $Z \neq 0$ at x_0. We take $X \in T_{x_0}(M)$ that is linearly independent from Z_{x_0}. From the preceding equation, we find that $h'(X, X) = 0$, that is, $h(X, X) = \phi h'(X, X) = 0$. Now let $\{X_1, \cdots, X_n\}$ be a basis in $T_{x_0}(M)$ such that $Z = X_1$. By taking X to be X_k, $k \geq 2$, we get the identity $h(X_k, X_k) = 0$. By taking $X = X_j + X_k$, $j \neq k$, and $j, k \geq 2$, we get $h(X_j, X_k) = 0$. From

$$h(X_1 + X_k, X_1 + X_k) = h(X_1, X_1) + 2h(X_1, X_k) = 0,$$
$$h(X_1 + 2X_k, X_1 + 2X_k) = h(X_1, X_1) + 4h(X_1, X_k) = 0,$$

we obtain $h(X_1, X_k) = 0$ for any k. We have thus shown that $h = 0$ at x_0 if $Z_{x_0} \neq 0$. Thus if $h_{x_0} \neq 0$, then $Z_{x_0} = 0$. If $x_0 \in M - V$, then there exists a sequence of points x_k where h is not 0 and hence Z is zero. Then, again, $Z_{x_0} = 0$, concluding the proof of Proposition 7.4.

Assume that f is not totally geodesic at x_0 and hence at any point in a neighborhood of x_0. For any $\nabla \in P$ and $\tilde{\nabla} \in \tilde{P}$, we take a transversal vector field ξ, determined up to a scalar, relative to which $f : M \to \tilde{M}$ is an affine immersion. We write

$$\tilde{\nabla}_X \xi = -f_*(SX) + \tau(X)\xi,$$

where S is the shape operator and τ the transversal connection form. If we change ξ to $\xi' = \phi\xi$, then S changes to ϕS and τ to $\tau' = \tau + d\phi$. It follows that $d\tau$ does not depend on the choice of ξ. The property that S is a scalar multiple of the identity is also independent of the choice of ξ. We may further verify that any projective change of $\tilde{\nabla}$ within $\tilde{P}$ alters neither $d\tau$ nor the said property of the shape operator.

Definition 7.3. A projective immersion f, assumed not to be totally geodesic at a point x_0, is said to be *umbilical* at x_0 if, for some choice of $\nabla \in P$, $\tilde{\nabla} \in \tilde{P}$ and ξ for which f is an affine immersion, S is a scalar multiple of the identity at x_0. If f is umbilical at every point, then we say that f is *umbilical*.

Definition 7.4. Let (M, P) and $(\tilde{M}, \tilde{P})$ be two manifolds with projective structures with $\dim \tilde{M} = \dim M + 1$. An immersion $f : M \to \tilde{M}$ is said to be *equiprojective* if the following condition is satisfied.

(A_1): for each point x_0 of M, there exist equiaffine connections $\nabla \in P$ around x_0 and $\tilde{\nabla} \in \tilde{P}$ around $f(x_0)$ such that f is an affine immersion relative to ∇ and $\tilde{\nabla}$.

Note that for a projective immersion we can always choose one of ∇ and $\tilde{\nabla}$ to be equiaffine. The new element in Definition 7.4 is that both can be chosen to be equiaffine. Recall that then a transversal vector field ξ can be chosen to be equiaffine (that is, $\tilde{\nabla}_X \xi$ is tangential for each X). Just like the case of Proposition 7.2, if (A_1) holds, then the following conditions also hold.

(B_1): for each point $x_0 \in M$ and for any equiaffine connection $\tilde{\nabla} \in \tilde{P}$ around $f(x_0)$ there is an equiaffine connection $\nabla \in P$ around x_0 such that f is an affine immersion;

(C_1): for each point $x_0 \in M$ and for any equiaffine connection $\nabla \in P$ around x_0, there is an equiaffine connection $\tilde{\nabla} \in \tilde{P}$ around $f(x_0)$ such that f is an affine immersion.

We have

Proposition 7.5. *A projective immersion* $f : (M, P) \to (\tilde{M}, \tilde{P})$ *is equiprojective if and only if* $d\tau = 0$.

Proof. The only-if-part is easy. To prove the if-part assume that $d\tau = 0$. For $x_0 \in M$, choose an equiaffine $\tilde{\nabla} \in \tilde{P}$ around $f(x_0)$ with a local volume element $\tilde{\omega}$ with $\tilde{\nabla}\tilde{\omega} = 0$. Take a local connection $\nabla \in P$ around x_0 such that f is an affine immersion with a transversal vector field ξ. Since $d\tau = 0$, we get a function ϕ around x_0 such that $\tau = -d\phi$. Then for $\xi' = \phi\xi$ we have

$$\tilde{\nabla}_X f_*(Y) = f_*(\nabla_X Y) + h(X, Y)\xi',$$

and $\tau' = 0$. Thus the local volume element ω given by

$$\omega(X_1, \ldots, X_n) = \tilde{\omega}(X_1, \ldots, X_n, \xi'), \quad \text{where } X_1, \ldots, X_n \in T_x(M),$$

is parallel relative to ∇. This means that ∇ is equiaffine and that f is equiprojective.

We now prove one of the major results concerning projective immersions.

Theorem 7.6. *Let* $f : (M, P) \to (\tilde{M}, \tilde{P})$ *be an equiprojective immersion, where* $\dim M = n \geq 3$, $\dim \tilde{M} = n + 1$. *Assume that* $(\tilde{M}, \tilde{P})$ *is flat. Then* (M, P) *is flat if and only if at each point* x_0, *one of the following conditions holds:*

(1) $S = \rho I$; (2) rank $h = 1$ *and* $S = \rho I$ *on* ker h; *or* (3) $h = 0$.

Proof. Suppose that $(\tilde{M}, \tilde{P})$ is flat. For x_0 we take equiaffine connections $\nabla \in (M, P)$ and $\tilde{\nabla} \in (\tilde{M}, \tilde{P})$, both having symmetric Ricci tensors, such that f is an affine immersion with an equiaffine transversal vector field ξ. Since $\tilde{\nabla}$ is projectively flat, we have

(7.9) $$\tilde{R}(X, Y)Z = \tilde{\gamma}(Y, Z)X - \tilde{\gamma}(X, Z)Y,$$

where $\tilde{\gamma}$ is the normalized Ricci tensor of $\tilde{\nabla}$ (refer to Note 1). The Gauss equation for the affine immersion f gives

(7.10) $$R(X, Y)Z = \tilde{\gamma}(Y, Z)X - \tilde{\gamma}(X, Z)Y + h(Y, Z)SX - h(X, Z)SY.$$

We find that the normalized Ricci tensor γ for ∇ is expressed by

(7.11) $$\gamma(Y, Z) = \tilde{\gamma}(Y, Z) + \frac{1}{n-1}[h(Y, Z) \operatorname{tr} S - h(SY, Z)].$$

If ∇ is also projectively flat, we have

(7.12) $$R(X, Y)Z = \gamma(Y, Z)X - \gamma(X, Z)Y.$$

Comparing (7.11) and (7.12) we arrive at

$$(7.13) \qquad (n-1)[h(Y,Z)SX - h(X,Z)SY]$$
$$= [(\operatorname{tr} S)h(Y,Z) - h(SY,Z)]X - [(\operatorname{tr} S)h(X,Z) - h(SX,Z)]Y.$$

Conversely, if this equation holds, it implies (7.12) and ∇ is projectively flat.

In this way, the proof of Theorem 7.6 is reduced to proving that the condition (7.13) is satisfied if and only if one of the conditions (1), (2), or (3) of Theorem 7.6 holds. This is pure linear algebra and is omitted. The reader may refer to [NP4].

The case where the rank of h is at most equal to 1 will be treated in the next section. As an application of the case (1) in Theorem 7.6, we prove the following result regarding nondegenerate affine hypersurfaces in an $(n+1)$-dimensional space form $\tilde{M}$ of nonzero constant sectional curvature with its Levi-Civita connection. We consider $\tilde{M}$ with its connection $\tilde{\nabla}$, which admits, of course, a parallel volume element (namely, the metric volume element of g). For any immersion $f : M^n \to \tilde{M}$, we may define the notion of nondegeneracy and, under that assumption, prove the unique existence of the affine normal field just as in the classical case where $\tilde{M}$ is the affine space $\mathbf{R}^{n+1}$. The procedure for finding the affine normal field remains the same (Section 3 of Chapter II). We now prove

Theorem 7.7. *Let $\tilde{M}$ be an $(n + 1)$-dimensional pseudo-Riemannian manifold with metric $\tilde{g}$ of constant sectional curvature $c \neq 0$ and its Levi-Civita connection $\tilde{\nabla}$, where $n \geq 3$. If there exists a nondegenerate Blaschke hypersurface M^n whose connection is flat, then $c < 0$. In particular, S^{n+1} (or $\mathbf{P}^{n+1}$) with its canonical connection of constant positive sectional curvature does not admit any nondegenerate flat affine hypersurface.*

Proof. Assume M^n is a nondegenerate Blaschke hypersurface with flat connection ∇. We shall show that the affine normal is perpendicular to M^n and that M^n is umbilical in the metric sense. First, we have the Gauss equation for the affine hypersurface M^n

$$(7.14) \qquad R(X,Y) = \tilde{\gamma}(Y,Z)X - \tilde{\gamma}(X,Z)Y + h(Y,Z)SX - h(X,Z)SY,$$

where $\tilde{\gamma}$ is the normalized Ricci tensor (equal to $c\tilde{g}$) of $(\tilde{M}, \tilde{\nabla})$ and h and S are the affine fundamental form and the affine shape operator, respectively. Now $R = 0$ by assumption; in particular, ∇ is projectively flat. The space $(\tilde{M}, \tilde{\nabla})$ being projectively flat, we see by Theorem 7.6 that $S = \rho I$, where ρ is a constant. The Gauss equation now reduces to

$$[c\tilde{g}(Y,Z) + \lambda h(Y,Z)]X + [c\tilde{g}(X,Z) + \lambda h(X,Z)]Y = 0.$$

For arbitrary X and Z tangent to M take Y that is linearly independent of X. Then we get $c\tilde{g}(X,Z) = -\rho\, h(X,Z)$. Thus $\rho \neq 0$ and $h = -(c/\rho)g_0$, where g_0 is the restriction of $\tilde{g}$ to M^n.

If we denote by ξ_0 a unit normal vector field for M^n (relative to $\tilde{g}$) and by h_0 the second fundamental form in the metric sense, we see that $h_0 = \lambda h$, where λ is a certain scalar function. But then we have $h_0 = -(\lambda c/\rho)g_0$. As is well-known, it follows that $k = -\lambda c/\rho$ is constant. This means that M^n is umbilical in the metric sense.

We now go back to the procedure for finding the affine normal field. We have $h_0 = kg_0$. If $\{X_1, \ldots, X_n\}$ is an orthonormal basis, then the absolute value of the determinant of the matrix $[h_0(X_i, X_j)]$ is equal to the constant $|k|^n$. This implies that the affine normal vector ξ is in the same direction as the unit normal vector ξ_0. Hence the Blaschke connection ∇ coincides with the Levi-Civita connection ∇_0 of g_0. The metric shape operator S_0 coincides with kI, because $h_0 = kg_0$. From the Gauss equation in the metric case, we obtain $c + k^2 = 0$, which implies $c < 0$.

Remark 7.2. For $n = 2$, there are many Blaschke immersions of a flat torus (T^2, ∇) into S^3; for example, all Clifford tori $S^1(r) \times S^1(s)$, $r^2 + s^2 = 1$, imbedded naturally in S^3 as $\{(x, y, u, v) \in \mathbf{R}^4 : x^2 + y^2 = r^2, u^2 + v^2 = s^2\}$.

Remark 7.3. Theorem 7.7 is an affine result related to the problems on isometric immersions between Riemannian or pseudo-Riemannian space forms. For example, it is known that there is no isometric immersion of a Euclidean space E^n into S^{n+1} of constant curvature 1, whereas E^n can be isometrically imbedded (as a horosphere) into the hyperbolic space H^{n+1} of constant curvature -1.

8. Hypersurfaces in $\mathbf{P}^{n+1}$ and their invariants

First, we shall briefly discuss hyperquadrics in $\mathbf{P}^{n+1}$. A *hyperquadric* in $\mathbf{P}^{n+1}$ is defined as the set of points whose homogeneous coordinates satisfy a quadratic equation $F(x^1, \ldots, x^{n+2}) = 0$. By a projective change of coordinates we may assume that

$$(8.1) \qquad F = (x^1)^2 \pm \cdots \pm (x^{r+2})^2,$$

where $0 < r \leq n$. If we consider only those hyperquadrics that are the images of an imbedding of an n-dimensional manifold, we must have $r = n$. We call them *nonsingular* hyperquadrics.

For example, in $\mathbf{P}^3$, nonsingular quadrics are given by

$$(8.2) \qquad (x^1)^2 + (x^2)^2 + (x^3)^2 - (x^4)^2 = 0$$

or

$$(8.3) \qquad (x^1)^2 + (x^2)^2 - (x^3)^2 - (x^4)^2 = 0$$

with respect to an appropriate projective coordinate system.

The intersection of the quadric (8.2) with the affine 3-space $\{(x^1, x^2, x^3, 1)\}$ appears as an ellipsoid $(x^1)^2 + (x^2)^2 + (x^3)^2 = 1$. The intersection with the affine 3-space $\{(x^1, x^2, 1, x^4)\}$ is a two-sheeted hyperboloid $(x_1)^2 + (x^2)^2 - (x^4)^2 = -1$.

If we let $u = x^3 + x^4$, $v = x^3 - x^4$, then (8.2) takes the form $(x^1)^2 + (x^2)^2 + uv = 0$ and the intersection with the affine 3-space $v = -1$ appears as an elliptic paraboloid $u = (x^1)^2 + (x^2)^2$. In a similar way, the intersection of (8.3) with $x^4 = 1$ gives a one-sheeted hyperboloid $(x^1)^2 + (x^2)^2 - (x^3)^2 = 1$. By using $u = x^2 + x^4$, $v = x^2 - x^4$, (8.3) takes the form $(x^1)^2 - (x^3)^2 + uv = 0$ and its intersection with $v = -1$ appears as a hyperbolic paraboloid $u = (x^1)^2 - (x^3)^2$.

In $\mathbf{P}^2$, a nonsingular quadric (usually called a *conic*) is given by

$$(8.4) \qquad\qquad (x^1)^2 + (x^2)^2 - (x^3)^2 = 0.$$

Its intersection with the affine 2-plane $x^3 = 1$, $x^2 = 1$, or $x^2 - x^3 = -1$ appears as an ellipse, a hyperbola, or a parabola.

The quadric Q defined by $(x^1)^2 + (x^2)^2 - (x^3)^2 - (x^4)^2 = 0$ in $\mathbf{P}^3$ has a conic $\underline{Q}: (x^1)^2 + (x^2)^2 - (x^3)^2 = 0$ as intersection with the projective plane $\mathbf{P}^2 : x^4 = 0$. It follows that Q is the cone with vertex $(0, 0, 0, 1)$ and $\underline{Q}$ as base and that $f : \underline{Q} \times \mathbf{R}^+ \to \mathbf{P}^3$ is an immersion whose image is the cone with vertex deleted.

We shall later consider how to characterize hyperquadrics among the hypersurfaces in $\mathbf{P}^{n+1}$.

Let $f : M^n \to \mathbf{P}^{n+1}$ be an immersion of an n-dimensional manifold into the projective space. We want to define the notion of rank of f at each point $x_0 \in M^n$. Let (U, D) be a chart for $\mathbf{P}^{n+1}$ where U is a neighborhood of $f(x_0)$. Take any transversal vector ξ around x_0 and define the affine fundamental form h relative to ξ and D. We know that h changes by a scalar factor for a different choice of ξ and the rank of h depends only on D. If we choose another chart (U', D') such that $f(x_0) \in U'$, then in $U \cap U'$ the two connections D and D' are projectively related. Thus the affine fundamental form based on D coincides with that based on D' by using the same transversal field. It follows that the notion of *rank* of f is well-defined. In particular, if f has rank n at each point, we say that f is nondegenerate. An immersion $f : M^n \to \mathbf{P}^{n+1}$ is said to be *totally geodesic at* x_0 if the affine fundamental form h vanishes, that is, has rank 0 at x_0. If f is totally geodesic at each point, we say that f is *totally geodesic*.

In order to study immersions of low rank, it is necessary to recall a result of [Fe]. Let $f : M^n \to S^{n+1}$ be an isometric immersion of an n-dimensional complete Riemannian manifold into the unit sphere S^{n+1} with its canonical Riemannian metric of constant curvature 1. Let $r(n)$ be the smallest even integer $\geq \frac{n}{2}$. Then the theorem of Ferus says that if the rank of f (which is equal to the rank of the metric second fundamental form or of the metric shape operator, also called the *type number*) is less than $r(n)$, then f is totally geodesic.

We now prove the projective version in [NP4] of the theorem of Ferus.

Theorem 8.1. *Let* $f : M^n \to \mathbf{P}^{n+1}$ *be an immersion of a connected, compact manifold,* $n \geq 2$. *If the rank of* f *is less than* $r(n)$ *everywhere, then* f *is totally geodesic.*

Proof. We introduce a Euclidean metric $\tilde{g}$ in $\mathbf{R}^{n+2}$ and consider the unit sphere S^{n+1}. The restriction to S^{n+1} of the projection $\mathbf{R}^{n+2} - \{0\} \to \mathbf{P}^{n+1}$ gives a two-fold covering $\pi_1 : S^{n+1} \to \mathbf{P}^{n+1}$. The canonical metric on S^{n+1} induced by the Euclidean metric $\tilde{g}$ defines a natural Riemannian metric, say g_1, on $\mathbf{P}^{n+1}$ such that π_1 is isometric. Observe that the rank of f at each point of M is the same as the rank of the second fundamental form of f with respect to the metric g_1. Let $g = f^* g_1$ and let $\pi : (\hat{M}, \hat{g}) \to (M, g)$ be the universal covering so that π is isometric. The immersion $f \circ \pi : \hat{M} \to \mathbf{P}^{n+1}$ can be lifted to an isometric immersion $\hat{f} : \hat{M} \to S^{n+1}$ such that $\pi_1 \circ \hat{f} = f \circ \pi$. It is obvious that the rank of $\hat{f}$ at $\hat{x} \in \hat{M}$ is equal to the rank of f at $x = \pi_1(\hat{x}) \in M$. By assumption, the rank of $\hat{f}$ is less than $r(n)$ at every point of $\hat{M}$. The theorem of Ferus quoted above now implies that $\hat{f}$ is totally geodesic. It follows that f is totally geodesic.

Let $(\tilde{M}, \tilde{P})$ be an $(n+1)$-dimensional manifold with a projective structure $\tilde{P}$. For an arbitrary immersion f of an n-manifold M into $(\tilde{M}, \tilde{P})$, what is the geometric significance of the choice of a transversal vector field ξ around a point $x_0 \in M$ or of a chart $(U, D) \in \tilde{P}$ around $f(x_0)$? Even when f is nondegenerate, the situation is not as simple as the affine case. Some preliminary observations will now be made. In order to simplify the notation, we shall simply write (U, D) for a chart in $\tilde{P}$ or (U', D') for an alternative choice.

For each point $x_0 \in M$, we take a neighborhood W of x_0 and a chart $(U, D) \in \tilde{P}$ such that $f(W) \subset U$. For the immersion $f : W \to (U, D)$, any choice of a transversal vector field ξ determines a connection ∇ and h on W such that for any vector fields X and Y on W

$$(8.5) \qquad D_X f_* Y = f_*(\nabla_X Y) + h(X, Y)\xi.$$

If we choose $(U', D') \in \tilde{P}$ instead of (U, D), then in $U \cap U'$ the connections D and D' are projectively equivalent:

$$(8.6) \qquad D'_X f_* Y = f_*(\nabla_X Y) + \bar{p}(f_* X)f_* Y + \bar{p}(f_* Y)f_* X$$

for a 1-form $\bar{p}$ on $U \cap U'$. Hence we get

$$(8.7) \qquad \nabla'_X Y = \nabla_X Y + \rho(X)Y + \rho(Y)X,$$

where $\rho(X) = \bar{p}(f_* X)$. Thus for the same ξ we see that $h = h'$ and that ∇ and ∇' are projectively equivalent.

Still keeping the same ξ, we see from

$$(8.8) \qquad D_X \xi = -f_*(SX) + \tau(X)\xi$$

and a similar formula for D' that the shape operators S and S' differ by a scalar multiple of the identity; to be more precise, the scalar is $\bar{p}(\xi)$. We conclude that

$$(8.9) \qquad \tau' = \tau + \rho.$$

If we assume that P consists of local affine connections with symmetric Ricci tensors (see Section 7), then ρ is a closed form and hence $d\tau = d\tau'$.

Now we keep the same chart $(U, D) \in \tilde{P}$ and consider changing ξ to another transversal vector field

$$(8.10) \qquad\qquad \xi' = \phi\xi + f_*(Z),$$

where Z is a vector field on M and ϕ a nonvanishing function. Then we have

$$(8.11) \qquad\qquad h' = \frac{1}{\phi}h,$$

$$(8.12) \qquad\qquad \nabla'_X Y = \nabla_X Y - h'(X, Y)Z.$$

We define the cubic form C relative to the chart (U, D) by

$$(8.13) \qquad C(X, Y, Z) = (\nabla_X h)(Y, Z) + \tau(X)h(Y, Z);$$

we say that C is divisible by h if

$$(8.14) \qquad C(X, Y, Z) = \mu(X)h(Y, Z) + \mu(Y)h(Z, X) + \mu(Z)h(X, Y)$$

for some 1-form μ (recall Definition 6.1). We have the following interesting fact.

Proposition 8.2. *The property that the cubic form C is divisible by h is independent of the choice of $D \in \tilde{P}$ or of ξ.*

Proof. Suppose for a point x_0 we choose $(U, D) \in \tilde{P}$ such that $f(W) \subset U$ for some neighborhood W of x_0. Then the property is independent of the choice of ξ as we saw in Lemma 6.3. Now change (U, D) to $(U', D') \in \tilde{P}$. Using the same ξ, we obtain $h' = h$. By virtue of (8.7) and (8.9) we may compute and obtain (setting $h = h'$)

$$C'(X, Y, Z) = C(X, Y, Z) - \rho(X)h(Y, Z) - \rho(Y)h(Z, X) - \rho(Z)h(X, Y).$$

Thus if C is divisible by h, then so is C'.

Now we can state

Theorem 8.3. *Let f be an immersion of an n-dimensional connected differentiable manifold M into $\mathbf{P}^{n+1}$ such that the rank of h is ≥ 2. Then $f(M)$ lies in a hyperquadric Q^n if and only if f has the property that C is divisible by h.*

Proof. Assume that C is divisible by h. Each point $x_0 \in M$ has a neighborhood W such that $f(W)$ is contained in an affine space $\mathbf{R}^{n+1} = \mathbf{P}^{n+1} - \Sigma$, where Σ is a certain projective hyperplane. The flat connection D in $\mathbf{R}^{n+1}$ belongs to the usual projective structure of $\mathbf{P}^{n+1}$. For $f : W \to (\mathbf{R}^{n+1}, D)$, we see that the rank of h is ≥ 2 and that C is divisible by h. By appealing to Theorem 6.6 we see that $f(W)$ lies in a hyperquadric in $\mathbf{R}^{n+1}$. Thus f is locally an immersion into a hyperquadric in $\mathbf{P}^{n+1}$; the connectivity argument shows that $f(M)$ lies in a hyperquadric. The converse also follows from Theorem 6.6.

We shall now indicate a few more invariants for a nondegenerate immersion $f : M \to \mathbf{P}^{n+1}$. We choose two Ricci-symmetric connections D and D' belonging to the projective structure P and let ω and ω' be volume elements that are parallel relative to D and D', respectively. The two volume elements are related by

$$(8.15) \qquad \omega' = \phi\omega$$

where ϕ is a nonvanishing function. The two connections are related by (8.7) with $\rho = \frac{1}{n+2} d \log \phi$, as is easily seen (see Appendix 1). Now let ξ and ξ' be the affine normal fields for the nondegenerate immersion f relative to the ambient equiaffine structures (D, ω) and (D', ω'), respectively. To find out their relationship, we use ξ as a tentative choice of transversal vector field for the immersion f relative to (D', ω') and go through the procedure for finding the affine normal field in Section 3 of Chapter II. Thus write

$$(8.16) \qquad D'_X f_*(Y) = f_*(\check{\nabla}_X Y) + \check{h}(X, Y)\xi.$$

We find that

$$(8.17) \qquad \check{\nabla}_X Y = \nabla_X Y + \rho(X)Y + \rho(Y)X,$$

$$(8.18) \qquad \check{h}(X, Y) = h(X, Y).$$

From

$$(8.19) \qquad D'_X \xi = -f_*(\check{S}X) + \check{\tau}(X)\xi$$

we get

$$(8.20) \qquad \check{\tau}(X) = \rho(X).$$

As for the induced volume element we have

$$(8.21) \qquad \check{\theta}(X_1, \dots, X_n) = \omega'(X_1, \dots, X_n, \xi).$$

We have

$$\theta(X_1, \dots, X_n) = \omega(X_1, \dots, X_n, \xi)$$

and hence

$$\check{\theta}(X_1, \dots, X_n) = \phi\theta(X_1, \dots, X_n).$$

Here we can assume that ϕ is positive by choosing, if necessary, $-\xi$ as a tentative transversal vector field. Now let $\{X_1,\ldots,X_n\}$ be a basis in $T_x(M)$ with $\check{\theta}(X_1,\ldots,X_n) = 1$. Then

$$\theta(\phi^{\frac{1}{n}}X_1,\ldots,\phi^{\frac{1}{n}}X_n) = 1,$$

$$h_{ij} = h(\phi^{\frac{1}{n}}X_i, \phi^{\frac{1}{n}}X_j) = \phi^{\frac{2}{n}}h(X_i, X_j).$$

Here we note that $\det [h_{ij}] = 1$ because h is the affine metric for M relative to D. We have also

$$\check{h}_{ij} = \check{h}(X_i, X_j) = h(X_i, X_j) = \phi^{-\frac{2}{n}}h_{ij}.$$

Since $\det [h_{ij}] = 1$, we have

$$\det {}_{\check{\theta}}\check{h} = (\phi^{-\frac{2}{n}})^n = \phi^{-2}.$$

It follows that the affine metric h' is given by

(8.22) $$h' = \phi^{\frac{2}{n+2}} h.$$

In order to find the affine normal field ξ' in the form $f_*Z + \phi^{-\frac{2}{n+2}}\xi$, we must determine the tangent vector field Z such that

$$X(\phi^{-\frac{2}{n+2}}) + h(X,Z) + \phi^{-\frac{2}{n+2}}\tau(X) = 0$$

for all X. Using $\rho = \frac{1}{n+2}d\log\phi$ and (8.20), the equation above becomes

$$h(X,Z) = \phi^{-\frac{2}{n+2}}\rho(X).$$

By introducing a vector field V by

(8.23) $$h(V,X) = \rho(X) \quad \text{for all} \quad X,$$

we get

$$Z = \phi^{-\frac{2}{n+2}}V$$

and the affine normal field ξ' is given by

(8.24) $$\xi' = \phi^{-\frac{2}{n+2}}(V + \xi).$$

We obtain the induced connection for ξ'

(8.25) $$\nabla'_X Y = \nabla_X Y + \rho(X)Y + \rho(Y)X - h(X,Y)V,$$

as can be verified by using (8.16), (8.17), (8.18), and (8.24).

Let us now recall the following general fact. If a metric h on a differentiable manifold M is changed conformally to $h' = \psi^2 h$ with a function $\psi > 0$, then the Levi-Civita connection ∇' for h' is related to the Levi-Civita connection ∇ for h exactly in the same way as (8.25) with $\rho = d \log \psi$. (See [NY].) If ψ is taken to be $\phi^{\frac{1}{n+2}}$, then $\rho = \frac{1}{n+2} d \log \phi$.

From this fact, we can make the following observation: If an equiaffine structure (D, ω) is changed projectively to another, the difference tensor K of the induced connection ∇ and the Levi-Civita connection for the affine metric h does not change. The cubic form C changes conformally with the same factor as the conformal change of the metric h. The Pick invariant $J = h(K, K)/n(n-1)$ changes by the reciprocal factor. We may say that the tensor K is an *invariant* object for a nondegenerate immersion $f : M \to \tilde{M}$.

For a nondegenerate hypersurface $f : M \to \mathbf{P}^{n+1}$, we have invariant objects and invariant properties (those objects and properties that do not depend on the choice of a transversal vector field ξ or on the choice of a chart belonging to $\mathbf{P}^{n+1}$) as follows.

Theorem 8.4. *The conformal class of h and that of the cubic form C are invariant objects. The property that C is divisible by h is invariant. The difference tensor field K between the induced connection ∇ and the Levi-Civita connection $\hat{\nabla}$ for h is invariant. Whether J is 0 or not is an invariant property. When J never vanishes, Jh is an invariant metric of M (called the projective metric). All invariant objects and properties are projectively invariant, that is, invariant under any projective transformation of $\mathbf{P}^{n+1}$.*

Remark 8.1. We may extend Theorems 11.3 and 11.4 of Chapter II to the case of a nondegenerate surface M^2 immersed in $\mathbf{P}^3$. If M^2 is ruled (see Definition 11.1 of Chapter II), then h is hyperbolic and J vanishes identically (each of these properties is invariant). Conversely, if J vanishes identically without C vanishing anywhere, then M^2 is a ruled surface in $\mathbf{P}^3$.

For these results, see [NP4], [S3, 4].

9. Complex affine geometry

This section treats holomorphic immersions of a complex n-manifold M^n into a complex affine space $\mathbf{C}^{n+1}$. The group for $\mathbf{C}^{n+1}$ is now the complex affine group $A(n+1, \mathbf{C})$, that is, $GL(n+1, \mathbf{C}) \cdot \mathbf{C}^n$. A natural question in this case is: what kind of induced structure do we have on M^n? In the following we shall exhibit two possibilities for induced connections on M^n: one is a holomorphic affine connection and the other is an affine Kähler connection. We begin with the basic information on affine connections on a complex manifold.

Let M^n be a complex manifold with complex structure tensor J. For each point $p \in M^n$, J_p is an endomorphism of the (real) tangent space $T_p = T_p(M^n)$ such that $J_p^2 = -I$, where I denotes the identity transformation. The complexification $T_p^{\mathbf{C}}$ is the set of all *complex vectors* $X + iY$, where

$X, Y \in T_p$. We may extend J_p as a complex endomorphism of the complex vector space $T_p^{\mathbf{C}}$. We may write

(9.1) $T_p^{\mathbf{C}} = T_p^{(1,0)} \oplus T_p^{(0,1)},$

where

$$T_p^{(1,0)} = \{X - iJX : X \in T_p\} \quad \text{and} \quad T_p^{(0,1)} = \{X + iJX : X \in T_p\}$$

are complex subspaces of $T_p^{\mathbf{C}}$. We have then

$$T_p^{(1,0)} = \{Z \in T_p^{\mathbf{C}} : JZ = iZ\} \quad \text{and} \quad T_p^{(0,1)} = \{Z \in T_p^{\mathbf{C}} : JZ = -iZ\}.$$

We can carry over these constructions to vector fields. Let $\mathfrak{X} = \mathfrak{X}(M)$ denote the space of real vector fields and $\mathfrak{X}^{\mathbf{C}} = \mathfrak{X}^{\mathbf{C}}(M)$ its complexification, which is decomposed as $\mathfrak{X}^{\mathbf{C}} = \mathfrak{X}^{(1,0)} \oplus \mathfrak{X}^{(0,1)}$, the direct sum of the space of vector fields of type $(1,0)$ and the space of vector fields of type $(0,1)$.

Definition 9.1. A vector field $Z \in \mathfrak{X}^{\mathbf{C}}$ is said to be *holomorphic* if Zf is holomorphic for every locally defined holomorphic function f on M. If $Z = \frac{1}{2}(X - iJX)$ for $X \in \mathfrak{X}$, then Z is holomorphic if and only if $[X, JY] = J[X, Y]$ for all $Y \in \mathfrak{X}$. This means that the 1-parameter group of local transformations generated by X preserves the complex structure. We can thus say that X is *real holomorphic*.

In terms of local coordinates, let $\{z^1, \ldots, z^n\}$ be a local complex coordinate system. Write $z^k = x^k + iy^k$, $1 \le k \le n$, where $x^k, y^k \in \mathbf{R}$. Then

$$J\frac{\partial}{\partial x^k} = \frac{\partial}{\partial y^k}, \quad J\frac{\partial}{\partial y^k} = -\frac{\partial}{\partial x^k}.$$

As is well-known, we have

(9.2) $\dfrac{\partial}{\partial z^k} = \dfrac{1}{2}\left(\dfrac{\partial}{\partial x^k} - i\dfrac{\partial}{\partial y^k}\right), \qquad \dfrac{\partial}{\partial \bar{z}^k} = \dfrac{1}{2}\left(\dfrac{\partial}{\partial x^k} + i\dfrac{\partial}{\partial y^k}\right).$

We see that $\mathfrak{X}^{(1,0)}$ is locally spanned by $\partial/\partial z^k$, $1 \le k \le n$, and that $Z \in \mathfrak{X}^{(1,0)}$ is holomorphic if and only if $Z = \sum_{k=1}^{n} f^k(\partial/\partial)z^k$, where f^k, $1 \le k \le n$, are holomorphic functions of $(z^1, \ldots, z^n)$.

Now for a complex manifold M^n we consider a torsion-free affine connection ∇ that is *compatible* with J, i.e., $\nabla J = 0$. It is known that a complex manifold M^n admits such a connection, which we shall call a *complex connection* for simplicity. (See [KN2, pp. 143–145].) The condition $\nabla J = 0$ is equivalent to $\nabla_X J = 0$, that is, $\nabla_X(JY) = J(\nabla_X Y)$ for all $X, Y \in \mathfrak{X}$. It follows that a vector field $X \in \mathfrak{X}$ is real holomorphic if and only if $\nabla_{JY} X = J\nabla_Y X$ for every $Y \in \mathfrak{X}$, as can be easily shown. The map ∇_X maps $\mathfrak{X}^{(1,0)}$ into itself and likewise for $\mathfrak{X}^{(0,1)}$. Thus we may consider ∇ as a $(1,0)$-connection that defines covariant differentiation

$$(9.3) \qquad (Z, W) \in \mathfrak{X}^{(1,0)} \times \mathfrak{X}^{(1,0)} \mapsto \nabla_Z W \in \mathfrak{X}^{(1,0)}.$$

Definition 9.2. We say that the $(1,0)$-connection is *holomorphic* if $\nabla_Z W$ is a holomorphic vector field for all holomorphic vector fields Z and W. As a real connection, ∇ is *real holomorphic* if $\nabla_X Y$ is holomorphic for all real holomorphic vector fields X and Y.

It is easy to prove

Proposition 9.1. *A torsion-free connection ∇ compatible with J is real holomorphic if and only if the corresponding $(1,0)$-connection is holomorphic.*

We have also

Proposition 9.2. *The connection ∇ is real holomorphic if and only if its curvature tensor R satisfies the condition*

$$(9.4) \qquad R(JX, Y) = JR(X, Y) \quad \text{for all} \quad X, Y \in \mathfrak{X}.$$

This condition is equivalent to

$$(9.5) \qquad R(Z, \bar{W}) = 0 \quad \text{for all} \quad Z, W \in \mathfrak{X}^{(1,0)}.$$

Proof. Assume that ∇ is real holomorphic. To prove the identity (9.4), it is sufficient to show

$$R(JX, Y)Z = JR(X, Y)Z$$

for all real holomorphic vector fields X, Y, and Z. We recall that $\nabla_{JU} V = J\nabla_U V$ for every $U \in \mathfrak{X}$ if V is real holomorphic. Note that $\nabla_Y Z$ is holomorphic. Hence $\nabla_{JX}(\nabla_Y Z) = J\nabla_X \nabla_Y Z$. Similarly, we have

$$\nabla_{JX} Z = J\nabla_X Z \quad \text{and} \quad \nabla_Y \nabla_{JX} Z = J\nabla_Y \nabla_X Z$$

as well as

$$\nabla_{[JX, Y]} Z = \nabla_{J[X, Y]} Z = J\nabla_{[X, Y]} Z.$$

Putting all this together, we obtain $R(JX, Y)Z = JR(X, Y)Z$. We omit the proof of the converse. It is easy to show that (9.4) and (9.5) are equivalent.

A Riemannian metric g on a complex manifold M^n is said to be *Hermitian* if $g(JX, JY) = g(X, Y)$ for all $X, Y \in \mathfrak{X}$. If furthermore its Levi-Civita connection ∇ is compatible with J, then g is called a *Kähler metric* (see [KN2, p. 149]). It is known that the Levi-Civita connection of a Kähler metric satisfies

$$(9.6) \qquad R(JX, JY) = R(X, Y) \quad \text{for} \quad X, Y \in \mathfrak{X}$$

and, equivalently,

$$(9.7) \qquad R(Z, W) = 0 \quad \text{for} \quad Z, W \in \mathfrak{X}^{(1,0)};$$

refer to Proposition 4.5 of [KN2, p. 149]. In view of this, we introduce the following notion.

Definition 9.3. A torsion-free affine connection ∇ compatible with the complex structure J is said to be *affine Kähler* if its curvature tensor satisfies the condition (9.6) above.

We remark that if ∇ is at the same time holomorphic and affine Kähler, then the curvature tensor vanishes identically and ∇ is locally flat.

Example 9.1. We give an affine Kähler connection that does not come from a Kähler metric on the 1-dimensional complex torus $T = \mathbf{C}/\mathbf{Z}^2$. We define the connection ∇ by

$$\nabla_{\frac{\partial}{\partial \bar{z}}} \frac{\partial}{\partial z} = \phi \frac{\partial}{\partial z},$$

where $\partial/\partial z$ is the standard complex vector field on $\mathbf{C}$ and ϕ is a smooth function. It is affine Kähler because of 1-dimensionality. The Ricci tensor is seen to be

$$\text{Ric}\left(\frac{\partial}{\partial z}, \frac{\partial}{\partial \bar{z}}\right) = -\frac{\partial \bar{\phi}}{\partial z};$$

hence, when $\phi(z) = \exp(2\pi i \,\text{Re}\, z)$ which is well-defined on T, the Ricci tensor is not real-valued. On the other hand, it is known that the Ricci tensor of a Kähler connection is real-valued (see the equation (24) on p. 158 of [KN2]). This implies that the connection ∇ above is not Kähler.

Remark 9.1. The notion of affine Kähler connection is in fact far more general than that of Kähler connection. We can construct affine Kähler connections on the Hopf surface, the quotient of $\mathbf{C}^2 - \{0\}$ by a discrete group generated by a dilatation. Refer to [NPo1]. Thus there are compact complex manifolds that admit an affine Kähler connection without admitting any Kähler metric.

Now we consider a holomorphic immersion f of a complex n-dimensional manifold M^n into $\mathbf{C}^{n+1}$. Let J denote the complex structure tensor for both M^n and $\mathbf{C}^{n+1}$; thus $f_* JX = J f_* X$ for $X \in \mathfrak{X}$. We shall obtain the complex versions of the formulas of Gauss and Weingarten in the real hypersurface theory in Chapter II. For this purpose, we choose a (real) vector field ξ along f that is transversal to the immersion. Then $J\xi$ is also transversal and linearly independent of ξ. Hence we have a decomposition of $T_{f(x)}(\mathbf{C}^{n+1})$ as the direct sum of $f_*(T_x(M))$, $\mathbf{R}\{\xi\}$, and $\mathbf{R}\{J\xi\}$. For all $X, Y \in \mathfrak{X}$, we have a decomposition

$$(9.8) \qquad D_X f_* Y = f_*(\nabla_X Y) + h(X, Y)\xi + k(X, Y)J\xi,$$

by which we define a torsion-free affine connection ∇ and symmetric tensors h and k. Since h and k are symmetric and f is holomorphic, we get $\nabla J = 0$ and

$$(9.9) \qquad k(X,Y) = -h(JX,Y),$$
$$h(JX,Y) = h(X,JY), \quad k(JX,Y) = k(X,JY).$$

We can also write

$$(9.10) \qquad D_X\xi = -f_*(AX) + \mu(X)\xi + \nu(X)J\xi,$$

thus defining a $(1,1)$-tensor A and two 1-forms μ and ν. Since D is compatible with J, we get

$$(9.11) \qquad D_X J\xi = -Jf_*(AX) - \nu(X)\xi + \mu(X)J\xi.$$

These formulas can be rewritten in terms of complex vector fields as follows. Set

$$(9.12) \qquad \zeta = \xi - iJ\xi, \quad S = A - iJA, \quad \tau = \mu + i\nu,$$

and extend h complex-bilinearly on $\mathfrak{X}^{\mathbb{C}} \times \mathfrak{X}^{\mathbb{C}}$. Then, for $Z = X - iJX$ and $W = Y - iJY$, we have

$$(9.13) \qquad h(Z,W) = 2(h(X,Y) + ik(X,Y)), \quad h(Z,\bar{W}) = 0.$$

From (9.8) we obtain

$$(9.14) \qquad D_Z f_* W = f_* \nabla_Z W + h(Z,W)\zeta, \quad . \quad D_Z f_* \bar{W} = f_* \nabla_Z \bar{W}$$

and their conjugates. The equations (9.10) and (9.11) are rewritten as

$$(9.15) \qquad D_Z\zeta = -f_*SZ + \tau(Z)\zeta, \quad D_Z\bar{\zeta} = -f_*\bar{S}Z + \bar{\tau}(Z)\bar{\zeta},$$

and their conjugates.

Since D is flat, we get the following equations of Gauss and Codazzi (for h) by straightforward computation.

Proposition 9.3. *For all vector fields* $Z, W, U \in \mathfrak{X}^{(1,0)}$ *we have*

$$(9.16a) \qquad R(Z,W)U = h(W,U)SZ - h(Z,U)SW,$$
$$(9.16b) \qquad R(Z,W)\bar{U} = 0,$$
$$(9.16c) \qquad R(Z,\bar{W})U = -h(Z,U)S\bar{W},$$
$$(9.17a) \qquad (\nabla_Z h)(W,U) + h(W,U)\tau(Z) = (\nabla_W h)(Z,U) + h(Z,U)\tau(W),$$
$$(9.17b) \qquad (\nabla_{\bar{W}} h)(Z,U) + h(Z,U)\tau(\bar{W}) = 0.$$

The equation of Codazzi for S and the equation of Ricci are as follows.

Proposition 9.4. *For all vector fields* Z, $W \in \mathfrak{X}^{(1,0)}$ *we have*

$$(9.18a) \qquad (\nabla_Z S)(W) - \tau(Z)SW = (\nabla_W S)(Z) - \tau(W)SZ,$$

$$(9.18b) \qquad (\nabla_Z \bar{S})(W) - \bar{\tau}(Z)\bar{S}W = (\nabla_W \bar{S})(Z) - \bar{\tau}(W)\bar{S}Z,$$

$$(9.18c) \qquad (\nabla_Z \bar{S})\bar{W} - \tau(Z)S\bar{W} = (\nabla_{\bar{W}} S)(Z) - \tau(\bar{W})SZ,$$

$$(9.19a) \qquad h(Z, SW) - h(W, SZ) = (\nabla_Z \tau)(W) - (\nabla_W \tau)(Z),$$

$$(9.19b) \qquad (\nabla_Z \bar{\tau})(W) - (\nabla_W \bar{\tau})(Z) = 0,$$

$$(9.19c) \qquad h(Z, S\bar{W}) = (\nabla_Z \tau)(\bar{W}) - (\nabla_{\bar{W}} \tau)(Z).$$

Definition 9.4. We say that the transversal $(1,0)$-vector field ζ is *holomorphic* or *anti-holomorphic* according as $D_{\bar{Z}}\zeta = 0$ or $D_Z\zeta = 0$ for all $Z \in \mathfrak{X}^{(1,0)}$.

From (9.16) we obtain

Theorem 9.5. *If the transversal* $(1,0)$-*vector field* ζ *is holomorphic, then the affine connection* ∇ *is holomorphic. If* ζ *is anti-holomorphic, then the affine connection* ∇ *is affine Kähler.*

The first natural extensions of Blaschke's theory for nondegenerate hypersurfaces to the case of complex hypersurfaces $f : M^n \to \mathbf{C}^{n+1}$ were given by K. Abe [Ab] and by Dillen, Vrancken and Verstraelen [DVV], [DVr1, 5]. Under an appropriate assumption similar to the nondegeneracy condition, they established the unique existence of a holomorphic transversal $(1,0)$-vector field ζ as an analogue of the affine normal field. The approach in [Ab], based on the complex affine transformation group of $\mathbf{C}^{n+1}$ with determinant 1, used the formalism of complex moving frames. [DVV], based on the group of complex affine transformations with determinant of absolute value 1, used the method of vector fields.

In practice, it seems simplest to proceed as follows. We say that a holomorphic immersion $f : M^n \to \mathbf{C}^{n+1}$ is *nondegenerate* if the form h on $\mathfrak{X}^{(1,0)} \times \mathfrak{X}^{(1,0)}$ is nondegenerate for any holomorphic transversal $(1,0)$-vector field ζ. This condition is independent of the choice of ζ, as in the real case. For a chosen ζ, define the induced complex volume element

$$\theta(Z_1, \ldots, Z_n) = \mathrm{Det}\,(f_*Z_1, \ldots, f_*Z_n, \zeta),$$

where Det is the complex determinant function on $\mathbf{C}^{n+1}$. It follows that θ is parallel relative to the induced $(1,0)$-connection ∇ if and only if the form τ in (9.15) is identically 0 (note that $\bar{\tau}$ vanishes because ζ is assumed holomorphic). Now the volume element θ coincides with the complex volume element induced by h if and only if

$$\det{}_\theta h = 1.$$

This is the case if and only if $\det [h(Z_j, Z_k)] = 1$ for any basis $\{Z_1, \ldots, Z_n\}$ such that $\theta(Z_1, \ldots, Z_n) = 1$. We now want to find ζ satisfying the two conditions

(9.20) $$\tau = 0 \quad \text{and} \quad \det {}_\theta h = 1.$$

For this purpose we can follow the procedure for finding the affine normal field (Section 3 of Chapter II), except that we take one branch ϕ of $(\det {}_\theta h)^{1/(n+2)}$ instead of $|\det {}_\theta h|^{1/(n+2)}$ as in the real case. For any $(n + 2)$-th root of unity ϵ, we have $\epsilon\phi, \ldots, \epsilon^{n+1}\phi$ as other choices. Using ϕ, let $\hat{\zeta} = \phi\zeta + Z$, where Z is determined by

$$\tau + \frac{1}{\phi} h(Z, \cdot) + d \log \phi = 0.$$

Example 9.2. A complex curve in $\mathbf{C}^2$ is a holomorphic immersion $f : \mathbf{C} \to \mathbf{C}^2$, which we can regard as the complex version of a plane curve discussed in Chapter I. We can define the notions of nondegenerate curve and affine curvature κ (which is a holomorphic function) in the same way. If κ is constant, then the curve is affinely congruent to $z_2 = z_1^2$ or $z_1^2 + z_2^2 = 1$. A complex curve can obviously be regarded as a surface $f : \mathbf{R}^2 = \mathbf{C}^1 \to \mathbf{R}^4 = \mathbf{C}^2$. We can characterize a complex curve among (real) nondegenerate surfaces in $\mathbf{R}^4$. See Note 2.

Example 9.3. The graph of $z_{n+1} = \frac{1}{2}(z_1^2 + \cdots + z_n^2)$, for which $\zeta = (0, \ldots, 0, 1)$ and the shape operator S is identically zero (improper affine sphere).

Example 9.4. $z_1^2 + \cdots + z_{n+1}^2 = 1$, for which $\zeta = (z_1, \ldots, z_{n+1})$ and $S = -I$ (proper affine sphere).

Example 9.5. $z_1 z_2 \cdots z_{n+1} = 1$, which is also a proper affine sphere.

[Ab] shows that if a complex nondegenerate hypersurface has vanishing cubic form, then it is affinely equivalent to Example 9.3 or 9.4 (that is, it is a quadric). He also followed the method of Guggenheimer [Gu] to classify equiaffinely homogeneous complex surfaces M^2 in $\mathbf{C}^3$. Their list can be completed by the following model.

Example 9.6. We give the complex version of the surface $z = xy + \log x$, $x > 0$, discussed in Sections 1 and 3 of Chapter III. Note that this equiaffinely homogeneous real surface can be parametrized by

(9.21) $$(u, v) \mapsto (e^u, v, e^u v + u).$$

Now this may be regarded as a holomorphic immersion of $\mathbf{C}^2$ into $\mathbf{C}^3$. By the procedure described above, we find that the holomorphic affine normal vector field ζ is $(0, \ldots, 0, 1)$ and hence $S = 0$. The induced connection can be described by

$$\nabla_u \partial_u = \partial_u, \quad \nabla_u \partial_v = 0, \quad \nabla_v \partial_v = 0.$$

The surface is homogeneous relative to the group of equiaffine complex affine transformations given by

$$\begin{bmatrix} 1 & be^a & 0 & a \\ 0 & e^a & 0 & 0 \\ 0 & 0 & e^{-a} & b \\ 0 & 0 & 0 & 1 \end{bmatrix} \quad \text{acting on} \quad \begin{bmatrix} z \\ x \\ y \\ 1 \end{bmatrix},$$

where $a, b \in \mathbf{C}$. In fact, the surface is the orbit of the point $(x, y, z) = (1, 0, 0)$. (We thank Alan Landman for suggesting this matrix form.)

We introduce the following notion.

Definition 9.5. A holomorphic immersion of M^n into $\mathbf{C}^{n+1}$ is said to be *affine Kähler* if there exists (around each point) an anti-holomorphic $(1, 0)$ transversal vector field ζ along f. In this case, the induced connection on M^n is affine Kähler in the sense of Definition 9.3.

This notion was originally introduced in [NPP] so that we may regard an isometric immersion of a Kähler manifold M^n into $\mathbf{C}^{n+1}$ (or any other Kähler manifold $\tilde{M}^{n+1}$) as an affine Kähler immersion. The geometry of complex hypersurfaces immersed in a Kähler space form (the complex Euclidean space $\mathbf{C}^{n+1}$, the complex projective space $\mathbf{P}^{n+1}(\mathbf{C})$, and the complex hyperbolic space $H^{n+1}(\mathbf{C})$) was first studied in [Sm].

For further results on affine Kähler immersions, see [NPP], [NPo2].

Finally, we shall explain yet another formalism dealing with complex hypersurfaces in a more general way as given by B. Opozda, see [O5, 6]. First, we give a preliminary explanation about complex structures and linearity over $\mathbf{C}$. For any finite-dimensional real vector space V with a complex structure, we can introduce complex scalar multiplication $(a + bi)x = ax + bJx$ for all $a, b \in \mathbf{R}$ and $x \in V$. In this way, V is considered as a complex vector space. A linear endomorphism S of V is said to be *complex* or *anti-complex* according as $S(Jx) = JS(x)$ or $S(Jx) = -JS(x)$ for all $x \in V$. A complex-valued function τ on V is said to be *complex* (or *linear over* $\mathbf{C}$) if $\tau(x + y) = \tau(x) + \tau(y)$ and $\tau(Jx) = i\tau(x)$ for all $x, y \in V$. We say that τ is *anti-complex* if $\tau(Jx) = -i\tau(x)$. A similar definition can be made for bilinearity over $\mathbf{C}$ of a complex-valued function h on $V \times V$.

Now consider a holomorphic immersion $f : M^n \to \mathbf{C}^{n+1}$ of a complex manifold. We use J for the complex structure of M^n as well as of $\mathbf{C}^{n+1}$. In the following, let $\mathfrak{X}$ denote the space of all real C^∞ vector fields on M^n. We consider a real transversal vector field ξ on M^n. (These tangent or transversal vector fields may be defined on some open subsets only.) We consider the decomposition at each point $x \in M^n$:

(9.22) $$T_{f(x)}(\mathbf{C}^{n+1}) = f_*(T_x(M^n)) + \mathbf{C}\{\xi\},$$

where the transversal subspace $\mathbf{C}\{\xi\}$ is real 2-dimensional. For $X, Y \in \mathfrak{X}$ we can write

(9.23) $$D_X f_*(Y) = f_*(\nabla_X Y) + h(X, Y)\xi,$$

(9.24) $$D_X \xi = -f_*(SX) + \tau(X)\xi.$$

Here we note that ∇ is a complex connection (that is, torsion-free and compatible with J), h is bilinear over $\mathbf{C}$, whereas S and τ are, in general, linear simply over $\mathbf{R}$.

The condition that $\zeta = \xi - iJ\xi$ is holomorphic, $D_{X+iJX}(\xi - iJ\xi) = 0$ for all $X \in \mathfrak{X}$, is equivalent to $D_{JX}\xi = JD_X\xi$. It follows that ξ is real holomorphic if and only if

(9.25a) $$D_{JX}\xi = JD_X\xi \quad \text{for all} \quad X \in \mathfrak{X}$$

holds. Similarly, ξ is real anti-holomorphic if and only if

(9.25b) $$D_{JX}\xi = -JD_X\xi \quad \text{for all} \quad X \in \mathfrak{X}.$$

Using these we get

Proposition 9.6.

(a) ξ *is real holomorphic if and only if S and τ are complex, that is, $SJ = JS$ and $\tau J = i\tau$.*

(b) ξ *is real anti-holomorphic if and only if S and τ are anti-complex, that is, $SJ = -JS$ and $\tau J = -i\tau$.*

(c) *If ξ is a unit normal vector field (relative to the usual Kähler metric g in $\mathbf{C}^{n+1}$), then S is anti-complex.*

Proof. We prove only (c). From $g(\xi, \xi) = 1$, we get

$$0 = g(D_X\xi, \xi) = g(-f_*(SX) + \tau(X)\xi, \xi) = \tau(X)g(\xi, \xi) = \tau(X)$$

for every tangent vector X. Thus $\tau = 0$. On the other hand, from $g(\xi, Y) = 0$ for each $Y \in \mathfrak{X}$, we get $g(D_X\xi, Y) + g(\xi, D_X Y) = 0$, which implies $g(SX, Y) = h(X, Y)$. Then $g(SJX, Y) = h(JX, Y) = ih(X, Y)$. Using the invariance of g under J, we have $g(JSX, Y) = -g(SX, JY) = -h(X, JY) = -ih(X, Y)$, thus proving that $SJ = -JS$.

Theorem 9.7. *Let $f : M^n \to \mathbf{C}^{n+1}$ be a holomorphic immersion and ξ a transversal vector field. If the corresponding S is anti-complex at each point, then the induced connection ∇ is affine Kähler. Conversely, if ∇ is affine Kähler and if the rank of the complex bilinear function h is greater than 1 at some point, then S is anti-complex.*

See [O5] for the proof of the converse as well as for more results based on the approach explained above.

NOTES

1. Affine immersions of general codimension

In the main text we have been mainly concerned with affine immersions of codimension 1. This note gives some remarks on the case of general codimension.

Let (M, ∇) and $(\tilde{M}, \tilde{\nabla})$ be two differentiable manifolds of dimension n and $n + p$ with torsion-free affine connections ∇ and $\tilde{\nabla}$, respectively.

An immersion $f : M \to \tilde{M}$ is called an *affine immersion* if there exists, around each point of M, a field $\mathcal{N}$ of transversal subspaces of dimension p, denoted by $x \mapsto N_x \subset T_{f(x)}(\tilde{M})$ and such that

(N1.1) $$T_{f(x)}(\tilde{M}) = f_*(T_x(M)) + N_x$$

holds and, for all vector fields X and Y on M, we have a decomposition

(N1.2) $$\tilde{\nabla}_X f_* Y = f_* \nabla_X Y + \alpha(X, Y),$$

where $\nabla_X Y \in T_x(M)$ and $\alpha(X, Y) \in N_x$ at each point x. We call N_x the *transversal* or *normal space* and α the *affine fundamental form* with the understanding that the choice in general is not unique. (See Section 1 of Chapter II.) If ξ is a vector field with values in $\mathcal{N}$, $\xi_x \in N_x$, then we write

(N1.3) $$\tilde{\nabla}_X \xi = -f_* S_\xi X + \nabla_X^\perp \xi,$$

where $S_\xi X \in T_x(M)$ and $\nabla_X^\perp \xi \in N_x$ at each point x. We call S_ξ the *shape operator* for ξ, and $\nabla^\perp$ the *normal connection*; the mapping $\xi \mapsto \nabla_X^\perp \xi$ is covariant differentiation relative to the normal connection. We may use $\mathcal{N}$ to denote the normal bundle rather than a field of normal spaces.

Just as in hypersurface theory, we get several basic equations that relate the curvature tensors $\tilde{R}$ of $(\tilde{M}, \tilde{\nabla})$ and R of (M, ∇). We define the covariant derivatives of α and S relative to a vector field X by

(N1.4) $$(\nabla_X \alpha)(Y, Z) = \nabla_X^\perp(\alpha(Y, Z)) - \alpha(\nabla_X Y, Z) - \alpha(Y, \nabla_X Z),$$

(N1.5) $$(\nabla_X S)_\xi(Y) = \nabla_X(S_\xi Y) - S_\xi(\nabla_X Y) - (S_{\nabla_X^\perp \xi})(Y),$$

and let the prefixes tan and nor denote the tangential and normal components of the vector in $T_{f(x)}(\tilde{M})$ according to the decomposition (N1.1). Then computation shows the following:

(N1.6) Gauss : $\quad \tan \tilde{R}(X,Y)Z = R(X,Y)Z + S_{\alpha(X,Z)}Y - S_{\alpha(Y,Z)}X,$

(N1.7) Codazzi : $\quad \operatorname{nor} \tilde{R}(X,Y)Z = (\nabla_X \alpha)(Y,Z) - (\nabla_Y \alpha)(X,Z),$

(N1.8) Codazzi : $\quad \tan \tilde{R}(X,Y)\xi = (\nabla_Y S)_\xi(X) - (\nabla_X S)_\xi(Y),$

(N1.9) Ricci : $\quad \operatorname{nor} \tilde{R}(X,Y)\xi = \alpha(S_\xi X, Y) - \alpha(X, S_\xi Y) + R^\perp(X,Y)\xi.$

Here X, Y, Z are vector fields on M and $R^\perp(X,Y)$ is the curvature tensor of the normal connection.

When the ambient affine connection $\tilde{\nabla}$ is Ricci-symmetric and projectively flat (see Section 3 of Chapter I as well as Appendix 1), we have

$$\tilde{R}(X,Y)Z = \tilde{\gamma}(Y,Z)X - \tilde{\gamma}(X,Z)Y$$

and

$$\tilde{R}(X,Y)\xi = \gamma(Y,\xi)X - \gamma(X,\xi)Y,$$

both of which are tangential. Hence we have simpler formulas

(N1.10) $\qquad \tilde{\gamma}(Y,Z)X - \tilde{\gamma}(X,Z)Y = R(X,Y)Z + S_{\alpha(X,Z)}Y - S_{\alpha(Y,Z)}X,$

(N1.11) $\qquad (\nabla_X \alpha)(Y,Z) = (\nabla_Y \alpha)(X,Z),$

(N1.12) $\qquad \tilde{\gamma}(Y,\xi)X - \tilde{\gamma}(X,\xi)Y = (\nabla_Y S)_\xi(X) - (\nabla_X S)_\xi(Y),$

(N1.13) $\qquad R^\perp(X,Y)\xi = \alpha(X, S_\xi Y) - \alpha(S_\xi X, Y).$

When the ambient affine connection $\tilde{\nabla}$ is flat, we have the equations that are generalizations of those in Theorem 3.3 of Chapter II,

(N1.14) $\qquad R(X,Y)Z = S_{\alpha(Y,Z)}X - S_{\alpha(X,Z)}Y,$

(N1.15) $\qquad (\nabla_X S)_\xi Y = (\nabla_Y S)_\xi X,$

as well as (N1.11) and (N1.13).

Now when $\tilde{M} = \mathbf{R}^{n+p}$ and $\tilde{\nabla} = D$ is the ordinary flat connection, we can define several notions following the discussions in Chapter II. Among others, we mention one property: the immersion is said to be *umbilical* if $S_\xi = \rho(\xi)I$ for a 1-form ρ. The notions such as graph immersion and centro-affine immersion can be easily generalized. Then we can state the following theorem.

Theorem N1.1. *Let $f : (M^n, \nabla) \to (\mathbf{R}^{n+p}, D)$ be an umbilical affine immersion, where $n \geq 2$. Then it is affinely equivalent to a graph immersion or to a centro-affine immersion.*

Given an affine immersion, the space generated by the values of α_x at each point $x \in M$ is called the *second osculating space*. The following theorem explains its role.

Theorem N1.2. *Let* $f : (M^n, \nabla) \rightarrow (\mathbf{R}^{n+p}, D)$ *be an affine immersion. Suppose* $\mathcal{N}_1$ *with fibers* $N_1(x)$, $x \in M$, *is a subbundle of the normal bundle* $\mathcal{N}$ *such that* (i) $N_1(x)$ *contains the range of* α_x *for every* $x \in M^n$ *and* (ii) $\mathcal{N}_1$ *is parallel relative to the normal connection. Then the image* $f(M^n)$ *is contained in a certain* $(n + q)$-*dimensional affine subspace of* $\mathbf{R}^{n+p}$, *where* $q = \dim N_1(x)$.

Refer to the paper [NP3] for proofs.

2. Surfaces in $\mathbf{R}^4$

In this note we summarize the results in [NV] concerning an immersion of a surface M^2 in $\mathbf{R}^4$. We shall first define the notion of nondegeneracy and construct the affine metric for a nondegenerate immersion.

Suppose we choose a field of normal (transversal) spaces $\mathcal{N} : x \rightarrow N_x$ for a given immersion $f : M^2 \rightarrow \mathbf{R}^4$. As in Note 1, we may consider $\mathcal{N}$ as the normal bundle $\bigcup_{x \in M^2} N_x$ over M^2. Thus

$$\mathbf{R}^4 = T_x(M^2) \oplus N_x,$$

where we identify $T_x(M^2)$ with $f_*(T_x(M^2))$. For any local basis $\{\xi_1, \xi_2\}$ of $\mathcal{N}$, we can write

$$
\begin{aligned}
(N2.1) \qquad D_X Y &= \nabla_X Y + h^1(X, Y)\xi_1 + h^2(X, Y)\xi_2, \\
D_X \xi_1 &= -S_1 X + \tau_1^1(X)\xi_1 + \tau_1^2(X)\xi_2, \\
D_X \xi_2 &= -S_2 X + \tau_2^1(X)\xi_1 + \tau_2^2(X)\xi_2.
\end{aligned}
$$

Then ∇ is a torsion-free affine connection on M^2, which depends only on $\mathcal{N}$ and not on the choice of local basis $\{\xi_1, \xi_2\}$. We call it the affine connection induced by $\mathcal{N}$. The other objects h^i, S_i, τ_j^i, $1 \leq i, j \leq 2$, are respectively the *affine fundamental forms*, the *shape operators* and the *normal connection forms*; they together satisfy the equations of Gauss, of Codazzi, and of Ricci – special cases of the equations (N1.6)–(N1.9) of Note 1. We also define a volume element by

$$(N2.2) \qquad\qquad \omega(X, Y) = |XY\xi_1\xi_2|,$$

where the right-hand side denotes the determinant. The form ω depends on the choice of local basis $\{\xi_1, \xi_2\}$. The cubic forms are defined by

$$
\begin{aligned}
(N2.3) \qquad C^1(X, Y, Z) &= (\nabla_X h^1)(Y, Z) + \tau_1^1(X)h^1(Y, Z) + \tau_2^1(X)h^2(Y, Z), \\
C^2(X, Y, Z) &= (\nabla_X h^2)(Y, Z) + \tau_1^2(X)h^1(Y, Z) + \tau_2^2(X)h^2(Y, Z).
\end{aligned}
$$

Let $u = \{X_1, X_2\}$ be a local frame field on M. We define a symmetric bilinear function G_u by

$$(N2.4) \qquad G_u(Y, Z) = \frac{1}{2}\{|X_1 \, X_2 \, D_Y X_1 \, D_Z X_2| + |X_1 \, X_2 \, D_Z X_1 \, D_Y X_2|\}.$$

Suppose $v = \{Y_1, Y_2\}$ is another local frame field such that

$$[Y_1 \ Y_2] = [X_1 \ X_2] \begin{bmatrix} a & c \\ b & d \end{bmatrix},$$

which we denote simply by

$$v = uA,$$

where A is the matrix appearing just above. It is now easy to verify that

$$(\text{N2.5}) \qquad G_v = (ad - bc)^2 G_u.$$

This means that G is a symmetric $(0, 2)$ tensor of weight 2 in classical terminology.

In general, for any symmetric bilinear function h we define

$$(\text{N2.6}) \qquad \det {}_u h = \begin{vmatrix} h(X_1, X_1) & h(X_1, X_2) \\ h(X_2, X_1) & h(X_2, X_2) \end{vmatrix}.$$

For $v = uA$ we have

$$(\text{N2.7}) \qquad \det {}_v h = (ad - bc)^2 \det {}_u h.$$

Thus we have by (N2.7) and (N2.5)

$$\det {}_v G_v = (ad - bc)^2 \det {}_u G_v = (ad - bc)^6 \det {}_u G_u.$$

This shows that whether G_u is nondegenerate or not is independent of the choice of local frame field u. We say that the immersion f (or the surface M^2) is *nondegenerate* if G_u is nondegenerate. In this case, the equation above implies that

$$(\text{N2.8}) \qquad g = \frac{G_u}{(\det {}_u G_u)^{1/3}}$$

is independent of the choice of u. We call it the *affine metric*.

The signature of the affine metric can be interpreted as follows. The metric g is definite if and only if $\lambda_1 h^1 + \lambda_2 h^2$ is nondegenerate for all $(\lambda_1, \lambda_2) \neq (0, 0)$. It is indefinite if and only if there exist exactly two ratios $(\lambda_1 : \lambda_2) \neq (\mu_1 : \mu_2)$ such that $\lambda_1 h^1 + \lambda_2 h^2$ and $\mu_1 h^1 + \mu_2 h^2$ are degenerate.

The following is a key lemma.

Lemma N2.1. *Assume that f is nondegenerate. Given a g-orthonormal frame field $\{X_1, X_2\}$, there exists a unique local basis $\{\xi_1, \xi_2\}$ relative to which*

$$h^1 = \begin{bmatrix} 1 & 0 \\ 0 & -\epsilon \end{bmatrix}, \qquad h^2 = \begin{bmatrix} 0 & 1 \\ 1 & 0 \end{bmatrix}.$$

Here $\epsilon = 1$ if g is definite and $\epsilon = -1$ otherwise.

Using this lemma, we define a metric $g_u^{\perp}$ in each normal space N_x by setting

$$g_u^{\perp}(\xi_1, \xi_1) = \epsilon, \quad g_u^{\perp}(\xi_1, \xi_2) = 0, \quad g_u^{\perp}(\xi_2, \xi_2) = 1.$$

We can, in fact, verify that the metric $g^{\perp}$ is independent of the choice of u.

We shall now explain the choice of normal spaces introduced by Burstin and Mayer [BM] for a nondegenerate surface $f : M^2 \to \mathbf{R}^4$. Let $\hat{\nabla}$ be the Levi-Civita connection for the affine metric g. For a given field of normal spaces $\mathcal{N}$, let ∇ be the affine connection it induces and let $K = \nabla - \hat{\nabla}$ be the difference tensor. Denoting by x the position vector representing f we take the Laplacian $\triangle_g x$ of x relative to the affine metric g. A simple computation shows that

$$\triangle_g x = \operatorname{tr}_g K,$$

which says in particular that $\triangle_g x$ is a tangent vector. (This is a contrast to the fact that $\triangle_g x$ is in the affine normal direction in the case of a nondegenerate surface in $\mathbf{R}^3$. See Theorem 6.5 of Chapter II.) We set for any vector fields X and Y on M^2

$$\eta(X, Y) = D_X Y - \hat{\nabla}_X Y - \frac{1}{2} g(X, Y) \triangle_g x.$$

One can show that the subspace spanned by $\eta(X, Y)$ is a 2-dimensional normal space at each point; we take it to be $\mathcal{N}_{BM}$ – the Burstin-Mayer normal field. From

$$D_X Y = \hat{\nabla}_X Y + \frac{1}{2} g(X, Y) \triangle_g x + \eta(X, Y),$$

we see that the affine connection induced by $\mathcal{N}_{BM}$ is given by

$$\nabla_X Y = \hat{\nabla}_X Y + \frac{1}{2} \triangle_g x.$$

In view of this we have

Theorem N2.2. *For the Burstin–Mayer normal field $\mathcal{N}_{BM}$ the following conditions are equivalent:*

(i) $\nabla = \hat{\nabla}$;

(ii) $\nabla \omega_g = 0$;

(iii) $\triangle_g x = 0$.

A nondegenerate surface M^2 in $\mathbf{R}^4$ is said to be *harmonic* if $\triangle_g x = 0$. J. Li [LiJ] has proved that if M^2 given as a graph $(u, v) \mapsto (u, v, \phi(u, v), \psi(u, v))$

has a positive-definite affine metric and is harmonic, then it is equiaffinely equivalent to the graph

$$(u,v) \mapsto (u, v, \frac{1}{2}(v^2 - u^2), uv).$$

We now explain a new approach taken in [NV] to define the canonical normal field for a nondegenerate surface M^2.

Theorem N2.3. *For a nondegenerate surface* $f : M^2 \to \mathbf{R}^4$, *there is a unique choice of normal field such that*

(i) $\nabla \omega_g = 0$ *(equiaffine)*;

(ii) $\nabla^{\perp} g^{\perp}$ *is symmetric in the sense that*

$$(\nabla^{\perp} g^{\perp})(X_1, \xi_2, \xi_1) = (\nabla^{\perp} g^{\perp})(X_2, \xi_1, \xi_1),$$
$$(\nabla^{\perp} g^{\perp})(X_2, \xi_1, \xi_2) = (\nabla^{\perp} g^{\perp})(X_1, \xi_2, \xi_2),$$

where X_1, X_2, ξ_1, *and* ξ_2 *are as in* Lemma N2.1.

Such a normal field is called the *canonical normal field*.

Remark N2.1. Condition (ii) is equivalent to the following:

(N2.9)
$$(\nabla g)(X_2, X_1, X_1) = -(\nabla g)(X_1, X_2, X_1),$$
$$(\nabla g)(X_1, X_2, X_2) = -(\nabla g)(X_2, X_1, X_2).$$

We have the following results.

Theorem N2.4. *If a normal field* $\mathcal{N}$ *for a nondegenerate surface has the property that* $\nabla g = 0$, *then it is the canonical normal field.*

Theorem N2.5. *Let* M^2 *be a nondegenerate surface with canonical normal field* $\mathcal{N}$. *Then the cubic forms* C_1 *and* C_2 *vanish if and only if* $\nabla g = 0$ *and* $\nabla^{\perp} g^{\perp} = 0$.

As an application we have

Theorem N2.6. *Let* M^2 *be a nondegenerate surface with canonical normal field* $\mathcal{N}$. *The cubic forms vanish if and only if either*

(i) g *is definite and* M^2 *is a holomorphic curve*;

or

(ii) g *is indefinite and* M^2 *is a product of two nondegenerate plane curves.*

Here a holomorphic curve is a holomorphic immersion $z \in M \mapsto f(z) \in \mathbf{C}^2$, where M is a domain in $\mathbf{C}$, for which the canonical normal field is constructed in [DVV]. A product of two nondegenerate plane curves is of the form

$$(u, v) \in I \times J \mapsto (\phi(u), \psi(v)) \in \mathbf{R}^2 \times \mathbf{R}^2 = \mathbf{R}^4,$$

where $u \in I \mapsto \phi(u) \in \mathbf{R}^2$ is a nondegenerate curve with affine arclength parameter u in an interval I with $|d\phi/du \ d^2\phi/du^2| = 1$ and similarly for $v \in J \mapsto \psi(v) \in \mathbf{R}^2$.

Remark N2.2. Let $\tilde{M}^4$ be a 4-dimensional manifold with an equiaffine structure $(D, \tilde{\omega})$. For a surface M^2 immersed in $\tilde{M}^4$ we may define ω in (N2.2) by using $\tilde{\omega}$ on the right-hand side. The definition of nondegeneracy and the construction of affine metric can be given in the same way so as to lead to Theorem N2.3. The situation here is analogous to that in Appendix 4 which extends the classical Blaschke theory from $\mathbf{R}^{n+1}$ to an arbitrary $\tilde{M}^{n+1}$ with an equiaffine structure.

Remark N2.3. The approach in [NV] has been further developed by Vrancken and others. See [MSV], [DMVV], and [SVr]. Refer to [S7] for a projective treatment of surfaces in $\mathbf{P}^4$.

3. Affine normal mappings

In this note we briefly discuss the normal mapping associated with an affine immersion $f : M^n \to \mathbf{R}^{n+1}$.

We fix a point in $\mathbf{R}^{n+1}$ as the origin. Then the *normal mapping* (*normal map*) denoted by ϕ is defined to be

$$\phi(x) = \xi_x \in \mathbf{R}^{n+1}$$

that is displaced so as to have the starting point at the origin. Let v denote the conormal mapping and $\bar{\nabla}$ the affine connection relative to the centro-affine immersion v; recall the notations used in Section 5 of Chapter II.

Since

$$\phi_*(X) = D_X \phi = D_X \xi = -f_*(SX),$$

the normal mapping ϕ is an immersion if and only if the shape operator S is nonsingular. Assume that this is the case. Then we can regard ϕ as a centro-affine immersion with $-\phi$ as a transversal vector field. Thus we can write

(N3.1) $$D_X(\phi_*(Y)) = \phi_*(\nabla'_X Y) + h'(X, Y)(-\phi)$$

by defining the torsion-free affine connection ∇' and the fundamental form h'. Then

(N3.2) $$\nabla'_X Y = S^{-1}(\nabla_X S)Y + \nabla_X Y$$

and the fundamental form is given by

(N3.3) $$h'(X, Y) = h(SX, Y).$$

Consequently, we have $h' = \bar{h}$, where $\bar{h}$ is the fundamental form of the conormal mapping. It is easy to verify the identity

(N3.4) $$Xh'(Y,Z) = h'(\nabla'_X Y, Z) + h'(Y, \bar{\nabla}_X Z).$$

This means that the two connections ∇' and $\bar{\nabla}$ are conjugate relative to the metric h'.

Analogously to the formula for $\triangle v$ in Theorem 6.3 of Chapter II, we can compute the formula for $\triangle \phi$ as follows.

Theorem N3.1. *Let $f : M^n \to \mathbf{R}^{n+1}$ be an equiaffine immersion. Then the normal mapping ϕ satisfies*

(N3.5) $$\triangle \phi = -f_*(\mathrm{tr}\,_h \nabla S) - f_*(S(\mathrm{tr}\,_h K)) - \mathrm{tr}\,(S)\phi,$$

where $\triangle$ is the Laplacian relative to the affine metric.

Proof. By the definition of ϕ, we get

$$D_X(\phi_*(Y)) = D_X(f_*(-SY))$$
$$= -f_*((\nabla_X S)(Y)) - f_*(S(\nabla_X Y)) - h(X, SY)\xi$$

and

$$\phi_*(\hat{\nabla}_X Y) = f_*(-S(\hat{\nabla}_X Y)) = -f_*(S(\nabla_X Y)) + f_*(SK(X,Y)).$$

Thus we have

$$\mathrm{Hess}\,_\phi(X,Y) = -f_*((\nabla_X S)(Y)) - f_*(SK(X,Y)) - h(X,SY)\xi,$$

which shows the formula.

The next result complements the characterizations of ellipsoids discussed in Section 8 of Chapter III.

Corollary N3.2. *An ovaloid $f : M^2 \to \mathbf{R}^3$ is an ellipsoid if and only if the Laplacian of the normal map ϕ is proportional to ϕ.*

Proof. For the Blaschke structure of the immersion, the formula (N3.5) says that the Laplacian of the normal map is proportional to ϕ if and only if $\mathrm{tr}\,_h \nabla S = 0$. Then Proposition 8.3 and Theorem 8.6 of Chapter III imply the result.

Corollary N3.2 was obtained in [NO2], which also contains a result on a hyperovaloid M^n in $\mathbf{R}^{n+1}$ and its normal and conormal images.

4. Affine Weierstrass formula

Let $f : M^2 \to \mathbf{R}^3$ be a nondegenerate affine minimal immersion with Blaschke structure. Under the assumption that the affine metric is definite, we shall provide a formula, called the affine Weierstrass formula, that represents the map f as an indefinite integral and gives rise to related formulas and applications. For simplicity, we further assume that the surface is simply connected.

Let $\{x, y\}$ be an isothermal coordinate system and set $X = \partial/\partial x$ and $Y = \partial/\partial y$; we have

$$h(X, X) = h(Y, Y) = \lambda, \text{ say,} \quad \text{and} \quad h(X, Y) = 0.$$

We introduce the complex coordinate $z = x + iy$ and complex vector fields

$$Z = \frac{1}{2}(X - iY) \quad \text{and} \quad \overline{Z} = \frac{1}{2}(X + iY).$$

Differentiation of a vector relative to x, y, or z is denoted by attaching these letters as subscript to the vector. In the following we frequently regard tensors such as h, C, $S, \ldots$, defined over $\mathbf{R}$, as tensors over $\mathbf{C}$ by naturally extending the field of coefficients. The connections D and ∇ also are regarded as connections acting on complex tangent vector fields. Thus we have the fundamental equations

(N4.1)
$$D_U f_* V = f_*(\nabla_U V) + h(U, V)\xi,$$
$$D_U \xi = -f_*(SU)$$

for complex tangent vector fields U and V; here ξ denotes the affine normal field. The Blaschke metric h has the property

(N4.2) $$h(Z, Z) = h(\overline{Z}, \overline{Z}) = 0, \qquad h(Z, \overline{Z}) = \frac{1}{2}\lambda.$$

The determinant function Det of $\mathbf{R}^3$ and that of the dual space $\mathbf{R}_3$ denoted by Det* are extended by linearity to functions on the complexifications $\mathbf{C}^3$ and $\mathbf{C}_3$, respectively. We denote by v the affine conormal field as before. Then we see

(N4.3) $$\mathrm{Det}\,(f_*Z, f_*\overline{Z}, \xi) = \frac{i}{2}\lambda \quad \text{and} \quad \mathrm{Det}^*(v, v_*Z, v_*\overline{Z}) = \frac{i}{2}\lambda.$$

The Laplace operator of the affine metric is $\triangle = (1/\lambda)(X^2 + Y^2) = (4/\lambda)Z\overline{Z}$. We know from Theorem 6.5 of Chapter II that

(N4.4) $$\xi = \frac{1}{2}\triangle f,$$

and the conormal vector field v satisfies the equation

$$\triangle v + 2Hv = 0,$$

refer to (6.12) of Chapter II. Since f is assumed to be affine minimal, the map v is an $\mathbf{R}_3$-valued harmonic function. Thus there exists a $\mathbf{C}_3$-valued holomorphic map, which we denote by w, such that v is the imaginary part:

$$v = \frac{i}{2}(\overline{w} - w).$$

Recall the identification of $\overset{2}{\wedge}\mathbf{R}_3$ with $\mathbf{R}^3$ given in Section 7 of Chapter II. We extend it to a natural identification of $\overset{2}{\wedge}\mathbf{C}_3$ with $\mathbf{C}^3$. Since ξ is determined by the property $v(\xi) = 1$ and $v_*X(\xi) = v_*Y(\xi) = 0$, we must have $\xi = av_x \wedge v_y$ for some scalar function a. Then the identity

$$1 = v(\xi) = av(v_x \wedge v_y) = a\mathrm{Det}^*(v, v_x, v_y) = a\lambda$$

implies

$$\xi = \frac{1}{\lambda}v_x \wedge v_y.$$

This is rewritten as

(N4.5) $$\xi = -\frac{2i}{\lambda}v_z \wedge v_{\overline{z}} = -\frac{i}{2\lambda}w_z \wedge \overline{w_z}.$$

We now prove the following formula:

Theorem N4.1. *Let $f : M^2 \to \mathbf{R}^3$ be a nondegenerate affine minimal immersion of a simply connected surface M with Blaschke structure. Then there exists a holomorphic map w of M into $\mathbf{C}_3$ relative to the complex coordinate z on M such that the immersion f is given by the integral*

(N4.6) $$f(z) = -\frac{i}{4}\left\{ w \wedge \overline{w} + \int^z (w \wedge dw - \overline{w} \wedge d\overline{w}) \right\}$$

up to a constant vector.

Proof. Let w be the holomorphic map defined above and consider the $\overset{2}{\wedge}\mathbf{C}_3$-valued 1-form

$$d(w \wedge \overline{w}) + w \wedge dw - \overline{w} \wedge d\overline{w}$$
$$= (w - \overline{w}) \wedge d(w + \overline{w})$$
$$= 4iv \wedge d\eta,$$

where $w = \eta + iv$. The Cauchy–Riemann equations $\eta_x = v_y$ and $\eta_y = -v_x$ show that this form is equal to $4iv \wedge (v_y dx - v_x dy)$. Then the Lelieuvre formula (7.15) of Chapter II implies the result.

The integral (N4.6) is called the *affine Weierstrass formula.*

We shall next rewrite various geometric quantities in terms of the map w. First, the second equation of (N4.3) is now

$$\frac{i}{2}\lambda = \text{Det}^*(\frac{i}{2}(\overline{w}-w), -\frac{i}{2}w_z, \frac{i}{2}\overline{w_z})$$

$$= \frac{i}{8}\text{Det}^*(\overline{w}-w, w_z, \overline{w_z});$$

that is,

(N4.7) $$\lambda = \frac{1}{4}\text{Det}^*(\overline{w}-w, w_z, \overline{w_z}).$$

This implies the following.

Theorem N4.2. *Let M be a Riemann surface and $w : M^2 \to \mathbf{C}^{n+1}$ a holomorphic map. Assume*

$$\text{Det}^*(\overline{w}-w, w_z, \overline{w_z}) \neq 0$$

relative to local complex coordinate z. Then the formula (N4.6) defines an affine minimal immersion with the Blaschke metric given by (N4.2) with λ defined in (N4.7).

Example N4.1. Let us consider the holomorphic immersion $w = (z, iz, z^2)$. By the formula (N4.6) the corresponding immersion turns out to be

$$f = (\frac{2}{3}x^3, \frac{2}{3}y^3, \frac{1}{2}(x^2 + y^2)),$$

where $z = x + iy$.

Next, by (N4.7), we have

(N4.8)
$$f_z = f_*Z = \frac{i}{4}(\overline{w}-w) \wedge w_z,$$

$$f_{\bar{z}} = f_*\overline{Z} = \frac{i}{4}(\overline{w}-w) \wedge \overline{w_z}.$$

We want to know the vector w_{zz}: since $\overline{w}-w$, w_z, and $\overline{w_z}$ are linearly independent, we can write

(N4.9) $$w_{zz} = pw_z + q\overline{w_z} + r(\overline{w}-w).$$

Then

Lemma N4.3. *We have*

(N4.10)
$$p = \frac{\lambda_z}{\lambda},$$

$$q = \frac{1}{4\lambda}\text{Det}^*(\overline{w}-w, w_z, w_{zz}),$$

$$r = \frac{1}{4\lambda}\text{Det}^*(w_z, \overline{w_z}, w_{zz}).$$

Proof. By differentiating (N4.7), we have

$$4\lambda_z = \mathrm{Det}^*(\overline{w} - w, w_{zz}, \overline{w}_z)$$
$$= p\mathrm{Det}^*(\overline{w} - w, w_z, \overline{w}_z)$$
$$= 4\lambda p,$$

hence we get the first identity. For the second identity, we have

$$\mathrm{Det}^*(\overline{w} - w, w_z, w_{zz}) = \mathrm{Det}^*(\overline{w} - w, w_z, pw_z + q\overline{w}_z + r(\overline{w} - w))$$
$$= q\mathrm{Det}^*(\overline{w} - w, w_z, \overline{w}_z)$$
$$= 4\lambda q,$$

and similarly for the third identity.

Lemma N4.4. *The connection* ∇ *is given as follows:*

$$(N4.11) \qquad \nabla_Z Z = pZ + q\overline{Z}, \quad \nabla_Z \overline{Z} = 0, \quad \nabla_{\overline{Z}} \overline{Z} = \overline{p}\overline{Z} + \overline{q}Z.$$

Proof. Because $h(Z, Z) = 0$, we have

$$f_* \nabla_Z Z = D_Z f_* Z = f_{zz}$$
$$= \frac{i}{4}(\overline{w} - w) \wedge w_{zz}$$
$$= \frac{i}{4}(\overline{w} - w) \wedge (pw_z + q\overline{w}_z)$$
$$= pf_* Z + qf_* \overline{Z},$$

which proves the first identity. The second identity follows from

$$f_* \nabla_Z \overline{Z} = D_Z f_* \overline{Z} - h(Z, \overline{Z})\xi$$
$$= f_{z\overline{z}} - (\lambda/2)\xi$$
$$= 0$$

because of (N4.4). The others follow by taking conjugates.

We can compute the cubic form C by (N4.2) and (N4.11):

$$C(Z, Z, Z) = \nabla_Z(h(Z, Z)) - 2h(\nabla_Z Z, Z)$$
$$= -2h(pZ + q\overline{Z}, Z)$$
$$= -\lambda q;$$

hence

$$(N4.12) \qquad C(Z, Z, Z) = -\frac{1}{4}\mathrm{Det}^*(\overline{w} - w, w_z, w_{zz}).$$

Note here that the apolarity implies

$$C(Z,Z,Z) = \frac{1}{2}\{C(X,X,X) + iC(Y,Y,Y)\}$$

and $C(Z,Z,\overline{Z}) = C(Z,\overline{Z},\overline{Z}) = 0$, $C(\overline{Z},\overline{Z},\overline{Z}) = \overline{C(Z,Z,Z)}$.

Finally, the shape operator is given by

(N4.13) $$SZ = 2ir\overline{Z} = \frac{i}{2\lambda}\mathrm{Det}^*(w_z, \overline{w_z}, w_{zz})\overline{Z}.$$

In fact, we have

$$\begin{aligned}
D_Z\xi &= -\frac{i}{2\lambda}w_{zz} \wedge \overline{w_z} + \frac{i\lambda_z}{2\lambda^2}w_z \wedge \overline{w_z} \quad \text{by (N4.4)} \\
&= -\frac{i}{2\lambda}(pw_z + r(\overline{w} - w)) \wedge \overline{w_z} + \frac{i\lambda_z}{2\lambda^2}w_z \wedge \overline{w_z} \\
&= -\frac{i}{2\lambda}r(\overline{w} - w) \wedge \overline{w_z} \quad \text{by (N4.10)} \\
&= -2ir\overline{Z};
\end{aligned}$$

hence the result follows.

On the other hand, the derivative of (N4.12) is

(N4.14) $$\overline{Z}(C(Z,Z,Z)) = -\frac{1}{4}\mathrm{Det}^*(\overline{w_z}, w_z, w_{zz}) = ih(SZ,Z).$$

Therefore we can state

Proposition N4.5. *The cubic form $C(Z,Z,Z)$ is holomorphic if and only if the surface is an improper affine minimal surface.*

Moreover we have the following theorem characterizing elliptic paraboloids in terms of the map w, which is considered to be a complex curve.

Theorem N4.6. *The affine minimal surface represented by the integral (N4.6) is an elliptic paraboloid if and only if the complex curve w in $\mathbf{C}^3$ lies on a complex line.*

Proof. Assume that the surface is an elliptic paraboloid. Then $C = 0$ and, say from (N4.13) and (N4.14), $S = 0$. Thus, from (N4.10), we see that w_{zz} is linearly dependent on w_z: $w_{zz} = \rho w_z$ for some function ρ. By the assumption that $w_z \neq 0$, ρ is holomorphic. Now it is easy to see that w has a form $w = \mathbf{a}u + \mathbf{b}$, where $\mathbf{a}$ and $\mathbf{b}$ are constant vectors and u is a holomorphic function.

Conversely, assume that w has a representation of this form. Then $w_z = \mathbf{a}u_z$ and $w_{zz} = \mathbf{a}u_{zz}$; this means that $C = 0$, that is, the given surface is an elliptic paraboloid.

Remark N4.1. The nondegeneracy condition $\mathrm{Det}^*(\overline{w} - w, w_z, \overline{w_z}) \neq 0$ for the w above is

$$u_z \overline{u_z} \mathrm{Det}^* (\overline{\mathbf{b}} - \mathbf{b}, \mathbf{a}, \overline{\mathbf{a}}) \neq 0.$$

Finally, we give a remark due to [Wan1] on the vanishing of the Pick invariant. The calculation above of the cubic form says that $C(Z,Z,Z)\,dz^3$ is an invariant $(3,0)$-form on the immersed surface. Since the Euler number of the bundle of $(3,0)$-forms on a 2-dimensional sphere is -6, any such form must have at least one zero (refer to, say [MS]). That is, we have

Theorem N4.7. *The Pick invariant on any ovaloid must vanish at least at one point.*

In particular, we have

Corollary N4.8. *Let* $f : M \to \mathbf{R}^3$ *be an ovaloid. If the Pick invariant is constant, then it is an ellipsoid.*

In comparison to the formula (1.7) of [Cal5, p. 206], we rewrite the equations (N4.12) and (N4.13) as follows:

(N4.15)
$$C(Z,Z,Z) = 2i \, \mathrm{Det}\,(\xi, f_z, f_{zz}),$$
$$h(SZ,Z) = -\lambda \mathrm{Det}\,(\xi, f_z, \xi_z).$$

5. Affine Bäcklund transformations

This is a continuation of Note 4. Here we review the transformation called the affine Bäcklund transformation following [Bl] and [Bu].

First, we treat affine minimal surfaces whose affine metrics are definite. In Note 4, we have seen that any affine minimal surface of this type has locally an integral representation

(N5.1)
$$f(z) = -\frac{i}{4}\left\{ w \wedge \overline{w} + \int^z (w \wedge dw - \overline{w} \wedge d\overline{w}) \right\}$$

up to a constant vector, where z is a complex coordinate and w is a holomorphic mapping into $\mathbf{C}^3$. If we replace w with $\check{w} = \sqrt{-1}w$, then we get another affine minimal immersion

$$\check{f}(z) = -\frac{i}{4}\left\{ \check{w} \wedge \overline{\check{w}} + \int^z (\check{w} \wedge d\check{w} - \overline{\check{w}} \wedge d\overline{\check{w}}) \right\},$$

that is

(N5.2)
$$\check{f}(z) = -\frac{i}{4}\left\{ w \wedge \overline{w} - \int^z (w \wedge dw - \overline{w} \wedge d\overline{w}) \right\},$$

provided that

$$\check{\lambda} = -\frac{1}{4}\text{Det}^{*}(w + \overline{w}, w_z, \overline{w_z})$$

is nonvanishing (see Theorem N4.2). The affine normal vector of $\check{f}$ is given by

$$\check{\xi} = -\frac{i}{2\check{\lambda}} w_z \wedge \overline{w_z}.$$

Now it is straightforward to see

$$f - \check{f} = -\frac{i}{2}w \wedge \overline{w},$$

$$f.Z = \frac{i}{4}(\overline{w} - w) \wedge w_z, \qquad f.\overline{Z} = \frac{i}{4}(\overline{w} - w) \wedge \overline{w_z},$$

$$\check{f}.Z = \frac{i}{4}(\overline{w} + w) \wedge w_z, \qquad \check{f}.\overline{Z} = \frac{i}{4}(\overline{w} + w) \wedge \overline{w_z}.$$

Therefore we have $(\overline{w}-w)(f+\check{f}) = 0$, $(\overline{w}-w)(f.Z) = 0$, and $(\overline{w}-w)(f.\overline{Z}) = 0$, which together imply that the vector $f + \check{f}$ is linearly dependent on $f.Z$ and $f.\overline{Z}$. Similarly, it is linearly dependent on $\check{f}.Z$ and $\check{f}.\overline{Z}$. Hence the vector $f(x)+\check{f}(x)$ that connects two points $f(x)$ and $-\check{f}(x)$ is tangent to both of the immersions f and $-\check{f}$ for any point x. Further, the affine normal vectors $\xi(x)$ and $\check{\xi}(x)$ are parallel for any point x. Such a correspondence of immersions is called the *Bäcklund transformation*.

Example N5.1. Let us reconsider Example N4.1 where $w = (z, iz, z^2)$. Then $\check{w} = (iz, -z, iz^2)$, which defines the immersion

$$\check{f} = (xy^2 + \frac{1}{3}x^3, x^2y + \frac{1}{3}y^3, \frac{1}{2}(x^2 + y^2)).$$

Compare this immersion with f in Example N4.1; $\check{f}$ is affinely congruent to f. In fact, it is the rotation of f relative to the third axis by an angle $\frac{1}{4}\pi$ up to dilatation.

We next consider an affine minimal surface with indefinite metric. Relative to the asymptotic coordinate system $\{x^1, x^2\}$ used in Section 7 of Chapter II, the conormal immersion v is harmonic:

$$v_{12} = 0.$$

This means that v defines a translation surface. We can set

$$v(x^1, x^2) = \lambda(x^1) + \mu(x^2).$$

Then

$$v \wedge v_1 dx^1 - v \wedge v_2 dx^2 = \lambda \wedge d\lambda - \mu \wedge d\mu + d(\mu \wedge \lambda).$$

Hence the equation (7.13) of Chapter II implies

$$(N5.3) \qquad f(x) = \mu \wedge \lambda + \int^x (\lambda \wedge d\lambda - \mu \wedge d\mu)$$

up to a constant vector and the induced metric is $2F dx^1 \wedge dx^2$, where

$$F = -\text{Det}^*(\lambda + \mu, \lambda_1, \mu_2) \neq 0.$$

Further, if

$$\check{F} = \text{Det}^*(\lambda - \mu, \lambda_1, \mu_2) \neq 0,$$

then the mapping

$$(N5.4) \qquad \check{f}(x) = -\mu \wedge \lambda + \int^x (\lambda \wedge d\lambda - \mu \wedge d\mu)$$

defines another affine minimal surface. This pair $\{f, \check{f}\}$ enjoys a nice property similar to what we saw in the definite case. We define

$$z = \frac{1}{2}(f + \check{f}) = \int^x (\lambda \wedge d\lambda - \mu \wedge d\mu),$$

which is again a translation surface. Assume $\text{Det}^*(\lambda, \lambda_1, \lambda_{11}) > 0$. Then, since $z_1 = \lambda \wedge \lambda_1$, $z_{11} = \lambda \wedge \lambda_{11}$, and $z_{111} = \lambda_1 \wedge \lambda_{11} + \lambda \wedge \lambda_{111}$, we can see

$$z_1 \wedge z_{11} = \text{Det}^*(\lambda, \lambda_1, \lambda_{11})\lambda,$$
$$\text{Det}(z_1, z_{11}, z_{111}) = (\text{Det}^*(\lambda, \lambda_1, \lambda_{11}))^2,$$

in view of (7.4) and (7.5) of Chapter II; in fact, for example,

$$\text{Det}(z_1, z_{11}, z_{111}) \cdot \text{Det}^*(\lambda, \lambda_1, \lambda_{11})$$
$$= \det \begin{bmatrix} 0 & 0 & \lambda(\lambda_1 \wedge \lambda_{11}) \\ 0 & \lambda_1(\lambda \wedge \lambda_{11}) & \lambda_1(\lambda \wedge \lambda_{111}) \\ \lambda_{11}(\lambda \wedge \lambda_1) & 0 & \lambda_{11}(\lambda \wedge \lambda_{111}) \end{bmatrix}$$
$$= \det \begin{bmatrix} 0 & 0 & \text{Det}^*(\lambda, \lambda_1, \lambda_{11}) \\ 0 & -\text{Det}^*(\lambda, \lambda_1, \lambda_{11}) & * \\ \text{Det}^*(\lambda, \lambda_1, \lambda_{11}) & 0 & * \end{bmatrix}$$
$$= (\text{Det}^*(\lambda, \lambda_1, \lambda_{11}))^3.$$

In particular, the mapping λ is recovered by the immersion z as

$$\lambda = \frac{z_1 \wedge z_{11}}{\sqrt{\text{Det}(z_1, z_{11}, z_{111})}},$$

provided that $\mathrm{Det}\,(z_1, z_{11}, z_{111}) \neq 0$. The mapping μ is similarly treated. Hence we can summarize the argument as follows:

Theorem N5.1. *Given a translation surface $z = u(x^1) + v(x^2)$ with the property that $\mathrm{Det}\,(u_1, u_{11}, u_{111}) > 0$ and $\mathrm{Det}\,(v_2, v_{22}, v_{222}) < 0$, the surfaces*

$$u(x^1) + v(x^2) \pm \frac{(u_1 \wedge u_{11}) \wedge (v_2 \wedge v_{22})}{\sqrt{\mathrm{Det}\,(u_1, u_{11}, u_{111})}\sqrt{\mathrm{Det}\,(v_2, v_{22}, v_{222})}}$$

define a pair of affine minimal surfaces.

The correspondence of $\check{f}$ to f is an example of Weingarten congruence of lines in the following sense. Let $\ell = \{\ell(x^1, x^2)\}$ be a family of lines given by

$$\ell(x^1, x^2) = \mathbf{R}\{f - \check{f}\} = \mathbf{R}\{\mu \wedge \lambda\}.$$

Since $f_1 = (\lambda + \mu) \wedge \lambda_1$ and $f_2 = \mu_2 \wedge (\lambda + \mu)$, we see that the line ℓ is tangent to the surface f and hence to the surface $\check{f}$. This means that two surfaces f and $\check{f}$ are the focal surfaces of the congruence of lines ℓ. The induced metrics $2F dx^1 dx^2$ and $2\check{F} dx^1 dx^2$ are obviously conformal to each other. Such a congruence of lines is said to be *Weingarten*; refer to [ChT]. Moreover, we note that the affine normal fields are parallel because they are given by $\frac{1}{2}\triangle f$ and $\frac{1}{2}\check{\triangle}\check{f}$, both scalar multiples of $\mu_2 \wedge \lambda_1$ and $-\mu_2 \wedge \lambda_1$, respectively. Conversely, the following theorem is known.

Theorem N5.2. ([ChT]). *Given a congruence of lines, assume that the two focal surfaces are immersed and the correspondence is Weingarten. If the affine normals of both surfaces are parallel at each pair of corresponding points, then the surfaces are both affine minimal.*

Refer to [Bu] for a somewhat more general treatment of the material in this note.

Example N5.2. *Enneper surface.* Define λ and μ by

$$\lambda = (x, 0, \frac{1}{2}x^2 + a) \quad \text{and} \quad \mu = (0, y, \frac{1}{2}y^2 + b).$$

Then it is easy to see that corresponding immersions are given by

$$f = (\frac{1}{2}x^2 y - \frac{1}{6}y^3 + cy, \frac{1}{2}xy^2 - \frac{1}{6}x^3 + cx, -xy),$$
$$\check{f} = (-\frac{1}{2}x^2 y - \frac{1}{6}y^3 - c'y, -\frac{1}{2}xy^2 - \frac{1}{6}x^3 + c'x, xy)$$

up to constant vectors, where $c = a + b$ and $c' = a - b$. The immersion f defines the Enneper surface and the middle translation surface is now the plane

$$f + \check{f} = (-\frac{1}{3}y^3 + (c - c')y, -\frac{1}{3}x^3 + (c + c')x, 0).$$

6. Formula for a variation of ovaloid with fixed enclosed volume

Let $f : M \to \mathbf{R}^{n+1}$ be an ovaloid. We define an n-form α by

(N6.1) $\qquad \alpha(X_1,\ldots,X_n) = \tilde\omega(f_*X_1,\ldots,f_*X_n, f(x) - p),$

where p is a fixed point in $\mathbf{R}^{n+1}$ and $\tilde\omega$ is the determinant function that we have fixed. Relative to the decomposition (5.8) of Chapter II,

$$f(x) - p = f_*Z_x + \rho(x)\xi_x,$$

we have

$$\alpha = \rho\theta$$

where ρ is the affine support function.

We consider the $(n+1)$-dimensional volume $B(M)$ of the convex body bounded by $f(M)$. It is given by

(N6.2) $\qquad B(M) = -\dfrac{1}{n+1}\displaystyle\int_M \alpha.$

This definition is independent of the choice of point p. To see this, choose two points p and q and define an $(n-1)$-form γ by

$$\gamma(X_1,\ldots,X_{n-1}) = \tilde\omega(f_*X_1,\ldots,f_*X_{n-1}, f(x), q - p).$$

Then it is easy to see

$$d\gamma = (-1)^{n-1}(\alpha_p - \alpha_q)$$

where α_p (resp. α_q) is the n-form associated to the point p (resp. q).

Let $F = \{f_t\}$ be a smooth variation of f as was defined in Section 11 of Chapter III. The variation vector field $V = F_*(\partial/\partial t)|_{t=0}$ is decomposed as in (11.2); that is,

(N6.3) $\qquad V = V_0 + v\xi.$

The $(n+1)$-dimensional volume $B(t)$ of the convex body bounded by $f_t(M)$ is by definition

$$B_t = -\dfrac{1}{n+1}\displaystyle\int_M \alpha_t,$$

where

$$\alpha_t(X_1,\ldots,X_n) = \tilde\omega(f_{t*}X_1,\ldots,f_{t*}X_n, f_t(x) - p).$$

Let $\{X_1, \ldots, X_n\}$ be the basis of vector fields that is constructed in the proof of Proposition 11.1 of Chapter III. Since every basis vector field commutes with $\partial/\partial t$, we see

$$\frac{\partial}{\partial t} f_{t*} X_i = D_{\frac{\partial}{\partial t}} f_{t*} X_i$$
$$= D_{X_i} f_{t*} \frac{\partial}{\partial t}$$
$$= D_{X_i} (V_0 + v\xi)$$
$$= \nabla_{X_i} V_0 - v S X_i + [h(X_i, V_0) + X_i(v)]\xi.$$

Hence

$$\frac{\partial}{\partial t} \alpha_t(X_1, \ldots, X_n)|_{t=0}$$

$$= \sum_i \tilde{\omega}(\ldots, \frac{\partial}{\partial t} t_{t*} X_i|_{t=0}, \ldots, f(x) - p)$$

$$+ \tilde{\omega}(f_* X_1, \ldots, f_* X_n, \frac{\partial}{\partial t} f_t(x)|_{t=0})$$

$$= \sum_i \tilde{\omega}(\ldots, \nabla_{X_i} V_0 - v S X_i + [h(X_i, V_0) + X_i(v)]\xi, \ldots, f_* Z + \rho\xi)$$

$$+ \tilde{\omega}(f_* X_1, \ldots, f_* X_n, V_0 + v\xi)$$

$$= \rho \operatorname{trace}\{X \mapsto \nabla_X V_0 - v S X\}\theta(X_1, \ldots, X_n)$$
$$- [h(Z, V_0) + Zv]\theta(X_1, \ldots, X_n) + v\theta(X_1, \ldots, X_n);$$

that is,

(N6.4) $$\frac{\partial}{\partial t} \alpha_t|_{t=0} = [\rho(\operatorname{div} V_0 - v \operatorname{tr} S) + V_0\rho - Zv + v]\theta$$

because $h(Z, V_0) = -V_0\rho$, as shown in (5.10) of Chapter II. Since $\rho \operatorname{div} V_0 + V_0\rho = \operatorname{div}(\rho V_0)$, we have

$$\frac{\partial}{\partial t} B_t|_{t=0} = -\frac{1}{n+1} \int_M v\theta + \frac{1}{n+1} \int_M (\rho v \operatorname{tr} S + Zv)\theta.$$

Now recall the definition of the $(n-1)$-form β_0 in (9.13) of Chapter III:

$$\beta_0(X_1, \ldots, X_{n-1}) = \theta(Z, X_1, \ldots, X_{n-1}),$$

which has the property

$$d\beta_0 = (H\rho + 1)\theta;$$

see (9.14) of Chapter III (refer also to Remark 6.1 of Chapter II). We easily get

$$d(v\beta_0) = \frac{1}{n} Z v \cdot \theta + v d\beta_0.$$

Therefore

$$d(v\beta_0) = \frac{1}{n}Zv + v(H\rho + 1)\theta,$$

which implies that

$$\int_M (\rho v \operatorname{tr} S + Zv)\theta = -n \int_M v\theta.$$

Finally, we have

(N6.5)
$$\frac{\partial}{\partial t}B_t|_{t=0} = -\int_M v\theta.$$

Let us consider a variation of an immersion of a compact manifold as an ovaloid that preserves the volume of the surrounding convex body. Then the extremal problem relative to the affine volume is solved by finding a constant λ, called the Lagrange multiplier, for which

$$\frac{\partial}{\partial t}A|_{t=0} + \lambda\frac{\partial}{\partial t}B|_{t=0} = 0.$$

By Proposition 11.1 of Chapter III we see that this condition is equivalent to

(N6.6)
$$\int_M \left(\frac{n(n+1)}{n+1}H + \lambda\right)v\theta = 0$$

for every variation. This implies the following.

Proposition N6.1 *Let* $f : M \to \mathbf{R}^{n+1}$ *be an ovaloid with Blaschke structure. Then* f *is a critical point of the volume functional A for any variation* f_t *with constant* $(n + 1)$-*volume for the convex body bounded by* $f_t(M)$ *if and only if the affine mean curvature is constant (hence* $f(M)$ *is an ellipsoid).*

For the last statement, see Theorem 9.7 of Chapter III. For the classical treatment of related topics, see Blaschke [Bl, pp. 198–204] and Deicke [De]. Refer also to [S2] and [MMi3].

7. Completeness and hyperbolic affine hyperspheres

Given an n-manifold M that is immersed into $\mathbf{R}^{n+1}$ as a nondegenerate hypersurface, we can associate at least three notions of completeness; namely,

(1) completeness relative to the affine metric,
(2) completeness relative to the Riemannian metric induced from a Euclidean metric in $\mathbf{R}^{n+1}$,
(3) completeness of the Blaschke connection.

For the completeness (2), note that the notion is independent of the choice of Euclidean metric compatible with the affine structure of $\mathbf{R}^{n+1}$. The relationships between these notions of completeness are rather complicated. In this note, we cite two examples due to [Schn1] and [N3] and the important results on hyperbolic affine hyperspheres due to [CY1] and [L5, 8].

Let us begin with

Example N7.1. [Schn1]. Let $f : M = \mathbf{R}^+ \times \mathbf{R} \to \mathbf{R}^3$ be the imbedding defined by

$$f(x, y) = \left(x, y, \frac{1}{2} \left(\frac{1}{x} + y^2 \right) \right).$$

The surface is complete relative to the induced metric. Because of the formulas in Section 3 of Chapter II, the affine metric is equal to

$$x^{-\frac{9}{4}} dx^2 + x^{\frac{3}{4}} dy^2.$$

Let $\gamma(t) = (t, 0)$ be the curve in M; it is divergent as $t \to \infty$. However, its length is bounded. Hence the affine metric is not complete. The Blaschke connection ∇ is given by

$$\nabla_{\partial_x}\partial_x = -\frac{3}{4x}\partial_x, \quad \nabla_{\partial_x}\partial_y = \nabla_{\partial_y}\partial_x = 0, \quad \nabla_{\partial_y}\partial_y = -\frac{3}{4}x^2\partial_x,$$

where $\partial_x = \partial/\partial_x$ and $\partial_y = \partial/\partial_y$. Therefore the curve $t \mapsto (t^4, 0)$ $(t > 0)$ is a geodesic with the affine parameter t. Since it is not extendable to $t = 0$, the affine connection is not complete.

This example shows that (2) does not imply (1) nor (3). The next example reveals a similar phenomenon when the ambient affine space carries a Lorentzian structure.

Example N7.2. [N3]. Let $f : M = \mathbf{R}^2 \to \mathbf{L}^3$ be the imbedding into the Lorentz–Minkowski space $\mathbf{L}^3$ with metric $dx^2 + dy^2 - dz^2$ which is defined by

$$f(x, y) = \left(\int_0^x \sqrt{1 + e^{2t}} dt, \ e^x \sinh y, \ e^x \cosh y \right).$$

The imbedded surface is spacelike and the induced Riemannian metric $dx^2 + e^{2x} dy^2$ is complete and of constant curvature -1. On the other hand, the affine metric is equal to

$$e^x(1 + e^{2x})^{-\frac{1}{2}} dx^2 + e^x(1 + e^{2x})^{\frac{1}{2}} dy^2$$

and is not complete; consider the curve $(t, 0)$ where $-t \in [0, \infty)$. However, in this example, the Blaschke connection is identical with the induced Riemannian connection and is complete.

For a locally strictly convex hyperbolic affine hypersphere, A.-M. Li has proved the following theorem in [L5, 8].

Theorem N7.1. *Every locally strictly convex hyperbolic affine hypersphere that is complete relative to the affine metric is complete relative to the induced Riemannian metric.*

As was explained in the Introduction, E. Calabi made significant progress on the classification of affine hyperspheres. In particular, the problem of determining global hyperbolic affine hyperspheres attracted the attention of several people. The answer to the case where the hypersphere is closed in the ambient affine space was given by Cheng and Yau [CY1], and by Gigena [Gi] as well as by Sasaki [S1]. This answer, however, is not fully satisfactory because it assumes an extrinsic condition that the hypersurface is closed in the ambient space or, in other words, the completeness relative to the Euclidean metric. In the course of the proof in [CY1], they showed that the closedness implies also the completeness of the affine metric. Theorem N7.1 above proves that the converse implication holds and thus brings a complete answer to the Calabi problem in the following form.

Theorem N7.2. *Every locally strictly convex hyperbolic affine hypersphere that is complete relative to the affine metric is closed in the affine space and is asymptotic to the boundary of a convex cone with vertex at the center of the hypersphere.*

In the paper [L7], Li has also extended one of the results by [CY1] as follows.

Theorem N7.3. *Let $f : M^n \to \mathbf{R}^{n+1}$ be a closed, locally strictly convex hypersurface. If the eigenvalues of the shape operator are bounded, then the affine metric is complete.*

This has a corollary that solves partially the affine Bernstein problem stated in the Introduction:

Theorem N7.4. *Let M be the graph of $x^3 = f(x^1, x^2)$, where f is a strictly convex function f defined on all of $\mathbf{R}^2(x^1, x^2)$. If it is affine minimal and if the eigenvalues of the shape operator are bounded, then the surface is affinely congruent to an elliptic paraboloid.*

Refer to [MMi2] for another kind of solution to the affine Bernstein problem.

8. Locally symmetric surfaces

This note supplements Section 4 of Chapter IV by discussing nondegenerate locally symmetric Blaschke surfaces. Let $f : M^2 \to \mathbf{R}^3$ be a nondegenerate locally symmetric immersion of a surface M. From the Gauss equation we see that, at each point, the image of S, im S, coincides with the image of R, im R, i.e. the subspace spanned by all vectors of the form $R(X, Y)Z$. Since $\nabla R = 0$, im R is parallel and hence S has constant rank. The structure of this immersion depends crucially on the rank of the shape operator S. When rank $S = 0$, that is, when $S = 0$, M^2 is by definition an improper affine sphere. So the following discussion is concerned with the case where rank $S = 2$ and the case where rank $S = 1$. The results will be summarized in Theorems N8.1–N8.3 below. Then we shall add some results on projectively flat surfaces.

Recall first the notation. The curvature tensor in the case where dim $= 2$ takes the form

(N8.1) $R(X,Y)Z = \text{Ric}(Y,Z)X - \text{Ric}(X,Z)Y,$

and its covariant derivative is given by

(N8.2) $(\nabla_U R)(X,Y)Z = (\nabla_U \text{Ric})(Y,Z)X - (\nabla_U \text{Ric})(X,Z)Y.$

The assumption that the connection is symmetric is equivalent to the condition $\nabla\text{Ric} = 0$. We recall that the Ricci tensor is given by

(N8.3) $\text{Ric}(X,Y) = \text{tr}\, S \cdot h(X,Y) - h(X,SY);$

see (2.5) of Chapter II.

We now consider the case where S is nondegenerate. Then Ric is also nondegenerate by (N8.3). Hence the condition $\nabla\text{Ric} = 0$ implies that Ric defines a nondegenerate Riemannian metric on M and ∇ is a metric connection. We put $g = \epsilon\text{Ric}$ where $\epsilon = -1$ when Ric is negative-definite and $\epsilon = 1$ otherwise. Then (N8.1) means that the Riemannian curvature of g is of constant curvature ϵ. The Cartan–Norden theorem in Section 3 of Chapter IV says that there exists an inner product $\langle\,,\,\rangle$ on $\mathbf{R}^3$ relative to which ξ is orthonormal to $f(M)$ and the immersion f of (M,g) is isometric. This means

Theorem N8.1. *A locally symmetric Blaschke surface whose shape operator is nondegenerate is locally affinely congruent to a surface of constant curvature in the 3-dimensional pseudo-Euclidean space.*

Example N8.1. The quadrics with center in $\mathbf{R}^3$ in Example 3.5 of Chapter II.

Example N8.2. Let $f : (u,v) \to \mathbf{R}^3$ be an immersion given by

$$f(u,v) = (p(u)\cos v, p(u)\sin v, q(u))$$

into the ordinary Euclidean space, where

$$p(u) = \frac{1}{c}e^{-cu}, \quad q(u) = \int_0^u \sqrt{1 - e^{-2ct}}\,dt.$$

Here c is a positive constant and the range of u is restricted to $u > 0$. This surface is known to be of constant Gaussian curvature $-c^2$. By computation, we see

$$S\partial_u = c\sqrt{c}\,e^{-cu}(1 - e^{-2cu})^{-\frac{1}{2}}\partial_u,$$
$$S\partial_v = -c\sqrt{c}\,e^{cu}(1 - e^{-2cu})^{\frac{1}{2}}\partial_v,$$

and

$$\nabla_{\partial_u}\partial_u = 0, \quad \nabla_{\partial_u}\partial_v = -c\partial_v, \quad \nabla_{\partial_v}\partial_v = \frac{1}{c}e^{-2cu}\partial_u,$$

where $\partial_u = f_*(\partial/\partial u)$ and $\partial_v = f_*(\partial/\partial v)$. Therefore $\nabla R = 0$ and $\det S = -c^3$.

Next we treat the case where rank $S = 1$, that is, dim im $S = 1$. We have two subcases: (a) ker $S \cap$ im $S = \emptyset$, which implies $S^2 \neq 0$, and (b) ker $S =$ im S, which implies $S^2 = 0$.

In subcase (a), set im $S = \mathbf{R}\{X\}$ and ker $S = \mathbf{R}\{Y\}$. Then

(N8.4) $$SX = \lambda X \quad \text{and} \quad SY = 0$$

for some λ, say > 0. Since $\lambda h(X, Y) = h(SX, Y) = h(X, SY) = 0$, we see $h(X, Y) = 0$. Put $h(X, X) = \epsilon_1$ and $h(Y, Y) = \epsilon_2$. By multiplying X and Y by scalars if necessary, we can assume that ϵ_1 and ϵ_2 are ± 1. As we know, the space im S is parallel. This means that we can write

$$\nabla_X X = \alpha X, \quad \nabla_Y X = \beta X,$$

for some scalar functions α and β. Now our task is to check all of the integrability conditions, namely the Gauss equation, the Codazzi equations, the apolarity condition and further the condition $\nabla R = 0$. We find that the coefficients α and β are given by

(N8.5) $$\alpha = -\frac{1}{2}X(\log \lambda), \qquad \beta = -\frac{1}{2}Y(\log \lambda),$$

and the connection ∇ is determined by the equations

(N8.6) $$\begin{aligned} \nabla_X X &= \alpha X, & \nabla_X Y &= 2\beta X - \alpha Y, \\ \nabla_Y X &= \beta X, & \nabla_Y Y &= -2\alpha\epsilon_1\epsilon_2 X - \beta X. \end{aligned}$$

The function λ must satisfy the equation

(N8.7) $$2X(\alpha)\epsilon + 6\alpha^2\epsilon + 6\beta^2 + 2Y(\beta) = \epsilon_2\lambda,$$

where $\epsilon = \epsilon_1\epsilon_2$. Since, in view of (N8.5), we can set $X = \lambda^{\frac{1}{2}}\partial/\partial u$ and $Y = \lambda^{\frac{1}{2}}\partial/\partial v$ for a local coordinate system $\{u, v\}$, the equation (N8.7) is simplified to

(N8.8) $$\epsilon\psi_{uu} + \psi_{vv} + \epsilon_2\delta\psi = 0,$$

where $\psi = |\lambda|^{-1}$ and $\delta = \text{sign}(\lambda)$. We can see that the affine normal vector satisfies the equation

(N8.9) $$\frac{d^2}{du^2}\xi + \delta\epsilon_1\xi = 0,$$

which implies that the vector ζ runs on a plane; in fact, it is on a sine curve or a hyperbolic sine curve.

Conversely, let ψ be a solution of (N8.8) and ζ a nontrivial solution of (N8.9). Then the procedure can be reversed to obtain an immersion of the 2-manifold into $\mathbf{R}^3$ as a locally symmetric nondegenerate surface; in other words, the Gauss and Weingarten equations defining the immersion are integrable. Their explicit form is given in the following theorem.

Theorem N8.2. *Let* $f : M^2 \to \mathbf{R}^3$ *be a locally symmetric Blaschke surface. Assume that the rank of the shape operator S is equal to 1 and that $S^2 \neq 0$. Then, around each point of M, there exist local coordinates (u, v) relative to which the immersion f is given by solving the system of equations*

$$\epsilon_1 f_{uu} - \epsilon_2 f_{vv} = 2\epsilon_1 (\log \psi)_u f_u,$$
$$f_{uv} = (\log \psi)_v f_u,$$

where the function ψ is a solution of the equation (N8.8) and ϵ_1, ϵ_2, and δ are equal to ± 1, and are determined by the signatures of h and S.

Example N8.3. We may assume $\epsilon_1 = 1$. Consider the case where the function ψ depends only on v. Then $\psi = C(v)$ is a solution where $C(v) = \cos v$ ($\epsilon_2 \delta = 1$) or $\cosh v$ ($\epsilon_2 \delta = -1$). In this case, it is easy to solve the systems (N8.14) and (N8.15) and the immersion has the form

$$f = \pm (C(v) C(u), C(v) S(u), v)$$

relative to an appropriate affine coordinate system of $\mathbf{R}^3$, where $S(u) = \sin u$ or $\sinh u$ depending on whether $\epsilon_2 \delta = 1$ or $\epsilon_2 \delta = -1$.

Subcase (b) where $S^2 = 0$ is treated similarly. Let $\ker S = \mathbf{R}\{X\}$. Then the condition $\ker S = \operatorname{im} S$ implies $SU = X$ for some U. This shows $h(X, X) = h(SU, SU) = h(S^2 U, U) = 0$. Choose Y so that $h(X, Y) = 1$ and $h(Y, Y) = 0$ and let $SY = \lambda X$ for a scalar λ, say > 0. Replacing X and Y by $\lambda^{-1/2} X$ and $\lambda^{1/2} Y$ respectively, we can set

(N8.10) $SY = \delta X$ $(\delta = \pm 1)$, $SX = 0$.

Since $\operatorname{im} S$ is parallel, we can set

$$\nabla_X X = \alpha X, \qquad \nabla_Y X = \beta X,$$

for some scalar functions α, β. Further it turns out that $\alpha = \beta = 0$ because of the integrability conditions and that

$$\nabla_X Y = 0, \qquad \nabla_Y Y = cX,$$

for a scalar function c satisfying the condition $X(c) = \delta$. Now, introducing local coordinates (u, v) for which $X = \partial/\partial u$ and $Y = \partial/\partial v$, we see that the function c and the immersion f are determined by the equations

$$c = -\delta u + c_1(v),$$
$$f = p(v) + u q(v),$$

where c_1 is an arbitrary function of v; moreover, the intermediate functions p and q are nontrivial solutions of the system of differential equations

(N8.11)
$$p'' = c_1(v)q \quad \text{and} \quad q'' + \delta q = 0.$$

This defines a ruled surface and we have the following theorem.

Theorem N8.3. *A locally symmetric Blaschke surface whose shape operator S has rank 1 and $S^2 = 0$ is ruled and can be given by solving the system* (N8.11).

We remark that the helicoidal surface in Remark 4.1 of Chapter IV belongs to this class. The results in this note are due to B. Opozda [O2], which should be referred to for more detail. See also [O4].

The study of locally symmetric Blaschke surfaces in $\mathbf{R}^3$ goes back to W. Ślebodziński [Sl1, 2], who took it up as a special case of the problem posed by E. Cartan [Car1]. Opozda [O7] has extended the results to the case of projectively flat surfaces and, in particular, has shown that, unlike locally symmetric connections of rank 1, Ricci-symmetric projectively flat connections of rank 1 on surfaces may not be realized on Blaschke surfaces.

Theorem N8.4. *Let ∇ be a projectively flat connection on a 2-manifold M. Assume that the distribution $\operatorname{im} R$ is of constant rank 1. Then, for every $x \in M$, there are a neighbourhood U of x and a nondegenerate affine immersion $f : U \to \mathbf{R}^3$ such that $\nabla|_U$ is the Blaschke connection induced by f if and only if the distribution $\operatorname{im} R$ is parallel relative to ∇.*

Example N8.4. Define a connection ∇ on $\mathbf{R}^2(u,v) - \{u = 0\}$ by

(N8.12)
$$\nabla_{\partial_u}\partial_u = 0, \quad \nabla_{\partial_v}\partial_u = \frac{1}{u}\partial_v, \quad \nabla_{\partial_v}\partial_v = u^2\partial_u.$$

This connection is seen to be projectively flat and Ricci-symmetric while $\operatorname{im} R$ is not parallel relative to ∇. Hence, ∇ is not realizable on any Blaschke surface.

Podestà [Po2] also found an interesting example of projectively flat Blaschke surfaces. C. Lee [Le1,2] made use of his study of affine rotation surfaces and produced many more projectively flat surfaces.

9. Centro-affine immersions of codimension 2

In this note, we discuss the notion of centro-affine immersion $M^n \to \mathbf{R}^{n+2} - \{0\}$ of codimension 2. In addition to its own interest, it has applications to projective differential geometry.

Let F be an immersion of an n-manifold M into $\mathbf{P}^{n+1}$ and $\pi : \mathbf{R}^{n+2} - \{0\} \to \mathbf{P}^{n+1}$ the natural projection. Then, locally, there is an immersion f of M into $\mathbf{R}^{n+2} - \{0\}$ such that $\pi \circ f = F$. We call f a *local lift* of F and write $F = [f]$. Another local lift g is written as $g = \phi f$ for some nonzero scalar function ϕ. Both immersions f and g are centro-affine immersions of codimension

2 in the sense we define below. Conversely, any centro-affine immersion f of an n-manifold M into $\mathbf{R}^{n+2}$ defines an immersion of M into $\mathbf{P}^{n+1}$. Thus the theory of centro-affine immersions of codimension 2 provides a new approach to the study of hypersurfaces in $\mathbf{P}^{n+1}$; this is our main concern in this note.

We use the following notation. By D we mean a flat affine connection of $\mathbf{R}^{n+2}$ and by η the radial vector field on $\mathbf{R}^{n+2} - \{0\}$: $\eta = \sum_{i=1}^{n+2} x^i \left(\partial / \partial x^i \right)$, where $\{x^1, \ldots, x^{n+2}\}$ is an affine coordinate system. The letter ω denotes a parallel volume form of $\mathbf{R}^{n+2}$ that is fixed once and for all. Let M be an n-manifold and f an immersion of M into $\mathbf{R}^{n+2} - \{0\}$. Assume f is transversal to η. We can choose, at least locally, a vector field ξ along f that is transversal to f such that ξ and η are linearly independent. At each $x \in M$ the tangent space $T_{f(x)}(\mathbf{R}^{n+2})$ is decomposed as the direct sum of the span $\mathbf{R}\{\eta\}$, the tangent space $f_*(T_x(M))$, and the span $\mathbf{R}\{\xi\}$. According to this decomposition, the vectors $D_X \eta$, $D_X f_* Y$, and $D_X \xi$, where X, Y are vector fields on M, have the following expressions:

$$
\begin{aligned}
D_X \eta &= f_* X, \\
\text{(N9.1)} \qquad D_X f_* Y &= T(X, Y)\eta + f_*(\nabla_X Y) + h(X, Y)\xi, \\
D_X \xi &= \rho(X)\eta - f_*(SX) + \tau(X)\xi.
\end{aligned}
$$

An n-form θ is defined by

$$
\text{(N9.2)} \qquad \theta(X_1, \ldots, X_n) = \omega(f_* X_1, \ldots, f_* X_n, \xi, \eta).
$$

Thus we have several objects associated with ξ. They have the following properties:

    (1) ∇ is a torsion-free affine connection on M;

    (2) h and T are symmetric tensors;

    (3) $\nabla_X \theta = \tau(X)\theta$.

The condition that D is flat is expressed by the following set of equations: Gauss equation (N9.3), Codazzi equations (N9.4), (N9.5), (N9.6), and Ricci equations (N9.7), (N9.8).

$$
\text{(N9.3)} \quad R(X, Y)Z = h(Y, Z)SX - h(X, Z)SY - T(Y, Z)X + T(X, Z)Y,
$$

$$
\text{(N9.4)} \quad (\nabla_X T)(Y, Z) + \rho(X)h(Y, Z) = (\nabla_Y T)(X, Z) + \rho(Y)h(X, Z),
$$

$$
\text{(N9.5)} \quad (\nabla_X h)(Y, Z) + \tau(X)h(Y, Z) = (\nabla_Y h)(X, Z) + \tau(Y)h(X, Z),
$$

$$
\text{(N9.6)} \quad (\nabla_X S)(Y) - \tau(X)SY + \rho(X)Y = (\nabla_Y S)(X) - \tau(Y)SX + \rho(Y)X,
$$

$$
\text{(N9.7)} \quad T(X, SY) - T(Y, SX) = (\nabla_X \rho)(Y) - (\nabla_Y \rho)(X)
$$
$$
+ \tau(Y)\rho(X) - \tau(X)\rho(Y),
$$

$$
\text{(N9.8)} \quad h(X, SY) - h(Y, SX) = (\nabla_X \tau)(Y) - (\nabla_Y \tau)(X) = d\tau(X, Y).
$$

These equations are obtained in a manner similar to the proof of Theorem 2.1 of Chapter II. From (N9.3) we have in particular

$$(\text{N9.9}) \qquad \text{Ric}(Y,Z) = \text{tr}\, S \cdot h(Y,Z) - h(SY,Z) - (n-1)T(Y,Z).$$

In comparison to the case of codimension 1, we have two additional invariants T and ρ. The vanishing of T means the following.

Proposition N9.1. *If T vanishes and* rank $h \geq 2$, *then the image of the immersion is contained in an affine hyperplane that does not go through 0 and the vector field ξ is tangent to this hyperplane.*

The proof is given as follows: If $T \equiv 0$, then (N9.4) says $\rho(X)h(Y,Z) = \rho(Y)h(X,Z)$. Then, by Lemma 6.2 of Chapter IV, $\rho = 0$ and, therefore, the distribution spanned by $f_*(T_x(M))$ and ξ_x, $x \in M$, is parallel relative to D; this implies the result.

Under the condition of this proposition, the immersion f considered as a mapping into the hyperplane is an affine immersion of M as a hypersurface; relative to the induced flat connection D' on the hyperplane, f satisfies

$$D'_X f_* Y = f_*(\nabla_X Y) + h(X,Y)\xi \quad \text{and} \quad D'_X \xi = -f_*(SX) + \tau(X)\xi;$$

that is, the immersion is reduced to the case of codimension 1.

On the other hand, in comparison to the case of a centro-affine hypersurface, we have two additional invariants h and τ. The vanishing of h has the following significance.

Proposition N9.2. *If h vanishes and $n \geq 2$, then the image of the immersion is included in a hyperplane through 0.*

In fact, if $h \equiv 0$, then the distribution spanned by $f_*(T_x(M))$ and $\eta_{f(x)}$, $x \in M$, is parallel relative to D. In this situation, the immersion f defines a centro-affine hypersurface immersion and the tensor T is the fundamental tensor of this immersion.

These two observations say that a centro-affine immersion of codimension 2 has the features that combine a centro-affine immersion of codimension 1 and an affine immersion also of codimension 1. This point of view will be helpful in the following discussions.

As is in the case of codimension 1, various objects depend on ξ. Examination of this dependence is carried out very similarly to those for Proposition 2.5 of Chapter II. In order to state the result, let ξ' be another choice of transversal vector field related to ξ by

$$\lambda\xi' = \xi + a\eta + f_* U,$$

where λ is a nonzero scalar function, a is also a scalar, and U is a tangent vector field. Let T', ∇', h', ρ', S', and τ' denote the quantities corresponding to ξ'. Then we see

(N9.10) $\nabla'_X Y = \nabla_X Y - h(X,Y)U,$

(N9.11) $T'(X,Y) = T(X,Y) - ah(X,Y),$

(N9.12) $h'(X,Y) = \lambda h(X,Y),$

(N9.13) $\tau'(X) = \tau(X) - X(\log \lambda) + h(X,U),$

(N9.14) $\lambda\rho'(X) = \rho(X) + X(a) + T(X,U) - ah(X,U) - a\tau(X),$

(N9.15) $\lambda S'X = SX + \tau(X)U - aX - \nabla_X U + h(X,U)U.$

Formula (N9.12) implies that the conformal class of h is independent of the choice of ξ. When the class of h is nondegenerate we say that the immersion is *nondegenerate*. If we assume that h is nondegenerate, then we can find, by using (N9.13), a vector field ξ such that $\tau = 0$; in this case, θ is ∇-parallel. We say that this choice of ξ defines (or that the pair $\{f,\xi\}$ is) an *equiaffine* immersion. Further, we can restrict the choice of ξ so that the form θ is equal to the volume form of the nondegenerate metric tensor h; such ξ is uniquely determined mod η up to sign. We call this pair $\{f,\xi\}$ a *Blaschke immersion* of codimension 2. Formulas (N9.11), (N9.12), and (N9.15) with $U = 0$ show

$$T'(X,Y) + h'(S'X,Y) = T(X,Y) + h(SX,Y) - 2ah(X,Y).$$

By determining the scalar function a we can assume that ξ is so chosen that

(N9.16) $\text{trace}_h\{(X,Y) \mapsto T(X,Y) + h(SX,Y)\} = 0.$

If this condition is satisfied, we say that ξ is *pre-normalized*. In particular, a pre-normalized Blaschke immersion $\{f,\xi\}$ is uniquely determined up to the sign of ξ.

Let us return to the general situation. The equation (N9.11) says that the condition $h|T$, namely, $T = \alpha h$ for some scalar function α, does not depend on the choice of ξ. Hence this is a centro-affine invariant property and we can restate Proposition N9.1 in an invariant form independent of the choice of ξ as follows:

Proposition N9.3. *Assume* rank $h \geq 2$. *Then the image of the immersion is contained in an affine hyperplane if and only if* $T = \alpha h$ *for some scalar function* α.

The relation of connections given in (N9.10) is the same as the equation (2.9) of Chapter II. From this fact we have an analogue of Proposition 2.8 of Chapter II.

Proposition N9.4. *Let* $n \geq 3$. *Assume* ∇ *is flat and* rank $h \geq 2$. *Then the image of the immersion lies on an affine hyperplane and the immersion turns out to be a graph immersion into this hyperplane.*

We define the cubic forms by

$$(N9.17) \qquad \begin{aligned} C(X,Y,Z) &= (\nabla_X h)(Y,Z) + \tau(X)h(Y,Z), \\ \delta(X,Y,Z) &= (\nabla_X T)(Y,Z) + \rho(X)h(Y,Z). \end{aligned}$$

They are symmetric in their arguments by the equations (N9.4) and (N9.5). We call C the *first cubic form* and δ the *second cubic form*. Under the change of ξ to $\xi' = \lambda^{-1}(\xi + a\eta + f_*U)$ the cubic forms C and δ transform according to the rules

$$(N9.18) \quad \lambda^{-1}C'(X,Y,Z)$$
$$= C(X,Y,Z) + h(X,Y)h(U,Z) + h(Y,Z)h(U,X) + h(Z,X)h(U,Y),$$
$$(N9.19) \quad \delta'(X,Y,Z) + a\lambda^{-1}C'(X,Y,Z)$$
$$= \delta(X,Y,Z) + h(X,Y)T(U,Z) + h(Y,Z)T(U,X) + h(Z,X)T(U,Y).$$

Now let us consider immersions into $\mathbf{P}^{n+1}$. Given an immersion F into $\mathbf{P}^{n+1}$, let f and g be two local lifts. We want to find how the invariants for f are related to those for g.

First, the relationship between f_* and g_* is given as follows. For a tangent vector field X, we have

$$D_X g = D_X(\phi f) = (X\phi)f + \phi D_X f$$

and

$$g_* X = (X\phi)\eta_{f(x)} + \phi f_* X = X(\log \phi)\eta_{g(x)} + \phi f_* X,$$

where $f_* X \in T_{f(x)}(\mathbf{R}^{n+2})$ is considered to be in $T_{g(x)}(\mathbf{R}^{n+2})$ by parallel translation. We write for the moment η_g for $\eta_{g(x)}$ and η for $\eta_{f(x)}$. Then

$$(N9.20) \qquad\qquad g_* X = \sigma(X)\eta_g + \phi f_* X$$

where $\sigma = d \log \phi$. The quantities for g are denoted with "$\bar{\ }$". Then the relationship between the invariants with respect to the immersions (f, ξ, η) and (g, ξ, η_g) is the following.

$$(N9.21) \qquad \bar{\nabla}_X Y = \nabla_X Y + \sigma(Y)X + \sigma(X)Y,$$
$$(N9.22) \qquad \bar{T}(X,Y) = \operatorname{Hess}^\nabla_{\log \phi}(X,Y) - \sigma(X)\sigma(Y) + T(X,Y),$$
$$(N9.23) \qquad \bar{h}(X,Y) = \phi h(X,Y),$$
$$(N9.24) \qquad \bar{\tau} = \tau,$$
$$(N9.25) \qquad \phi\bar{\rho}(X) = \rho(X) + \sigma(SX),$$
$$(N9.26) \qquad \phi\bar{S} = S,$$

where $\operatorname{Hess}^\nabla_{\log \phi}$ is the Hessian of $\log \phi$ relative to ∇. Moreover, we can see that

$$(N9.27) \qquad\qquad\qquad \bar{\theta} = \phi^{n+1}\theta.$$

Hence, the conformal class of h is preserved and, if f is equiaffine, i.e. $\tau = 0$, then g is also equiaffine relative to the same ξ.

Combining the discussions above, we can conclude that the following properties are projectively invariant (refer to Theorem 8.4 of Chapter IV):

(1) the conformal class of h is a projective invariant;

(2) the property that S is diagonal is projectively invariant;

(3) the property that $h|C$ is projectively invariant.

Thus we have reformulated the treatment in Section 8 of Chapter IV in terms of codimension-2 centro-affine immersions. Relative to the invariant properties above, we can draw more general geometrical conclusions. Let us first discuss the property (3).

A *quadratic cone* is defined to be a cone with the origin as vertex over a quadratic hypersurface in an affine hyperplane that does not pass through the origin. Then we have

Theorem N9.5. *Assume that the immersion* $M \to \mathbf{R}^{n+2} - \{0\}$ *satisfies that* rank $h \geq 2$, $\nabla h = 0$, *and* $n \geq 2$. *Then the image lies on a quadratic cone.*

The proof is given similarly to Theorem 4.5 of Chapter II. As a corollary we have

Corollary N9.6. (1) *Assume that the immersion* $M \to \mathbf{R}^{n+2} - \{0\}$ *satisfies the condition* $h|C$. *Then the image lies on a quadratic cone.*

(2) *Let* $F : M \to \mathbf{P}^{n+1}$ *be an immersion and* f *an arbitrary local lift. If the immersion* f *satisfies the condition* $h|C$, *then the image of* F *lies on a quadratic hypersurface.*

The last statement is a new version of Theorem 8.3 of Chapter IV. If we lay emphasis on the invariant T instead of h, an analogous property to $\nabla h = 0$ is $\nabla T = 0$. We can prove

Theorem N9.7. *Assume that the immersion* $M \to \mathbf{R}^{n+2} - \{0\}$ *satisfies that* rank $h \geq 2$, $\nabla T = 0$, *and* $n \geq 2$. *Then the image lies on a quadratic hypersurface or on an affine hyperplane.*

We next consider the invariant property (2). The immersion f is said to be *umbilical* relative to ξ if $S = \nu I$ for some scalar function ν. Its geometrical implication is found in the following.

Theorem N9.8. *Assume the immersion* $f : M \to \mathbf{R}^{n+2} - \{0\}$ *is umbilical. Then each 2-dimensional linear subspace spanned by* η_x *and* ξ_x *contains a fixed line through the origin; in other words, each projective line through* $[f(x)]$ *in* $\mathbf{P}^{n+1}$ *in the direction of* ξ_x *passes through a fixed point.*

So far we have been interested in the invariants T and h. We shall next be concerned with the connection ∇. Recall the definition of projective flatness of a torsion-free affine connection for which the Ricci tensor is not necessarily symmetric; refer to Appendix 1. The Weyl curvature tensor of such a connection is defined to be

$$W(X,Y)Z = R(X,Y)Z - [P(Y,Z)X - P(X,Z)Y] - (P(X,Y) - P(Y,X))Z$$

where

$$P(X, Y) = \frac{1}{n^2 - 1}(n\text{Ric}(X, Y) + \text{Ric}(Y, X)).$$

The projective flatness for the case when $n \geq 3$ is equivalent to the vanishing of the tensor W. In the present case, we have

(N9.28) $$W(X, Y)Z = W_1(X, Y)Z + W_2(X, Y)Z$$

where

(N9.29) $$W_1(X, Y)Z = h(Y, Z)S^o X - h(X, Z)S^o Y$$
$$+ \frac{1}{n - 1}\{h(S^o Y, Z)X - h(S^o X, Z)Y\},$$

(N9.30) $$W_2(X, Y)Z = \frac{1}{n^2 - 1}\left[\{h(S^o Z, Y) - h(S^o Y, Z)\}X\right.$$
$$\left. - \{h(S^o Z, X) - h(S^o X, Z)\}Y\right]$$
$$+ \frac{1}{n + 1}\{h(S^o Y, X) - h(S^o X, Y)\}Z,$$

and $S^o = S - (\text{tr }S/n)I$ denotes the traceless part of S. Note that, if ∇ has symmetric Ricci tensor, $W_2 = 0$ and hence $W = W_1$. With this preparation, we can state an analogous theorem to Theorem 7.6 of Chapter IV as follows.

Theorem N9.9. *Let $n \geq 3$. The connection ∇ is projectively flat if and only if one of the following conditions holds: (1) $h = 0$; (2) rank $h = 1$ and $S^o = v \cdot I$ on ker h; or (3) $S^o = 0$.*

We remark that, in comparison to Theorem 7.6, the Ricci-symmetry of the connection is not required. The verification of the if-part is done without much effort. To prove the only-if-part, we rely on the algebraic fact that $S^o = 0$ follows from the condition $W = 0$ provided rank $h \geq 2$.

Now we want to formulate the uniqueness theorem referring to Theorem 2.6 of Chapter II. Consider two centro-affine immersions $f^i : M \to \mathbf{R}^{n+2} - \{0\}$, $i = 1, 2$, with transversal vector fields ξ^i. Then we say that f^1 and f^2 are *affinely* (resp. *projectively*) *equivalent* if $f^1 = Af^2$ (resp. if $[f^1] = [Af^2]$) for a general linear transformation A in $GL(n + 2, \mathbf{R})$. The problem is writing these equivalence relations in terms of the connection ∇ and the fundamental form h. The key property is the transformation rules that we have seen in (N9.10–15) and in (N9.21–26).

Consider a pair (∇, h) that arises from an immersion $f : M \to \mathbf{R}^{n+2} - \{0\}$ together with a transversal vector field ξ. In the set of all such pairs (∇, h) associated to all immersions $M \to \mathbf{R}^{n+2} - \{0\}$, we define an equivalence relation: $(\nabla, h) \sim_a (\nabla', h')$ if there exist a vector field U and a function $\lambda \neq 0$ on M such that

(N9.31) $$\nabla'_X Y = \nabla_X Y - h(X, Y)U \quad \text{and} \quad h' = \lambda h.$$

Combining (N9.10) and (N9.12) we see that, given $f : M \to \mathbf{R}^{n+2} - \{0\}$, the equivalence class $[(\nabla, h)]_a$ is defined independently of the choice of ξ. Conversely, we have

Theorem N9.10. *Let $n \geq 3$. Two centro-affine immersions $f^1, f^2 : M \to \mathbf{R}^{n+2} - \{0\}$ of rank ≥ 2 are affinely equivalent if and only if the equivalence classes $[(\nabla^1, h^1)]_a$ and $[(\nabla^2, h^2)]_a$ for f^1 and f^2 coincide.*

We sketch the proof. The equations (N9.10) and (N9.12) enable us to assume $\nabla = \nabla'$ and $h = h'$; just change the choice of ξ appropriately. Then the projective curvature tensors W and W' of ∇ and ∇' become the same. Set $A = S - S'$, the difference of the shape operators. Then the tensor $W(X,Y)Z$ where S^o is replaced by the traceless part of A in the expression (N9.28) vanishes identically. Hence the same algebraic fact mentioned for Theorem N9.10 works because we have assumed rank $h \geq 2$. This means that $A = aI$ for a scalar function a. On the other hand, for $B = T - T'$, we have the identity

$$ah(Y,Z)X - ah(X,Z)Y - B(Y,Z)X + B(X,Z)Y = 0$$

in view of the Gauss equation (N9.3). This implies $B(Y,Z) = ah(Y,Z)$. Now take $\xi' - a\eta$ as the transversal vector field for ∇'; then, T' and S' are changed to $T' + ah$ and $S' + aI$ because of (N9.11) and (N9.15). Hence, with this choice of ξ', we have $T = T'$ and $S = S'$. Next, we can deduce $\tau = \tau'$ and $\rho = \rho'$ from (N9.4) and (N9.5) by using Lemma 6.2 of Chapter IV together with the condition rank $h \geq 2$. This implies the result; for the rest, refer to the argument in Proposition 1.3 of Chapter II.

As for projective equivalence we proceed as follows. Given two immersions $F^1, F^2 : M \to \mathbf{P}^{n+1}$, we shall say that F^1 and F^2 are *projectively equivalent* if there is a projective transformation $\tilde{A}$ of $\mathbf{P}^{n+1}$ such that $F^2 = \tilde{A}F^1$. In this case, any local lift f^1 of F^1 and any local lift f^2 of F^2 are projectively equivalent in the sense we defined above.

Given an immersion $F : M \to \mathbf{P}^{n+1}$, we consider a pair (∇, h) that arises from the choice of a local lift together with a transversal vector field. In the set of all such pairs (∇, h) associated to all immersions $M \to \mathbf{P}^{n+1}$, we define an equivalence relation: $(\nabla, h) \sim_p (\nabla', h')$ if there exist a closed 1-form σ, a vector field U, and a function $\lambda \neq 0$ on M such that

$$(N9.32) \quad \nabla'_X Y = \nabla_X Y + \sigma(X)Y + \sigma(Y)X - h(X,Y)U \quad \text{and} \quad h' = \lambda h.$$

It is easily checked by (N9.11) and (N9.21) that, given $F : M \to \mathbf{P}^{n+1}$, we get an equivalence class $[(\nabla, h)]_p$ independently of the choice of $\{f, \xi\}$ representing F. We define the rank of F as the rank of h. Conversely, we have

Theorem N9.11. *Let $n \geq 3$. Two immersions F^1 and $F^2 : M \to \mathbf{P}^{n+1}$ of rank ≥ 2 are projectively equivalent if and only if the equivalence classes $[(\nabla^1, h^1)]_p$ and $[(\nabla^2, h^2)]_p$ for F^1 and F^2 coincide.*

We finally remark on duality. The dual mapping is naturally defined as follows. Let $\mathbf{R}_{n+2}$ denote the dual vector space of $\mathbf{R}^{n+2}$ and η^* the radial vector field of $\mathbf{R}_{n+2}$. We define two mappings v and w from M into $\mathbf{R}_{n+2}$ as follows. To each point x, we let $v(x)$ and $w(x)$ be linear functions on $T_{f(x)}(\mathbf{R}^{n+2})$, naturally identified with $\mathbf{R}^{n+2}$, such that

$$v(x)(\xi_{f(x)}) = 1, \quad v(x)(\eta_{f(x)}) = 0, \text{ and } v(x)(f_*X) = 0 \text{ for all } X \in T_x(M);$$
$$w(x)(\xi_{f(x)}) = 0, \quad w(x)(\eta_{f(x)}) = 1, \text{ and } w(x)(f_*X) = 0 \text{ for all } X \in T_x(M).$$

Similarly to Proposition 5.1 of Chapter II, we have

$$
\begin{aligned}
(D_X v)(\xi) &= -\tau(X), & (D_X w)(\xi) &= -\rho(X), \\
(D_X v)(\eta) &= 0, & (D_X w)(\eta) &= 0, \\
(D_X v)(f_*Y) &= -h(X,Y), & (D_X w)(f_*Y) &= -T(X,Y).
\end{aligned}
$$

From this we can assert that the mapping v defines an immersion if h is nondegenerate and that the vector field η^* is transversal to the mapping v. The vector field w is also transversal to the mapping v. Since by definition the two vector fields v and w are linearly independent, the mapping v defines a centro-affine immersion of M. The pair $\{v, w\}$ is called the *dual mapping* of $\{f, \xi\}$. The fundamental equations for $\{v, w\}$ can be derived similarly to Proposition 5.4 of Chapter II; we shall omit them here. We remark, however, that, if $\tau = 0$, then the connection ∇ and the induced connection ∇^* relative to $\{v, w\}$ are conjugate to each other:

$$X(h(Y,Z)) = h(\nabla_X Y, Z) + h(Y, \nabla_X^* Z).$$

Now it is easy to define the dual immersion of an immersion into $\mathbf{P}^{n+1}$. Let F be an immersion of M into the projective space $\mathbf{P}^{n+1}$ and f a local lift of F: $[f] = F$. The dual mapping v is associated with a transversal vector field ξ. The dual mapping v' associated with another choice of vector field $\xi' = (\xi + a\eta + f_*U)/\lambda$ differs from v by $v' = \lambda v$. Hence, $[v] = [v']$ as mappings into the dual projective space $\mathbf{P}_{n+1}$. Let g be another choice of local lift of F; then $g = \phi f$ for a nonzero scalar function ϕ. In this case, the dual mapping v_g for $\{g, \xi\}$ is equal to v. So the dual immersion F^* of F can be defined by $F^* = [v]$. We see that $(F^*)^* = F$.

The last notion we introduce in this note is selfduality. We say that the pair $\{f, \xi\}$ is *affinely selfdual* if $f = Av$ for a linear isomorphism A of $\mathbf{R}_{n+2}$ with $\mathbf{R}^{n+2}$ and that the immersion $F = [f]$ is *selfdual* if $F = AF^*$ for a projective linear isomorphism A of $\mathbf{P}_{n+1}$ with $\mathbf{P}^{n+1}$. Then, an analogous theorem to Theorem 5.6 of Chapter II is the following:

Theorem N9.12. *The image of an affinely selfdual (nondegenerate) centro-affine immersion lies in a quadratic cone. The image of a selfdual (nondegenerate) projective immersion is part of a nondegenerate quadratic hypersurface.*

Remark N9.1. For more detail and proofs, refer to [NS5]. In Note 11, we shall again discuss nondegenerate immersions of hypersurfaces into $\mathbf{P}^{n+1}$ and how to make use of the results in this note.

10. Projective minimal surfaces in $\mathbf{P}^3$

In Theorem 8.4 of Chapter IV, we have defined the projective metric for immersed nondegenerate hypersurfaces in $\mathbf{P}^{n+1}$. This note gives some remarks on the notion of projective minimality associated to this metric.

Let $f : M^n \to \mathbf{P}^{n+1}$ be a nondegenerate immersion. We choose a Ricci-symmetric affine connection D from an arbitrary atlas belonging to the standard projective structure on $\mathbf{P}^{n+1}$. Relative to a fixed D-parallel volume form ω, let ξ be the affine normal field. Then, in Section 8 of Chapter IV, we have shown that the symmetric 2-form Jh is an absolute invariant; that is, it is independent of the choice of (D, ω). Hence the n-form $\pi = |J|^{\frac{n}{2}}\theta$ defines a pseudo-volume form on the immersed manifold M; it is a volume form in the usual sense if h is definite and J does not vanish. Thus the integral of π over M^n defines a volume functional associated to f. We say that the immersion f is *projective minimal* if this functional is critical relative to any variation of f. Note that this definition is valid if M^n is orientable, even when h is indefinite and J vanishes somewhere.

When $n = 2$, the situation is much simplified. Computation shows that the immersion is projective minimal if and only if the affine mean curvature satisfies the following differential equation:

$$(\text{N}10.1) \qquad \qquad \triangle H = \|S - HI\|_h^2.$$

Refer to [S4] for the deduction of this equation and for the case of general dimension. The following proposition is a direct consequence of this equation.

Proposition N10.1. *Any nondegenerate affine sphere is projective minimal.*

This fact was first claimed by H. Behnke, see [Bl, p. 248]. The equation (N10.1) gives also the following global result.

Theorem N10.2. *If the immersed surface is compact, strictly convex, and projective minimal, then it is an ellipsoid.*

The class of projective minimal surfaces has nice geometrical properties; see [S6] and references cited therein.

11. Projectively homogeneous surfaces in $\mathbf{P}^3$

This note supplements the treatment in Sections 7 and 8 of Chapter IV that are mainly concerned with the notion of projective immersions and applications to hypersurfaces in $\mathbf{P}^{n+1}$. We shall sketch how to associate invariants to a hypersurface immersed in $\mathbf{P}^{n+1}$ with nonvanishing Pick invariant. As one important application, we sketch the classification of projectively homogeneous surfaces in $\mathbf{P}^3$. To get a concrete classification list, the method of codimension-2 centro-affine immersions in Note 9 is useful.

Let $\tilde{M}$ be an $(n + 1)$-dimensional differentiable manifold with a flat projective structure P and f an immersion of a manifold M into $\tilde{M}$. Recall that a flat projective structure is given by an atlas of projectively flat local affine connections that are projectively equivalent in each common domain. We fix for the moment a torsion-free affine connection D defined on an open set U belonging to the structure P, and consider M immersed in U. Let ξ be an arbitrary field of vectors transversal to M. For any vector fields on M we write

(N11.1)
$$D_X Y = \nabla_X Y + h(X, Y)\xi,$$
$$D_X \xi = -SX + \tau(X)\xi,$$

thus defining as usual an affine connection ∇, a symmetric tensor h of type $(0, 2)$, a tensor S of type $(1, 1)$ called the shape operator, and a 1-form τ on M. Let R^D and R denote the curvature tensors of D and ∇, respectively, and let $\gamma^D(U, V) = \frac{1}{n}\mathrm{Ric}^D(U, V)$ and $\gamma(X, Y) = \frac{1}{n-1}\mathrm{Ric}(X, Y)$ denote the respective normalized Ricci tensors.

The fundamental equations (N1.10)–(N1.13) are now written as follows:

(N11.2) $R(X, Y)Z = \gamma^D(Y, Z)X$
$$- \gamma^D(X, Z)Y + h(Y, Z)SX - h(X, Z)SY,$$

(N11.3) $(\nabla_X h)(Y, Z) + \tau(X)h(Y, Z) = (\nabla_Y h)(X, Z) + \tau(Y)h(X, Z),$

(N11.4) $(\nabla_Y S)(X) - \tau(Y)SX - \gamma^D(Y, \xi)X$
$$= (\nabla_X S)(Y) - \tau(X)SY - \gamma^D(X, \xi)Y,$$

(N11.5) $h(SX, Y) + (\nabla_X \tau)Y = h(X, SY) + (\nabla_Y \tau)X.$

From the first identity we have

(N11.6) $\gamma(X, Y) = \gamma^D(X, Y) + \dfrac{1}{n-1}\{\mathrm{tr}\, S \cdot h(X, Y) - h(SX, Y)\}.$

We define the cubic form by setting

(N11.7) $C(X, Y, Z) = (\nabla_X h)(Y, Z) + \tau(X)h(Y, Z)$

and the *shape form* by

(N11.8) $\mathscr{S}(X, Y) = h(SX, Y) - \gamma^D(X, Y).$

Assuming that the fundamental form h is nondegenerate, we define a tensor K_X by

(N11.9) $h(K_X Y, Z) = -\dfrac{1}{2}C(X, Y, Z).$

Definitions and formulas above all depend on the choice of a transversal vector field ξ and a connection D. A convenient choice is described in the next proposition.

Proposition N11.1. *There exist a pair* (D, ξ) *and a volume form* ω *that satisfy the following conditions:*

(N11.10) $D\omega = 0, \qquad \tau = 0, \qquad \text{tr}\,(K_X) = 0, \qquad \text{tr}\,_h\mathscr{S} = 0.$

If $(\overline{D}, \overline{\omega}, \overline{\xi})$ *is another choice, then there exists a nonvanishing function* σ *such that this triple is related to* (D, ω, ξ) *by the equations*

(N11.11)
$$\overline{D}_X Y = D_X Y + \varphi(X)Y + \varphi(Y)X,$$
$$\overline{\omega} = \sigma\omega,$$
$$\overline{\xi} = \frac{1}{\lambda}(U + \xi),$$

where $\lambda = c\sigma^{2/(n+2)}$ *for a constant* c, φ *is a 1-form on* M *determined by* $\varphi = \frac{1}{n+2}d\log\sigma$, *the vector field* U *is determined by* $h(U, X) = \varphi(X)$, *and the value of* φ *at* ξ *is given by* $\varphi(\xi) = -\frac{1}{2}\varphi(U)$. *Furthermore, the invariants* C, $\mathscr{S}$, *and* γ^D *change as follows:*

(N11.12) $\overline{C}(X, Y, Z) = \lambda C(X, Y, Z),$
(N11.13) $\overline{\mathscr{S}}(X, Y) = \mathscr{S}(X, Y) + C(U, X, Y),$

(N11.14) $\lambda\overline{\gamma^D}(X, \overline{\xi}) = \gamma^D(X, \xi) - \mathscr{S}(U, X) - \frac{1}{2}C(U, U, X).$

To summarize, the set $\{C, \mathscr{S}, \gamma^D\}$ *together is covariant relative to the triple* (D, ω, ξ).

The Fubini–Pick invariant J is defined by

$$J = \frac{1}{n(n-1)}h(K, K)$$

and it changes as follows:

(N11.15) $\lambda\overline{J} = J.$

Hence, if $J \neq 0$, we can eliminate the ambiguity in the choice of function σ in the previous proposition; that is we have

Proposition N11.2. *Assume* $J \neq 0$ *and let* J_0 *be an arbitrary nonzero constant. Then there exists a triple* (D, ω, ξ) *with the properties in* (N11.10) *such that* J *is equal to* J_0. *Such a triple is uniquely determined.*

Remark N11.1. The conditions (N11.10) correspond to the Blaschke prenormalization (N9.16) in Note 9.

The last proposition enables us to get explicit absolute invariants; we exhibit this by considering a concrete problem of classifying projectively homogeneous surfaces in $\mathbf{P}^3$. To be precise, we recall the notion of projective homogeneity. A projectively homogeneous hypersurface M^* in $\mathbf{P}^{n+1}$ is a space in the form G/H, where G is a Lie group of projective transformations of $\mathbf{P}^{n+1}$ and H a closed subgroup. Let M be an open subset of G/H and x_0 an arbitrary point of M. Then there is an open neighborhood V of x_0 in M and a differentiable map $\phi : V \to G$ such that $\phi_x(x_0) = x$. We consider the map $\Phi : V \times V \to M^*$ given by $\Phi(x, y) = \phi_x(y) \in M^*$. Since $\Phi(x_0, x_0) = x_0$, there exists a neighborhood, say U, of x_0 such that $\Phi(U, U)$ is contained in V. This means that for any $x \in U$, ϕ_x maps U into V and takes x_0 to x.

Assume now $n = 2$. Having chosen such V and U, we proceed to choose an orthonormal basis $\{X_1, X_2\}$ that has the property

(N11.16)
$$h(X_i, X_j) = \epsilon_i \delta_{ij}, \quad \text{where} \quad \epsilon_1 = 1 \quad \text{and} \quad \epsilon_2 = \epsilon = \pm 1,$$
$$C_{111} = -2, \quad C_{122} = 2\epsilon, \quad C_{112} = C_{222} = 0,$$

where $C_{ijk} = C(X_i, X_j, X_k)$. The existence of such a basis is shown by the discussion of the cubic form in Section 11 of Chapter II. An important fact is that the choice of such $\{X_1, X_2\}$ is finite because of Proposition N11.2. For each $x \in U$ the projective transformation ϕ_x leaves the triple (D, ω, ξ) and hence h, C, and S invariant. Thus the pull-backs by ϕ_x of $\{X_1, X_2\}$ around x differ from $\{X_1, X_2\}$ around x_0 up to finite possibilities. Since ϕ_x and its differential depend continuously on x, we may assume, by taking U smaller if necessary, that the pull-backs coincide with $\{X_1, X_2\}$. Since ∇ is invariant under ϕ_x, we see that the components of ∇ and S relative to the basis $\{X_1, X_2\}$ are constants. The possibility of such constants can be determined by the integrability conditions (N11.2)–(N11.5). To obtain an analytic expression for each surface corresponding to each allowable set of constants, we pass from $\mathbf{P}^3$ to $\mathbf{R}^4$ by the method described in Note 9.

Instead of giving detail in the determination of allowable constants, we cite one typical example of a homogeneous surface called the Enriques surface. In the inhomogeneous coordinates (x, y, z) of $\mathbf{P}^3$, this surface is defined by the equation

[1]
$$(y - x^2)^3 = k(z + \frac{2}{3}x^3 - xy)^2,$$

where k is a constant. It is homogeneous under the group of matrices of the form

$$\begin{bmatrix} 1 & u & u^2 & \frac{1}{3}u^3 \\ & v & 2uv & u^2v \\ & & v^2 & uv^2 \\ & & & v^3 \end{bmatrix}, \quad v \neq 0,$$

which acts from the right on the homogeneous coordinate vector $[1, x, y, z]$. Note that the surface is degenerate when $k = 0$, and $J = 0$ for $k = -9/4$. We can regard the surface when $k = \infty$ as a Cayley surface (see Section 6 of Chapter III).

Remark N11.2. The remaining homogeneous surfaces that are nondegenerate with nonvanishing Fubini–Pick invariant are listed below:

[2] $z = y^2 + \epsilon x^k$;

[3] $z = y^2 + \epsilon e^x$;

[4] $z = y^2 + \epsilon x \log x$;

[5] $y^2 + \epsilon z^2 = x^{1-2/k}$ where $k > 2$;

[6] $y^2 - \epsilon z^2 = x^{1-2/k}$ where $0 < k < 2$;

[7] $x = \exp(\epsilon y^2 - z)$;

[8] $y^2 - \epsilon z^2 = e^x$;

[9] $(y^2 - \epsilon z^2) = (x^2 + 1)\exp(-\dfrac{4}{k}\arctan x)$;

[10] $z = x^k y^\ell$;

[11] $z = \log x + k \log y$;

[12] $z = (x^2 + y^2)^{k/2}\exp\left(\ell \arctan \dfrac{x}{y}\right)$;

[13] $z = \dfrac{1}{2}\log(x^2 + y^2) + \ell \arctan \dfrac{x}{y}$;

[14] $\arctan z = \dfrac{k}{2}\log \dfrac{x^2 + y^2}{1 + z^2} + \ell \arctan \dfrac{x}{y}$.

(k and ℓ are constant parameters and $\epsilon = \pm 1$.) Refer to [NS3] for this classification. The classification of projectively homogeneous surfaces when $J = 0$ needs a different consideration.

APPENDICES

1. Torsion, Ricci tensor, and projective invariants

In the text, we mainly consider torsion-free affine connections and often assume that the Ricci tensor is symmetric (that is, the connection is locally equiaffine). In this appendix, we briefly indicate how we can handle more general cases, particularly, when we discuss pregeodesics and projective equivalence.

1. Let ∇ be an affine connection on a manifold M^n. The torsion tensor is, of course, defined by

$$(A1.1) \qquad T(X, Y) = \nabla_X Y - \nabla_Y X - [X, Y].$$

In index notation we have

$$(A1.1a) \qquad T_{ij}^k = \Gamma_{ij}^k - \Gamma_{ji}^k,$$

where Γ_{ij}^k are the Christoffel symbols given by

$$(A1.2) \qquad \nabla_{\frac{\partial}{\partial x^i}} \frac{\partial}{\partial x^j} = \sum_{k=1}^n \Gamma_{ij}^k \frac{\partial}{\partial x^k}.$$

When the torsion is not 0, we can define a new connection $\hat{\nabla}$ with zero torsion and with the same pregeodesics as ∇ by setting

$$(A1.3) \qquad \hat{\nabla}_X Y = \nabla_X Y - \frac{1}{2} T(X, Y).$$

In index notation we have

$$(A1.3a) \qquad \hat{\Gamma}_{ij}^k = \frac{1}{2}(\Gamma_{ij}^k + \Gamma_{ji}^k).$$

2. Two torsion-free affine connections ∇ and $\bar{\nabla}$ have the same pregeodesics if they are projectively equivalent, that is, if there is a 1-form ρ such that

(A1.4) $$\bar{\nabla}_X Y = \nabla_X Y + \rho(X)Y + \rho(Y)X.$$

(In Definition 3.3 of Chapter I, we considered only locally equiaffine connections.) We shall indicate the proof for the converse, namely,

Proposition A1.1. *If two torsion-free affine connections have the same pregeodesics, they are projectively equivalent.*

Proof. Let $K(X,Y) = \bar{\nabla}_X Y - \nabla_X Y$. From the assumption of zero torsion, we see that $K(X,Y) = K(Y,X)$. We shall show that there is a 1-form ρ such that $K(X,X) = 2\rho(X)X$. Then by polarization we have $K(X,Y) = \rho(X)Y + \rho(Y)X$.

We first show that for each $x \in M$ there exists a function ρ on $T_x(M)$ such that $K(X,X) = 2\rho(X)X$ for each $X \in T_x(M)$. Let $X \neq 0 \in T_x(M)$ and let x_t be a geodesic with affine parameter for ∇ such that $x_0 = x$ and $\dot{x}_0 = X$. By assumption, the curve x_t is also a pregeodesic for $\bar{\nabla}$ so that $\bar{\nabla}_t \dot{x}_t = \phi(t)\dot{x}_t$. Thus $K(\dot{x}_t, \dot{x}_t) = \bar{\nabla}_t \dot{x}_t - \nabla_t \dot{x}_t = \phi(t)\dot{x}_t$. Evaluating this at $t = 0$, we obtain $K(X,X) = 2\rho X$, where ρ is a scalar that depends on X. We define ρ to be 0 on the zero tangent vector. We now prove the following (see Tanaka [Ta]).

Lemma A1.2. *Let $K : V \times V \to V$ be a symmetric bilinear function, where V is an n-dimensional real vector space. Suppose ρ is a function on V such that $K(X,X) = 2\rho(X)X$ for all $X \in V$, then ρ is a linear function.*

Proof. From $2t^2\rho(X)X = t^2 K(X,X) = K(tX,tX) = 2t\rho(tX)X$ for any $t \neq 0$, we get $\rho(tX) = t\rho(X)$. Next we show that $\rho(X+Y) = \rho(X) + \rho(Y)$ for all $X,Y \in T_x(M)$. This is obvious if X and Y are linearly dependent from what we have done. Assume that they are linearly independent. We have

$$K(X+Y, X+Y) = 2\rho(X+Y)(X+Y).$$

The left-hand side is equal to $K(X,X) + 2K(X,Y) + K(Y,Y) = 2\rho(X)X + 2\rho(Y)Y + 2K(X,Y)$. It follows that $K(X,Y)$ is a linear combination of X and Y. Let $K(X,Y) = aX + bY$ and write

$$K(sX+tY, sX+tY) = 2\rho(sX+tY)(sX+tY).$$

The left-hand side is equal to $s^2 K(X,X) + 2st K(X,Y) + t^2 K(Y,Y) = 2s^2\rho(X)X + 2st K(X,Y) + 2t^2\rho(Y)Y$, where the middle term is equal to $2st(aX+bY)$.

Comparing this with the right-hand side and using linear independence, we obtain

$$\rho(sX+tY)s = s^2\rho(X) + ast,$$

$$\rho(sX+tY)t = t^2\rho(Y) + bst.$$

Multiply the first equation by t and the second by s. We obtain

$$\rho(X)s^2t + ast^2 = \rho(Y)t^2s + bs^2t,$$

for all s, t. Hence $\rho(X) = b, \rho(Y) = a$, that is, $K(X, Y) = \rho(Y)X + \rho(X)Y$. Clearly, we get $\rho(X + Y) = \rho(X) + \rho(Y)$. This proves the lemma.

Finally, to complete the proof of Proposition A1.1, we show that the 1-form ρ is differentiable as follows. From $K(X, Y) = \rho(X)Y + \rho(Y)X$, we get trace $\{X \mapsto K(X, Y)\} = n\rho(Y) + \rho(Y) = (n + 1)\rho(Y)$.

3. If two affine connections ∇ and ∇' have the same pregeodesics, then the torsion-free connections $\hat{\nabla}$ and $\hat{\nabla}'$ obtained from ∇ and ∇', respectively, in the manner of (A1.3) have the same pregeodesics. Thus they are projectively equivalent, and thus there is a 1-form such that

$$\nabla'_X Y - \frac{1}{2}T'(X, Y) = \nabla_X Y - \frac{1}{2}T(X, Y) + \rho(X)Y + \rho(Y)X.$$

Conversely, if this relation holds, then ∇ and ∇' have the same pregeodesics.

4. Suppose that two torsion-free affine connections ∇ and $\bar{\nabla}$ are related by (A1.4). For any volume element ω, let τ and $\bar{\tau}$ be two 1-forms such that

(A1.5) $$\nabla_X \omega = \tau(X)\omega \quad \text{and} \quad \bar{\nabla}_X \omega = \bar{\tau}(X)\omega,$$

as in the proof of Proposition 3.1 of Chapter I, where we saw that ∇ (resp.$\bar{\nabla}$) has symmetric Ricci tensor if and only if $d\tau = 0$ (resp. $d\bar{\tau} = 0$). Now we can easily compute that

$$\tau - \bar{\tau} = (n + 1)\rho, \quad \text{hence} \quad d\tau - d\bar{\tau} = (n + 1)d\rho.$$

Suppose ∇ has symmetric Ricci tensor. Then $\bar{\nabla}$ has symmetric Ricci tensor if and only if $d\rho = 0$. Or, to put it differently, assume that $d\rho = 0$. Then ∇ has symmetric Ricci tensor if and only if $\bar{\nabla}$ does.

Another application of (A1.5) is the following. If ∇ is a torsion-free affine connection with nonsymmetric Ricci tensor, we can make a projective change to an affine connection with symmetric Ricci tensor. Take any volume element ω and define τ by $\nabla_X \omega = \tau(X)\omega$. Let $\rho = \frac{1}{n+1}\tau$ and let $\bar{\nabla}_X Y = \nabla_X Y + \rho(X) + \rho(Y)$. Then $\bar{\tau} = \tau - (n + 1)\rho = 0$. Thus $\bar{\nabla}\omega = 0$ and $\bar{\nabla}$ has symmetric Ricci tensor. We note that

(A1.6) $$d\rho(Y, Z) = \frac{1}{n + 1} \text{tr} R(Z, Y).$$

5. As long as we stay in the category of torsion-free affine connections with symmetric Ricci tensors, we consider only projective changes (A1.4) with

$d\rho = 0$. As in Section 3 of Chapter I, the Weyl projective curvature tensor W is defined by

(A1.7) $W(X,Y)Z = R(X,Y)Z - [\gamma(Y,Z)X - \gamma(X,Z)Y],$

where γ is the normalized Ricci tensor $\frac{1}{n-1}\mathrm{Ric}$. The invariance of the Weyl projective tensor by a projective change (with $d\rho = 0$) can be verified by using

(A1.8) $\bar{R}(X,Y)Z = R(X,Y)Z + [(\nabla_X\rho)(Z) - \rho(X)\rho(Z)]Y$
$\qquad\qquad\qquad\qquad - [(\nabla_Y\rho)(Z) - \rho(Y)\rho(Z)]X$

and

(A1.9) $\bar{\gamma}(Y,Z) = \gamma(Y,Z) + [\rho(Y)\rho(Z) - (\nabla_Y\rho)Z].$

For dim $M \geq 3$, the vanishing of W is a necessary and sufficient condition for ∇ to be projectively flat. Note that $W = 0$ implies

(A1.10) $(\nabla_X\gamma)(Y,Z) = (\nabla_Y\gamma)(X,Z).$

For $n = 2$, W is always 0, and (A1.10) is a necessary and sufficient condition for ∇ to be projectively flat.

When $W = 0$ and (A1.10) hold, we prove that ∇ is projectively flat in the following way. It is sufficient to show that there is a local 1-form ρ satisfying

(A1.11) $(\nabla_Y\rho)(Z) = \gamma(Y,Z) + \rho(Y)\rho(Z).$

We see that, for dim $M \geq 3$, (A1.11) and (A1.8) imply $\bar{R} = W = 0$. The same conclusion holds for dim$M = 2$. Finally, we note that the complete integrability condition for solving the system of partial differential equations (A1.11) is $W = 0$ and (A1.10); the proof is omitted.

6. Now let ∇ be a torsion-free affine connection on M^n whose Ricci tensor is not symmetric. In this case, how should one define the projective curvature tensor? By the second part of 4, we can find a 1-form ρ such that $\bar{\nabla}_X Y = \nabla_X Y + \rho(X)Y + \rho(Y)X$ has symmetric Ricci tensor. By computation we get in place of (A1.8) and (A1.9)

(A1.12) $\bar{R}(X,Y)Z = R(X,Y)Z + (d\rho)(X,Y)Z$
$\qquad\qquad + [(\nabla_X\rho)(Z) - \rho(X)\rho(Z)]Y - [(\nabla_Y\rho)(Z) - \rho(Y)\rho(Z)]X$

and

(A1.13) $\bar{\gamma}(Y,Z) = \gamma(Y,Z) + \dfrac{1}{n-1}d\rho(Z,Y) + [\rho(Y)\rho(Z) - (\nabla_Y\rho)Z].$

The Weyl projective curvature tensor, say $\bar{W}$, for the Ricci-symmetric connection $\bar{\nabla}$ can be computed by using (A1.12),(A1.13) and (A1.6). We have

(A1.14) $\bar{W}(X,Y)Z = R(X,Y)Z - \dfrac{1}{n+1}\operatorname{tr} R(X,Y)Z$

$- [\gamma(Y,Z)X - \gamma(X,Z)Y] - \dfrac{1}{n^2-1}[\operatorname{tr} R(Y,Z)X - \operatorname{tr} R(X,Z)Y].$

We define the Weyl projective curvature tensor of the original connection ∇ with nonsymmetric Ricci tensor to be $\bar{W}$ given by (A1.14).

To give it another expression, we introduce a tensor P by

(A1.15) $P(X,Y) = \gamma(X,Y) + \dfrac{1}{n^2-1}\operatorname{tr} R(X,Y)$

$= \dfrac{1}{n^2-1}[n\operatorname{Ric}(X,Y) + \operatorname{Ric}(Y,X)],$

the second equality holding by virtue of $\operatorname{Ric}(Y,X) = \operatorname{Ric}(X,Y) + \operatorname{tr} R(X,Y)$. It is worth noting that

(A1.16) $P(X,Y) - P(Y,X) = \dfrac{1}{n+1}[\operatorname{Ric}(Y,X) - \operatorname{Ric}(X,Y)]$

$= \dfrac{1}{n+1}\operatorname{tr} R(X,Y).$

If Ric were symmetric, then P would be just the normalized Ricci tensor γ. The tensor $\bar{W}$ for ∇ in (A1.14) can be written in the form

(A1.17) $\bar{W}(X,Y)Z = R(X,Y)Z - [P(X,Y) - P(Y,X)]Z$
$- [P(Y,Z)X - P(X,Z)Y].$

From the way we derived this projective curvature tensor, it is obviously invariant under any projective change of ∇. For $n \geq 3$, the vanishing of $\bar{W}$ is a necessary and sufficient condition for ∇ to be projectively flat. For $n = 2$, we have always $\bar{W} = 0$, and

(A1.18) $(\nabla_X P)(Y,Z) = (\nabla_Y P)(X,Z)$

is a necessary and sufficient condition for the projective flatness of ∇. Indeed, (A1.18) is the complete integrability for the equation

(A1.19) $(\nabla_Y \rho)(Z) - \rho(Y)\rho(Z) = P(Y,Z).$

If we have a 1-form ρ, then from (A1.13) and (A1.15) we obtain $\bar{\gamma} = 0$ and hence $\bar{R} = 0$. Conversely, if $\bar{R} = 0$ and thus $\bar{\gamma} = 0$, then (A1.19) has a solution ρ.

For the material in this Appendix, see Eisenhart [Ei].

2. Metric, volume, divergence, Laplacian

This appendix is supplementary to Sections 3 and 4 of Chapter I, Sections 3, 6 and 7 of Chapter II, and Section 10 of Chapter III.

 Let M be an oriented manifold with a nondegenerate metric g. Relative to a coordinate system $\{x^1, \ldots, x^n\}$ we express the components of g by

$$(A2.1) \qquad g_{ij} = g(\partial_i, \partial_j), \quad \text{where} \quad \partial_i = \frac{\partial}{\partial x^i},$$

and define the volume element $\omega = \omega_g$ by

$$(A2.2) \qquad \omega(\partial_1, \ldots, \partial_n) = \sqrt{G}, \quad \text{where} \quad G = |\det [g_{ij}]|,$$

when $\{x^1, \ldots, x^n\}$ is positively oriented. The formula defines a globally defined volume element. If we denote by ∇ the Levi-Civita connection for the metric g, then

$$(A2.3) \qquad \nabla \omega = 0.$$

For any vector field Z on M, the divergence of Z is defined by

$$(A2.4) \qquad \operatorname{div} Z = \operatorname{trace} \{X \mapsto \nabla_X Z\}.$$

In terms of a positively oriented coordinate system $\{x^1, \ldots, x^n\}$, we have

$$(A2.5) \qquad \operatorname{div}(\partial_k) = \sum_{i=1}^{n} \Gamma^i_{ik},$$

where Γ^i_{jk} are the Christoffel symbols for the connection ∇: $\nabla_{\partial_j} \partial_k = \sum_i \Gamma^i_{jk} \partial_i$.

 For any vector field ∂_k we have

$$(A2.6) \qquad \partial_k \, \omega(\partial_1, \ldots, \partial_n) = \operatorname{div}(\partial_k) \, \omega(\partial_1, \ldots, \partial_n).$$

This follows from (A2.3) and (A2.4):

$$\partial_k \, \omega(\partial_1, \ldots, \partial_n) = (\nabla_{\partial_k} \omega)(\partial_1, \ldots, \partial_n) + \sum_i \omega(\partial_1, \ldots, \nabla_{\partial_k} \partial_i, \ldots, \partial_n)$$

$$= \sum_i \omega(\partial_1, \ldots, \nabla_{\partial_i} \partial_k, \ldots, \partial_n)$$

$$= \operatorname{div}(\partial_k) \, \omega(\partial_1, \ldots, \partial_n).$$

Combining (A2.6) with (A2.2) we obtain

$$(A2.7) \qquad \operatorname{div}(\partial_k) = \frac{\partial \log \sqrt{G}}{\partial x^k}.$$

Using $\nabla_X(\phi Z) = (X\phi)Z + \phi\nabla_X Z$ we can derive

$$(A2.8) \qquad \operatorname{div}(\phi Z) = (Z\phi) + \phi \operatorname{div} Z.$$

Now we can prove the following formula:

$$(A2.9) \qquad \operatorname{div} Z = \frac{1}{\sqrt{G}} \sum_k \frac{\partial \sqrt{G} Z^k}{\partial x^k},$$

where $Z = \sum_k Z^k \partial_k$. To prove this, we apply (A2.8) and (A2.7) and get

$$\begin{aligned}
\operatorname{div} Z &= \sum_k \operatorname{div}(Z^k \partial_k) \\
&= \sum_k \left[\frac{\partial Z^k}{\partial x^k} + Z^k(\operatorname{div} \partial_k) \right] \\
&= \sum_k \left[\frac{\partial Z^k}{\partial x^k} + Z^k \frac{\partial \log \sqrt{G}}{\partial x^k} \right],
\end{aligned}$$

which is easily seen to be equal to the right-hand side of (A2.9).

Finally, we have the following formula for the Laplacian Δ relative to g applied to a function f

$$(A2.10) \qquad \Delta f = \frac{1}{\sqrt{G}} \sum_k \frac{\partial}{\partial x^k} \left(\sqrt{G} \sum_j g^{kj} f_j \right).$$

To prove this, recall the definition $\Delta f = \operatorname{div} \operatorname{grad} f$. If $Z = \operatorname{grad} f$, then $Z^k = \sum_j g^{kj} f_j$. Substituting this in (A2.9) we obtain (A2.10).

We mention the special case $n = 2$. First, suppose the given metric g on a 2-dimensional differentiable manifold is positive-definite. Let $\{x^1, x^2\}$ be an isothermal coordinate system, that is,

$$g_{11} = g_{22} = E \; (> 0), \quad g_{12} = g_{21} = 0.$$

Then

$$g^{11} = g^{22} = \frac{1}{E}, \quad g^{12} = g^{21} = 0.$$

From (A2.10) we obtain

(A2.11) $$\Delta f = \frac{1}{E}(f_{11} + f_{22}).$$

Now assume g is indefinite. Let $\{x^1, x^2\}$ be an asymptotic coordinate system, that is,

$$g_{11} = g_{22} = 0, \quad g_{12} = g_{21} = F \ (> 0).$$

Then

$$g^{11} = g^{22} = 0, \quad g^{12} = g^{21} = \frac{1}{F}.$$

From (A2.10) we get

(A2.12) $$\Delta f = \frac{2}{F} f_{12}.$$

3. Change of immersions and transversal vector fields

We fix a parallel volume element in $\mathbf{R}^{n+1}$ given by Det. Let $f : M^n \to \mathbf{R}^{n+1}$ be an immersion of an n-manifold into $\mathbf{R}^{n+1}$ provided with an equiaffine transversal vector field ξ. We denote by (∇, θ, h, S) the induced connection, the induced volume element, the affine fundamental form, and the affine shape operator. We shall obtain formulas that show how the induced structure changes when we change

(A3.1) f to $\tilde{f} = af$ and ξ to $\tilde{\xi} = b\xi,$

where a, b are arbitrary nonzero real numbers. For each $x \in M^n$, we consider $\tilde{\xi}_x = b\xi_x$ as transversal vector at the point $\tilde{f}(x) = af(x)$.
From $D_X \xi = f_*(-SX)$, it follows that

$$D_X \tilde{\xi} = b D_X \xi = b f_*(-SX) = f_*(-bSX) = \tilde{f}_*(-\frac{b}{a}SX) = -\tilde{f}_*(\tilde{S}X).$$

This implies that $\tilde{\xi}$ is an equiaffine transversal vector field and we have

(A3.2) $$\tilde{S} = \frac{b}{a}S.$$

From

$$D_X \tilde{f}_*(Y) = a D_X f_* \cdot Y = a f_*(\nabla_X Y) + \frac{a}{b} h(X, Y) b\xi$$
$$= \tilde{f}_*(\tilde{\nabla}_X Y) + \tilde{h}(X, Y)\tilde{\xi}$$
$$= a f_*(\tilde{\nabla}_X Y) + b\tilde{h}(X, Y)\xi,$$

we get

(A3.3)
$$\tilde{\nabla} = \nabla,$$

(A3.4)
$$\tilde{h} = \frac{a}{b}h.$$

For the induced volume elements we get

$$\tilde{\theta}(X_1,\ldots,X_n) = \mathrm{Det}\,(\tilde{f}_*X_1,\ldots,\tilde{f}_*X_n,\xi) = a^n b\,\mathrm{Det}\,(f_*X_1,\ldots,f_*X_n,\xi)$$
$$= a^n b\,\theta(X_1,\ldots,X_n),$$

thus

(A3.5)
$$\tilde{\theta} = a^n\,b\theta.$$

If f is nondegenerate, so is $\tilde{f}$. In this case, the metric volume elements ω, $\tilde{\omega}$ for h, $\tilde{h}$ are related by

$$\tilde{\omega}(X_1,\ldots,X_n) = |\det\,[\tilde{h}(X_i,X_j)]|^{\frac{1}{2}} = \left(\frac{a}{b}\right)^{\frac{n}{2}} \omega(X_1,\ldots,X_n)$$

and thus

(A3.6)
$$\tilde{\omega} = \left(\frac{a}{b}\right)^{\frac{n}{2}} \omega.$$

In the nondegenerate case, suppose ξ is the affine normal field for f so that $\theta = \omega$. For $\tilde{\xi}$ to be the affine normal field for $\tilde{f}$, up to sign, it is necessary and sufficient to have $\tilde{\theta} = \pm\tilde{\omega}$. From (A3.4) and (A3.5) we see that this is the case if and only if

(A3.7)
$$|a^n b^{n+2}| = 1.$$

4. Blaschke immersions into a general ambient manifold

In Chapter II we have developed the theory of Blaschke immersions into $\mathbf{R}^{n+1}$. Here we shall indicate how this idea goes over to the case of an immersion $f : M^n \to \tilde{M}^{n+1}$, where $\tilde{M}^{n+1}$ is equipped with an equiaffine structure $(\tilde{\nabla}, \tilde{\omega})$ (see Definition 3.2 of Chapter I).

For the given immersion $f : M^n \to \tilde{M}^{n+1}$ we choose locally an arbitrary transversal field ξ. It is easy to follow the development in Sections 1, 2, and 3 of Chapter II. More specifically, by Propositions 1.1 and 1.2 we have the induced connection ∇, the affine fundamental form h, the shape operator S and the transversal connection form τ. We can define the induced volume element θ by

$$\theta(X_1,\ldots,X_n) = \tilde{\omega}(f_*(X_1),\ldots,f_*(X_n),\xi),$$

and get Proposition 1.4. Of course, we cannot expect to have the same formulas as those in Theorem 2.1 and we do not worry about it now. As in the classical case, we say that ξ is *equiaffine* if $\tau = 0$. We find that the conformal class of h is independent of the choice of ξ. When it is of rank n everywhere, we say that f is *nondegenerate*.

For a nondegenerate immersion $f : M^n \to \tilde{M}^{n+1}$, we have Theorem 3.1 just in the same way as for the case of $\mathbf{R}^{n+1}$. The procedure for finding the affine normal also holds (simply replacing det by $\tilde{\omega}$). Once the affine normal field is obtained, the natural equiaffine structure (∇, θ) is determined on M^n together with the affine metric h whose volume element coincides with θ. This general approach to the equiaffine structure on a nondegenerate hypersurface was presented in the Münster Conference in 1982 and appeared in print in [N3]. It was also suggested in [Ba].

Now in this general theory, what happens to the geometry of the hypersurface M^n if the volume element $\tilde{\omega}$ on $\tilde{M}^{n+1}$ is changed to another volume element $\bar{\omega}$ parallel relative to $\tilde{\nabla}$? In this case, we have $\bar{\omega} = c^{n+1}\tilde{\omega}$ with a certain positive constant c. To find the affine normal field relative to $\bar{\omega}$, we take $\hat{\xi} = \xi/c^{n+1}$ as a tentative transversal field. Then we have

$$\hat{\theta}(X_1, \ldots, X_n) = \bar{\omega}(f_* X_1, \ldots, f_* X_n, \hat{\xi})$$
$$= c^{n+1}\tilde{\omega}(f_* X_1, \ldots, f_* X_n, c^{-(n+1)}\xi)$$
$$= \tilde{\omega}(f_* X_1, \ldots, f_* X_n, \xi)$$
$$= \theta(X_1, \ldots, X_n),$$

that is, $\hat{\theta} = \theta$ is the tentative volume element on M. It also follows from Proposition 2.5 that the tentative affine fundamental form is $\hat{h} = c^{n+1}h$.

Let $\{X_1, \ldots, X_n\}$ be a unimodular basis in $T_x(M) : \theta(X_1, \ldots, X_n) = 1$. Then we have $|\det [h(X_i, X_j)]| = 1$. On the other hand, we get

$$|\det [\hat{h}(X_i, X_j)]| = |\det [c^{n+1}h(X_i, X_j)]|$$
$$= c^{(n+1)n}|\det [h(X_i, X_j)]| = c^{(n+1)n}.$$

It follows that the affine normal field $\bar{\xi}$ for $f : M \to \tilde{M}^{n+1}$ relative to the volume element $\bar{\omega}$ in $\tilde{M}^{n+1}$ is

$$[c^{(n+1)n}]^{1/(n+2)}\hat{\xi} = c^{(n+1)n/(n+2)}c^{-(n+1)}\xi = c^{-2(n+1)/(n+2)}\xi.$$

Hence $\bar{\xi}$ is just a constant scalar multiple of ξ. It follows that the new induced connection $\bar{\nabla}$ is the same as the old one ∇. The new volume element $\bar{\theta}$ on M is

$$\bar{\theta}(X_1, \ldots, X_n) = \bar{\omega}(f_* X_1, \ldots, f_* X_n, c^{-2(n+1)/(n+2)}\xi)$$
$$= c^{n+1}c^{-2(n+1)/(n+2)}\tilde{\omega}(f_* X_1, \ldots, f_* X_n, \xi)$$
$$= c^{n(n+1)/(n+2)}\theta(X_1, \ldots, X_n),$$

that is, $\bar{\theta} = c^{n(n+1)/(n+2)}\theta$.

We can further determine that the new affine metric $\bar{h}$ and the new affine shape operator $\bar{S}$ are given by

$$\bar{h} = c^{2(n+1)/(n+2)}h \quad \text{and} \quad \bar{S} = c^{-2(n+1)/(n+2)}S.$$

To sum up, we have

Proposition A4.1. *If we change the fixed parallel volume element $\tilde{\omega}$ in $\tilde{M}^{n+1}$ to $\bar{\omega} = c^{n+1}\tilde{\omega}$, then the affine normal field and the Blaschke structure change as follows:*

$$\bar{\xi} = c^{-2(n+1)/(n+2)}\xi, \quad \bar{\theta} = c^{n(n+1)/(n+2)}\theta,$$
$$\bar{h} = c^{2(n+1)/(n+2)}h, \quad \bar{S} = c^{-2(n+1)/(n+2)}S.$$

In particular, the induced connection ∇ remains the same, that is, they are uniquely determined by the ambient equiaffine connection $\tilde{\nabla}$; the same can be said about the property that S is a scalar multiple of the identity. We may also say that they remain invariant under any automorphism of $(\tilde{M}^{n+1}, \tilde{\nabla})$.

In the special case where $\tilde{M}^{n+1} = \mathbf{R}^{n+1}$ with the usual flat connection $\tilde{\nabla}$, any nondegenerate hypersurface M^n immersed in $\mathbf{R}^{n+1}$ has the natural equiaffine connection ∇ independently of the choice of a $\tilde{\nabla}$-invariant volume element (that is, a determinant function) on $\mathbf{R}^{n+1}$. The affine normals (see Definition 3.1 of Chapter II) are also well-determined. It seems that Cartan preferred this point of view; see [Car1].

Bibliography

[Ab] K. Abe, Affine differential geometry of complex hypersurfaces, *Geometry and Topology of Submanifolds, III*, ed. by L. Verstraelen and A. West, World Scientific, Singapore, 1991, 1–31.

[Am1] S. Amari, *Differential geometric methods in statistics*, Lecture Notes in Statistics 28, Springer, New York, 1985.

[Am2] S. Amari, Differential geometry of a parametric family of invertible linear systems – Riemannian metric, dual affine connections, and divergence, *Math. Systems Theory*, 20(1987), 53–82.

[Ba] W. Barthel, Zur Affingeometrie auf Mannigfaltigkeiten, *Jahresber. Deutsch. Math.-Verein.*, 68(1966), 13–44.

[BF] C. Blanc and F. Fiala, Le type d'une surface et sa courbure totale, *Comm. Math. Helv.*, 14(1941/42), 230–233.

[Bl] W. Blaschke, *Vorlesungen über Differentialgeometrie II, Affine Differentialgeometrie*, Springer, Berlin, 1923.

[BM] C. Burstin and W. Mayer, Die Geometrie zweifach ausgedehnter Mannigfaltigkeiten F_2 im affinen Raum $\mathbf{R}_4$, *Math. Z.*, 27(1927), 373–407.

[BNS] N. Bokan, K. Nomizu, and U. Simon, Affine hypersurfaces with parallel cubic forms, *Tôhoku Math. J.*, 42(1990), 101–108.

[BS] W. Burau and U. Simon, Blaschke Beitrage zur affinen Differentialgeometrie, *Ges. Werke W. Blaschke*, Thales, Essen, 1985, Band 4, 11–34.

[Bu] S. G. Buyske, An algebraic representation of the affine Bäcklund transformation, *Geom. Dedicata*, 44(1992), 7–16.

[Cal1] E. Calabi, Improper affine hypersurfaces of convex type and a generalization of a theorem of Jörgens, *Michigan Math. J.*, 5(1958), 105–126.

[Cal2] E. Calabi, Complete affine hyperspheres I, *Sympos. Math.*, 10(1977), 19–38.

[Cal3] E. Calabi, Géométrie différentielle affine des hypersurfaces, *Séminaire Bourbaki* 1980/81, No. 573.

[Cal4] E. Calabi, Hypersurfaces with maximal affinely invariant area, *Amer. J. Math.*, 104(1982), 91–126.

[Cal5] E. Calabi, Convex affine maximal surfaces, *Results in Math.*, 13(1988), 199–223.

[Cal6] E. Calabi, Affine differential geometry and holomorphic curves, *Global Differential Geometry and Global Analysis,* Proceedings Berlin, 1990, Lecture Notes in Math. 1422, Springer, Berlin (1990), 15–21.

[Car1] E. Cartan, Sur la connexion affine des surfaces, *C. R. Acad. Sci. Paris*, 178(1924), 292–295.

[Car2] E. Cartan, Sur la connexion affine des surfaces développables, *C. R. Acad. Sci. Paris*, 178(1924), 449–451.

[Ce] T.E. Cecil, Focal points and support functions in affine differential geometry, to appear in Geom. Dedicata.

[CH] S.Cohn-Vossen and D. Hilbert, *Geometry and the Imagination,* Chelsea Publ. Co., New York 1952; *Anschauliche Geometrie,* Springer, Berlin, 1932.

[Ch1] S.S. Chern, Curves and surfaces in Euclidean space, *Studies in Global Geometry and Analysis, Studies in Math.*, ed. by MAA, 4(1967), 16–56.

[Ch2] S.S. Chern, The mathematical works of Wilhelm Blaschke, *Abh. Math. Sem.*, Univ. Hamburg 39(1973), 1–9.

[Ch3] S.S. Chern, Affine minimal hypersurfaces, Proc. *US–Japan Seminar on Minimal Submanifolds and Geodesics*, Kaigai, Tokyo, 1978, 17–30.

[ChT] S.S. Chern and C.L. Terng, An analogue of Bäcklund's theorem in affine geometry, *Rocky Mountain J. Math.*, 10(1980), 105–124.

[CR] T.E. Cecil and P.J. Ryan, Tight and taut immersions of manifolds, Pitman Publishing Limited, London, 1985.

[CY1] S.Y. Cheng and S.T. Yau, Complete affine hypersurfaces, Part I. The completeness of affine metrics, *Comm. Pure Appl. Math.*, 39(1986), 839–866.

[CY2] S.Y. Cheng and S.T. Yau, Differential equations on Riemannian manifolds and their geometric applications, *Comm. Pure Appl. Math.*, 28(1975), 333–354.

[De] A. Deicke, Über die Finsler-Räume mit $A_i = 0$, *Arch. Math.*, 4(1953), 45–51.

[Di1] F. Dillen, The complex version of a theorem by Berwald, *Soochow J. Math.*, 14(1988), 41–50.

[Di2] F. Dillen, Locally symmetric complex affine hypersurfaces, *J. Geom.*, 33(1988), 27–38.

[Di3] F. Dillen, Equivalence theorems in affine differential geometry, *Geom. Dedicata*, 32(1989), 81–92.

[Di4] F. Dillen, The classification of hypersurfaces of a real space form with parallel higher order fundamental form, *Soochow J. Math.*, 18(1992), 321-338.

[DMMSV] F. Dillen, A. Martínez, F. Milán, F.G. Santos, and L. Vrancken, On the Pick invariant, the affine mean curvature and the Gauss curvature of affine surfaces, *Results in Math.*, 20(1991), 622–642.

[DMVV] F. Dillen, G. Mys, L. Verstraelen, and L. Vrancken, The affine mean curvature vector for surfaces in $\mathbf{R}^4$, to appear in Math. Nachr. 164(1994).

[DNV] F. Dillen, K. Nomizu, and L. Vrancken, Conjugate connections and Radon's theorem in affine differential geometry, *Monatsh. Math.*, 109(1990), 221–235.

[DVe] F. Dillen and L. Verstraelen, Real and complex locally symmetric affine hypersurfaces, *Results in Math.*, 13(1988), 420–432.

[DVr1] F. Dillen and L. Vrancken, Complex affine hypersurfaces of C^{n+1}, Part I, *Bull. Soc. Math. Belg. Ser. B*, 40(1988), 245–271.

[DVr2] F. Dillen and L. Vrancken, Complex affine hypersurfaces of C^{n+1}, Part II, *Bull. Soc. Math. Belg. Ser. B*, 41(1989), 1–27.

[DVr3] F. Dillen and L. Vrancken, 3-dimensional affine hypersurfaces in $\mathbf{R}^4$ with parallel cubic form, *Nagoya Math. J.*, 124(1991), 41–53.

[DVr4] F. Dillen and L. Vrancken, Generalized Cayley surfaces, *Global Differential Geometry and Global Analysis*, Proceedings, Berlin 1990, Lecture Notes in Math. 1481, Springer, Berlin (1991), 36–47.

[DVV] F. Dillen, L. Verstraelen, and L. Vrancken, Complex affine differential geometry, Atti Acad. *Peloritana Pericolanti Cl. Sci. Fis. Mat. Nat.*, 66(1988), 231–260.

[Ei] L. P. Eisenhart, *Non-Riemannian Geometry*, Amer. Math. Soc. Colloq. Publ. 8, 1927.

[Fe] D. Ferus, On isometric immersions between hyperbolic spaces, *Math. Ann.*, 205(1973), 193–200.

[Fl] H. Flanders, Local theory of affine hypersurfaces, *J. Analyse Math.*, 15 (1965), 353–387.

[Ga] L. Gårding, An inequality for hyperbolic polynomials, *J. Math. Mech.*, 8(1959), 957–965.

[Gi] S. Gigena, On a conjecture by E. Calabi, *Geom. Dedicata*, 11(1981), 387–396.

[Gr1] L. Graves, Codimension one isometric immersions between Lorentz spaces, *Trans. Amer. Math. Soc.*, 252(1979), 367–392.

[Gr2] L. Graves, On codimension one isometric immersions between indefinite space forms, *Tsukuba J. Math.*, 3(1979), 17–29.

[Gu] H. Guggenheimer, *Differential Geometry*, McGraw-Hill, New York 1963.

[Hi] M.W. Hirsch, *Differential Topology*, Springer, New York, 1976.

[HLP] G.H. Hardy, J.E. Littlewood, and G. Pólya, *Inequalities*, Second Edition, Cambridge Univ. Press, Cambridge 1952.

[HN] P. Hartman and L. Nirenberg, On spherical image maps where Jacobians do not change sign, *Amer. J. Math.*, 81(1959), 901–920.

[Ho] H. Hopf, *Differential Geometry in the Large*, Lecture Notes in Math.1000, Springer, Berlin (1983).

[HS] C.C. Hsiung and J.K. Shahin, Affine differential geometry of closed hypersurfaces, *Proc. London Math. Soc.*, 17(1967), 715–735.

[IT] M. Ise and M. Takeuchi, *Lie Groups*, I–II, Translations of Mathematical Monographs, vol. 85, Amer. Math. Soc., Providence, RI (1991).

[Kl1] W. Klingenberg, Zur affinen Differentialgeometrie I: über p-dimensionale Minimalflächen und Sphären im n-dimensionalen Raum, *Math. Z.*, 54(1951), 65–80; II: über zweidimensionale Flächen im vierdimensionalen Raum, ibid., 184–216.

[Kl2] W. Klingenberg, Über das Einspannungsproblem in der projektiven und affinen Differentialgeometrie, *Math. Z.*, 55(1952), 321–345.

[KN1] S. Kobayashi and K. Nomizu, *Foundations of Differential Geometry*, vol. I, John Wiley and Sons, New York, 1963.

[KN2] S. Kobayashi and K. Nomizu, *Foundations of Differential Geometry*, vol. II, John Wiley and Sons, New York, 1969.

[Ko] O. Kobayashi, Maximal surfaces in the 3-dimensional Minkowski space L^3, *Tokyo J. Math.*, 6(1983), 297–309.

[Ku1] T. Kurose, Two results in the affine hypersurface theory, *J. Math. Soc. Japan*, 41(1989), 539–548.

[Ku2] T. Kurose, Dual connections and affine geometry, *Math. Z.*, 203(1990), 115–121.

[Ku3] T. Kurose, On the Minkowski problem in affine geometry, *Results in Math.*, 20(1991), 643–649.

[Ku4] T. Kurose, Dual connections and projective geometry, preprint.

[Ku5] T. Kurose, Dual surfaces in 3-dimensional affine space, preprint.

[L1] A.-M. Li, Uniqueness theorems in affine differential geometry, Part I, *Results in Math.*, 13(1988), 283–307.

[L2] A.-M. Li, Variational formulas for higher affine mean curvature, *Results in Math.*, 13(1988), 318–326.

[L3] A.-M. Li, Affine maximal surfaces and harmonic functions, *Global Differential Geometry and Global Analysis,* Proceedings, Berlin 1990, Lecture Notes in Math. 1369, Springer, Berlin (1989), 142–151.

[L4] A.-M. Li, Some theorems in affine differential geometry, *Acta Math. Sinica*, New Series 5(1989), 345–354.

[L5] A.-M. Li, Calabi conjecture on hyperbolic affine hyperspheres, *Math. Z.*, 203(1990), 483–491.

[L6] A.-M. Li, A characterization of ellipsoids, *Results in Math.*, 20(1991), 657–659.

[L7] A.-M. Li, Affine completeness and euclidean completeness, *Global Differential Geometry and Global Analysis,* Proceedings, Berlin

1990, Lecture Notes in Math. 1481, Springer, Berlin (1991), 116–126.

[L8] A.-M. Li, Calabi conjecture on hyperbolic affine hyperspheres (2), *Math. Ann.*, 293(1992), 485–493.

[Le1] C. Lee, Generalized affine rotation surfaces, Ph.D. Thesis 1993, Brown University.

[Le2] C. Lee, Generalized affine rotation surfaces, to appear in Proc. Amer. Math. Soc.

[LiJ] J. Li, Harmonic surfaces in affine 4-space, preprint.

[LNW] A.-M. Li, K. Nomizu, and Wang, A generalization of Lelieuvre's formula, *Results in Math.*, 20(1991), 682–690.

[LP] A.-M. Li and G. Penn, Uniqueness theorems in affine differential geometry, Part II, *Results in Math.*, 13(1988), 308–317.

[LSZ] A.-M. Li, U. Simon, and G. Zhao, *Global Affine Differential Geometry of Hypersurfaces*, W. de Gruyter, Berlin–New York, 1993.

[LW] A.-M. Li and C.-P. Wang, Canonical centroaffine hypersurfaces in $\mathbf{R}^{n+1}$, *Results in Math.*, 20(1991), 660–681.

[Mi] F. Milán, Pick invariant and affine Gauss–Kronecker curvature, *Geom. Dedicata*, 45(1993), 41–47.

[MMi1] A. Martínez and F. Milán, Convex affine surfaces with constant affine mean curvature, *Global Differential Geometry and Global Analysis*, Proceedings, Berlin 1990, Lecture Notes in Math. 1481, Springer, Berlin (1991), 139–144.

[MMi2] A. Martínez and F. Milán, On the affine Bernstein problem, *Geom. Dedicata*, 37(1991), 295–302.

[MMi3] A. Martínez and F. Milán, Affine isoperimetric problems and surfaces with constant affine mean curvature, *manuscripta math.*, 75(1992), 35-41.

[MMi4] A. Martínez and F. Milán, On affine-maximal ruled surfaces, *Math. Z.*, 208(1991), 635-644.

[MN] M. A. Magid and K. Nomizu, On affine surfaces whose cubic forms are parallel relative to the affine metric, *Proc. Japan Acad.*, Ser. A 65(1989), 215–218.

[MR1] M. A. Magid and P. J. Ryan, Flat affine spheres in $\mathbf{R}^3$, *Geom. Dedicata*, 33-3(1990), 277–288.

[MR2] M. A. Magid and P. Ryan, Affine 3-spheres with constant affine curvature, *Trans. Amer. Math. Soc.*, 330(1992), 887–901.

[MS] J. Milnor and J. Stasheff, *Characteristic Classes*, Ann. Math. Studies 76, Princeton Univ. Press, 1974.

[MSV] M.A. Magid, C. Scharlach and L. Vrancken, *Affine umbilical surfaces in* $\mathbf{R}^4$, preprint 1993.

[N1] K. Nomizu, On hypersurfaces satisfying a certain condition on the curvature tensor, *Tôhoku Math. J.*, 20(1968), 46–59.

[N2] K. Nomizu, Isometric immersions of the hyperbolic plane into the hyperbolic space, *Math. Ann.*, 205(1973), 181–192.

[N3] K. Nomizu, On completeness in affine differential geometry, *Geom. Dedicata*, 20(1986), 43–49.

[N4] K. Nomizu, Introduction to Affine Differential Geometry, Part I, MPI preprint MPI 88-37; revised 1989, Brown University.

[N5] K. Nomizu, A survey of recent results in affine differential geometry, *Geometry and Topology of Submanifolds, III*, ed. by L. Verstraelen and A. West, World Scientific, Singapore, 1991, 227–256.

[N6] K. Nomizu, Geometry of surfaces in codimension two, to appear in *Geometry and Topology of Submanifolds, V*, ed. by F. Dillen and L. Verstraelen, World Scientific, Singapore, 1993, 215–224.

[NO1] K. Nomizu and B. Opozda, Integral formulas for affine surfaces and rigidity theorems of Cohn-Vossen type, *Geometry and Topology of Submanifolds IV*, ed. by F. Dillen and L. Verstraelen, World Scientific, Singapore, 1992, 133–142.

[NO2] K. Nomizu and B. Opozda, On normal and conormal maps for affine hypersurfaces, *Tôhoku Math. J.*, 44(1992), 425–431.

[NO3] K. Nomizu and B. Opozda, Locally symmetric connections on possibly degenerate affine hypersurfaces, *Bul. Pol. Acad. Sci.*, 40(1992), 143–150.

[NO4] K. Nomizu and B. Opozda, On affine hypersurfaces with parallel nullity, *J. Math. Soc. Japan*, 44(1992), 693–699.

[NP1] K. Nomizu and U. Pinkall, On a certain class of homogeneous projectively flat manifolds, *Tôhoku Math. J.*, 39(1987), 407–427.

[NP2] K. Nomizu and U. Pinkall, On the geometry of affine immersions, *Math. Z.*, 195(1987), 165–178.

[NP3] K. Nomizu and U. Pinkall, Cubic form theorem for affine immersions, *Results in Math.*, 13(1988), 338–362.

[NP4] K. Nomizu and U. Pinkall, On the geometry of projective immersions, *J. Math. Soc. Japan*, 41(1989), 607–623.

[NP5] K. Nomizu and U. Pinkall, Cayley surfaces in affine differential geometry, *Tôhoku Math. J.*, 42(1990), 101–108.

[NPo1] K. Nomizu and F. Podestà, On affine Kähler structures, *Bull. Soc. Math. Belg., Ser. B*, 41(1989), 275–281.

[NPo2] K. Nomizu and F. Podestà, On the Cartan–Norden theorem for affine Kähler immersions, *Nagoya Math. J.*, 121(1991), 127–135.

[NPP] K. Nomizu, U. Pinkall, and F. Podestà, On the geometry of affine Kähler immersions, *Nagoya Math. J.*, 120(1990), 205–222.

[NR] R. Niebergall and P. J. Ryan, Affine isoparametric hypersurfaces, preprint 1992.

[NS1] K. Nomizu and T. Sasaki, On a theorem of Chern and Terng on affine surfaces, *manuscripta math.*, 66(1990), 303–307.

[NS2] K. Nomizu and T. Sasaki, A new model of unimodular-affinely homogeneous surfaces, *manuscripta math.*, 73(1991), 39–44.

[NS3] K. Nomizu and T. Sasaki, On the classification of projectively homogeneous surfaces, *Results in Math.*, 20(1991), 698–724.

[NS4] K. Nomizu and T. Sasaki, On certain quartic forms for affine surfaces, *Tôhoku Math. J.*, 44(1992), 25–33.

[NS5] K. Nomizu and T. Sasaki, Centroaffine immersions of codimension two and projective hypersurface theory, *Nagoya Math. J.*, 132(1993), 63-90.

[NSi] K. Nomizu and U. Simon, Notes on conjugate connections, *Geometry and Topology of Submanifolds, IV*, ed. by F. Dillen and L. Verstraelen, World Scientific, Singapore, 1992, 152–172.

[NV] K. Nomizu and L. Vrancken, A new equiaffine theory for surfaces in $\mathbf{R}^4$, *Internat. J. Math.*, 4(1993), 127–165.

[NY] K. Nomizu and K. Yano, Some results in the equivalence problem in Riemannian geometry, *Math. Z.*, 97(1967), 29–37.

[No] Norden, Über die innere Geometrie zweiter Art auf einer Hyperfläche des affinen Raumes, Anhang in [Schr].

[O1] B. Opozda, Some extensions of Radon's theorem, *Global Differential Geometry and Global Analysis*, Proceedings, Berlin 1990, Lecture Notes in Math. 1481, Springer, Berlin (1991), 185–191.

[O2] B. Opozda, Locally symmetric connections on surfaces, *Results in Math.*, 20(1991), 725–743.

[O3] B. Opozda, Some equivalence theorems in affine hypersurface theory, *Monatsh. Math.*, 113(1992), 245-254.

[O4] B. Opozda, Some relations between Riemannian and affine geometry, *Geom. Dedicata*, 47(1993), 225-236..

[O5] B. Opozda, Equivalence theorems for complex affine hypersurfaces, MPI 91-81.

[O6] B. Opozda, Fundamental theorems for complex hypersurfaces, *Kobe Math. J.*, 10(1993), 133–146.

[O7] B. Opozda, A class of projectively flat surfaces, preprint 1992.

[O8] B. Opozda, On some properties of the curvature and Ricci tensors in complex affine geometry, preprint 1992.

[O'N] B. O'Neill, *Semi-Riemannian Geometry with applications to relativity*, Academic Press, New York, 1983.

[OVe] B. Opozda and L. Verstraelen, On a new curvature tensor in affine differential geometry, *Geometry and Topology of Submanifolds, II*, ed. by Boyom *et al.*, World Scientific, Singapore, 1990, 271–293.

[Po1] F. Podestà, Affine kähler hypersurfaces satisfying certain conditions on the curvature tensor, *Results in Math.*, 19(1991), 148–156.

[Po2] F. Podestà, Projectively flat surfaces in $\mathbf{A}^3$, to appear in Proc. Amer. Math. Soc.

[Pog] A.V. Pogorelov, On the improper affine hypersurfaces, *Geom. Dedicata*, 1(1972), 33-46.

[R] J. Radon, Die Grundgleichungen der affinen Flächentheorie, *Leipziger Berichte*, 70(1918), 91–107.

[S1] T. Sasaki, Hyperbolic affine hyperspheres, *Nagoya Math. J.*, 77(1980), 107–123.

[S2]   T. Sasaki, On affine isoperimetric inequality for a strongly convex closed hypersurface in the unimodular affine space A^{n+1}, *Kumamoto J. Sci.*, (Math.) 16(1984), 23–38.

[S3]   T. Sasaki, On the projective geometry of hypersurfaces, MPI preprint 86-7, 1986.

[S4]   T. Sasaki, On a projectively minimal hypersurface in the unimodular affine space, *Geom. Dedicata*, 23(1987), 237–251.

[S5]   T. Sasaki, On the characteristic function of a strictly convex domain and the Fubini–Pick invariant, *Results in Math.*, 13(1988), 367–378.

[S6]   T. Sasaki, Projective Differential Geometry and Linear Homogeneous Differential Equations, Lecture Notes at Brown University, 1989.

[S7]   T. Sasaki, On the Veronese embedding and related system of differential equations, *Global Differential Geometry and Global Analysis,* Proceedings, Berlin 1990, Lecture Notes in Math. 1481, Springer, Berlin (1991), 210–247.

[Sac] R. Sacksteder, On hypersurfaces with nonnegative sectional curvatures, *Amer. J. Math.*, 82(1960), 609–630.

[Schn1] R. Schneider, Zur affinen Differentialgeometrie im Grossen, I, *Math. Z.*, 101(1967), 375–406.

[Schn2] R. Schneider, Zur affinen Differentialgeometrie im Grossen, II, *Math. Z.*, 102(1967), 1–8.

[Schr] P. A. Schirokov and A. P. Schirokov, *Affine Differentialgeometrie*, Teubner, Leibzig, 1962.

[Si1]   U. Simon, Zur Relativgeometrie: symmetrische Zusammenhänge auf Hyperflächen, *Math. Z.*, 106(1968), 35–46.

[Si2]   U. Simon, Kongruenzsätze der affinen Differentialgeometrie, *Math. Z.*, 120(1971), 365–368.

[Si3]   U. Simon, Characterisierungen von Relativsphären und Ellipsoiden, *Arch. Math.*, 24(1973), 100–104.

[Si4]   U. Simon, Zur Entwicklung der affinen Differentialgeometrie nach Blaschke, *Ges. Werke W. Blaschke*, Thales, Essen, 1985, Band 4, 35–88.

[Si5]   U. Simon, The fundamental theorem in affine hypersurface theory, *Geom. Dedicata*, 26(1988), 125–137.

[Si6]   U. Simon, Local classification of twodimensional affine spheres with constant curvature metric, *Differential Geom. Appl.*, 1 (1991), 123–132.

[Si7]   U. Simon, Global uniqueness for ovaloids in Euclidean and affine differential geometry, *Tôhoku Math. J.*, 44(1992), 327–334.

[SiSV] U. Simon, A. Schwenk-Schellschmidt, and H. Viesel, *Introduction to the Affine Differential Geometry of Hypersurfaces*, Lecture Notes, Science Univ. Tokyo, Tokyo, 1991.

[SiW] U. Simon and C.-P. Wang, Local theory of affine 2-spheres, *Proc. Symp. Pure Math.*, 54-3(1993), ed. by R.E. Greene and S.T. Yau, 585–598.

[Sl1] W. Ślebodziński, Sur quelques problèmes de la théorie des surfaces de l'espace affine, *Prace Mat. Fiz.*, 46(1939), 291–345.

[Sl2] W. Ślebodziński, Sur la réalisation d'une variété à connexion affine par une surface plongée dans un espace affine, *C. R. Acad. Sci. Paris*, 204(1937), 1536–1538.

[Sm] B. Smyth, Differential geometry of complex hypersurfaces, *Ann. of Math.*, 85(1967), 246–266.

[Sp] M. Spivak, *A Comprehensive Introduction to Differential Geometry*, I–V, Publish or Perish, Berkeley, California, 1970, 1975.

[Su] B. Su, *Affine Differential Geometry*, Science Press, Beijing 1983; Gordon and Breach, New York, 1983.

[SVr] C. Scharlach and L. Vrancken, Affine transversal planes for surfaces in $\mathbf{R}^4$, to appear in *Geometry and Topology of Submanifolds, V*, ed. by F. Dillen and L. Verstraelen, World Scientific, Singapore, 1993, 249–253.

[Ta] N. Tanaka, Projective connections and projective transformations, *Nagoya Math. J.*, 11(1957), 1–24.

[Tz] G. Tzitzéica, Sur une nouvelle classe de surfaces, *Rend. Circ. Mat. Palermo*, 25(1908), 180–187; 28(1909), 210–216.

[VeVe] P. Verheyen and L. Verstraelen, Locally symmetric affine hypersurfaces, *Proc. Amer. Math. Soc.*, 93(1985), 101–105.

[VeVr] L. Verstraelen and L. Vrancken, Affine variation formulas and affine minimal surfaces, *Michigan Math. J.*, 36(1989), 77–93.

[VLS] L. Vrancken, A.-M. Li, and U. Simon, Affine spheres with constant affine sectional curvature, *Math. Z.*, 206(1991), 651–658.

[Vr1] L. Vrancken, Affine higher order parallel hypersurfaces, *Ann. Fac. Sci. Toulouse Math.*, (5) 9(1988), 341–353.

[Vr2] L. Vrancken, Affine hypersurfaces with parallel cubic forms, *Tôhoku Math. J.*, 42(1990), 101–108.

[Vr3] L. Vrancken, Affine surfaces with constant affine curvatures, *Geom. Dedicata*, 33(1990), 177–194.

[Vr4] L. Vrancken, Affine surfaces with higher order parallel cubic form, *Tôhoku Math. J.*, 43(1991), 127–139.

[Vr5] L. Vrancken, Affine hypersurfaces with constant affine sectional curvatures, *Geometry and Topology of Submanifolds, IV*, ed. by F. Dillen and L. Verstraelen, World Scientific, Singapore, 1992, 17–29.

[Wal] R. Walter, Centroaffine differential geometry: submanifolds of codimension 2, *Results in Math.*, 13(1988), 386–402.

[Wan1] C.-P. Wang, Some examples of complete hyperbolic affine 2-spheres in $\mathbf{R}^3$, *Global Differential Geometry and Global Analysis*, Proceedings, Berlin 1990, Lecture Notes in Math. 1481, Springer, Berlin (1991), 272–280.

[Wan2] C.-P. Wang, Canonical equiaffine hypersurfaces in $\mathbf{R}^{n+1}$, preprint 1991.

[Wei] K.H. Weise, Der Berührungstensor zweier Flächen und die Affingeometrie der F_p im A_n, *Math. Z.*, 43(1938), 469–480; ibid. 44(1939), 161–184.

[Wet] B. Wettstein, Congruence and existence of differentiable maps, Thesis 1978, ETH.

[Wi] E. J. Wilczynski, On a certain class of self-projective surfaces, *Trans. Amer. Math. Soc.*, 14(1913), 421–443.

[Yu] Yu Jian-Hui, Affine hyperspheres with constant sectional curvature in $\mathbf{A}^4$, *J. Sichuan Univ., Nat. Sci.*, Ed., 27(1990), No.4.

List of Symbols

1. For an affine space

$\mathbf{R}^{n+1}$ the standard $(n+1)$-dimensional affine space
Det   the standard determinant function
Det* the dual determinant function on the dual space $\mathbf{R}_{n+1}$
$GL(n+1, \mathbf{R})$ the general linear group
$\mathfrak{gl}(n, \mathbf{R})$ the Lie algebra of the group $GL(n+1, \mathbf{R})$
$SL(n, \mathbf{R})$ the special linear group
$\mathfrak{sl}(n, \mathbf{R})$ the Lie algebra of the group $SL(n, \mathbf{R})$
$SO(n)$ the rotation group
$\mathfrak{o}(n)$ the Lie algebra of the group $SO(n)$
$A(n, \mathbf{R})$ the group of all affine transformations of $\mathbf{R}^n$
$SA(n, \mathbf{R})$ the group of all equiaffine (=unimodular affine) transformations

2. For a differentiable manifold M^n

$T_x(M)$ the tangent space of M at x
$T_x^*(M)$ the cotangent space of M at $x \in M$
$T(M)$ the tangent bundle of M
$T^*(M)$ the cotangent bundle of M
$\mathfrak{F}(M)$ the set of all differentiable functions on M
$\mathfrak{X}(M)$ the set of all differentiable vector fields on M
$\{x^1, \ldots, x^n\}, \{u, v\}$ etc.   local coordinate systems
$\{\partial/\partial x^1, \ldots, \partial/\partial x^n\} = \{\partial_1, \ldots, \partial_n\}, \{\partial_u, \partial_v\}$, etc.
       coordinate vector fields
$\dot{x}_t$ the tangent vector of a curve $x(t)$
g, h nondegenerate metrics = pseudo-Riemannian metrics
$g_{ij} = g(\partial/\partial x^i, \partial/\partial x^j)$ the components of the metric g
g^* the dual metric of g

$g^{ij} = g^*(dx^i, dx^j)$ the components of g^*

$f : M^n \to \tilde{M}^{n+1}$ a differentiable map (or immersion) of M^n into $\tilde{M}^{n+1}$

3. For an affine connection ∇ on a differentiable manifold M^n

$\nabla_X Y$ the covariant derivative of Y relative to X

Γ_{ij}^k Christoffel symbols

T the torsion tensor

R the curvature tensor

Ric the Ricci tensor

$\gamma = \frac{1}{n-1}$ Ric the normalized Ricci tensor

W the Weyl projective curvature tensor

4. For an immersion $f : M^n \to \mathbf{R}^{n+1}$

D the flat affine connection in $\mathbf{R}^{n+1}$

ξ a transversal vector field
(or the affine normal field if f is a Blaschke immersion)

∇ the induced connection (of the Blaschke immersion)

h the affine fundamental form (the affine metric if f is Blaschke)

S the affine shape operator

τ the transversal connection form

θ the induced volume element

ω_h the volume element of h (if h is nondegenerate)

$\hat{\nabla}$ the Levi-Civita connection for h

K the difference tensor $\nabla - \hat{\nabla}$

C the cubic form $C(X, Y, Z) = (\nabla_X h)(Y, Z) + \tau(X)h(Y, Z)$

ν the conormal map

ϕ the normal map

$K = \det S$ the affine Gauss–Kronecker curvature

$H = \operatorname{tr} S$ the affine mean curvature

$J = \frac{1}{n(n-1)} h(K, K)$ the Pick invariant

5. Others

$\mathbf{P}^{n+1}$ the $(n+1)$-dimensional real projective space

P a projective structure

(M, J) a complex manifold with complex structure J

$T_p^{\mathbf{C}}$ the complexification of the tangent space $T_p(M)$

$T_p^{1,0}, T_p^{0,1}$ the $(1,0)$-, $(0,1)$-tangent space

$\mathfrak{X}^{\mathbf{C}}$ the set of all complex vector fields

$\mathfrak{X}^{1,0}$ the set of all vector fields of type $(1,0)$

Index

Each bold-faced number indicates the page where the definition is given.

affine arclength, **3**
affine congruence, **102**, 113, 118–9, 125, 154–5, 165, 217–8
affine distance, **62**, 65–6
affine equivalence, 40, 143, 227–8
affine hyperplane, 223–6
affine hypersphere, **43**, 53, 63, 80–85, 124, 230
  center of, **43**
  homogeneous hyperbolic, 113
  hyperbolic, 217
  improper, **43**, 91, 161, 193, 208
  locally strictly convex, 124
  proper, **43**, 63, 94, 193
  with constant sectional curvature, 113–9
affine immersion, **27**, **29**, **196**
  Blaschke, **42**
  holomorphic, **192**
  nondegenerate, **36**, **199**, **224**, **244**
affine mean curvature, **44**, 66–7, 78, 132, 136, 142, 215, 230
  p-th, **135**, 137
affine metric, **42**, 67, 215–7
  flat, 113
  of surface in $\mathbf{R}^4$, **199**
affine minimal immersion, 142–4, 205–6, 212, 217
affine normal, **42**
affine normal field, *see* normal field
affine parameter, **12**
affine rotation surface, 221
affine space, **7**
  affine coordinate system in, **8**
  associated vector space, **8**
  dual, **57**

  standard, **8**
affine sphere, *see also* affine hypersphere
  of constant curvature, 113, 118
  with flat affine metric, 113
  ruled improper, 116
  ruled proper, 118
affine subspace, **9**
affine support function, *see* affine distance, **62**, 65–6
allowable change of coordinates, 1
allowable chart, 175
ambient manifold, 243
apolarity = apolarity condition, **42**, **51**–2, 75
atlas, **175**

B-scroll, 157, **158**
base manifold, **22**
Bernstein theorem (problem), 143
  affine, xi, 217
Berwald, 53, 169
Bianchi identities, **13**
Blaschke connection, **42**
Blaschke hypersurface, **42**
Blaschke immersion, **42**, 224
Blaschke normal field, **42**
Blaschke structure, **42**
Blaschke–Schneider theorem, 138
Burstin–Mayer normal field, **200**

Calabi, 48, 125, 217
Cartan, 158, 221, 245
Cartan–Norden theorem, 158
Cauchy–Riemann equations, 129
Cayley surface, 89, **91**, 93, 119, 233

centro-affine (hyper)surface, **15**, 37
  fundamental form of, **15**
  of codimension 2, 221
change
  conformal, 71, 187
  of immersion, 242
  of transversal vector field, **35**, 41, 223, 244
  of volume element, 244
Christoffel symbols, **11**, 20, 240
Clifford tori, 181
Cohn–Vossen theorem, 165
compatibility, **21**, 75, **188**
completeness, 215
  of affine connection, 170, 215
  of affine metric, 215–7
  of foliation, 148
complex curve, 193
complex linearity, **194**
complex vector, **187**
complexification, **188**
congruence of lines, **212**
conic, **182**
connection
  affine, **11**
  affine Kähler, **190**, 195
  bi-invariant, 109, 175
  Blaschke, **42**
  complex, **188**
  conjugate, **20**, 75
  dual, 22, **24**
  equiaffine, **14**
  flat affine, **13**
  holomorphic, **189**
  induced, **15**, **28**
  invariant affine, 110
  Kähler, 190
  Levi–Civita, **19**, 240
  linear, **24**
  metric, **19**
  normal, **196**
  projectively flat, **17**
  real holomorphic, **189**
conormal map, **57**
conormal vector, **57**
contravariant degree, **12**

convex cone, 217
coordinate system
  affine, **8**
  asymptotic, 72, 115, 210, 242
  flat, 92
  isothermal, 73, 128, 241
  null, 157
cotangent bundle, **23**
covariant degree, **12**
covariant derivative, **9**
covariant differentiation, **11**
cubic form, **21**, **34**, 50–3, **198**, 208
  divisible by h, **172**–3, 184, 187
  first, codimension 2, 225
  parallel, 93, 119, 122
  second, codimension 2, 225
curvature
  affine, **3**
  affine Gaussian = affine
    Gauss–Kronecker, **44**, 132, 168
  Euclidean Gauss, 48
  Euclidean Gauss–Kronecker, 63
  scalar, **78**
curvature operator, **24**
curvature tensor field, **13**
  of the normal connection, 197
  Weyl projective, **17**, 238, **239**
cylinder
  affine, **152**
  over a plane curve, 155–6
  over a timelike curve, 156
  proper affine, **154**–5
  representation, **150**

determinant, 1, 15, 45
  relative to a volume element, **40**
difference tensor, **50**
distribution, 27, 148
divergence, **65**, **240**
divisibility, *see* cubic form
dual immersion (mapping), **229**

ellipse, **5**, 6, 182
ellipsoid, 125, 129, 137, 203, 209, 230
elliptic paraboloid, **45**, 143–4, 208, 217
Enneper surface, **212**

Enriques surface, **233**
equation
  of Codazzi for *h*, **32**, 35, **42**, 71, **191**, **197**, **222**
  of Codazzi for *S*, **33**, 35, **42**, **192**, **197**, **222**
  of Gauss, **32**, **42**, **191**, **197**, **222**
  of Ricci, **33**, 35, **42**, **192**, **197**, **222**
equiaffine connection, **14**
  locally, **14**
  projectively flat, 125–6
equiaffine structure, **14**
equiaffinely homogeneous hypersurface, 92–3, 99–102

fiber, **22**
focal surface, 212
Fubini–Pick invariant, **232** *see also* Pick invariant
fundamental form
  affine, **28–9**, **196**, 198
  centro-affine, **15**
  first, **28**, 48
  projective, **177**
  second, **28**, 48
fundamental theorem, **73**

Gauss formula, **29**, 58, 148
generalized null cubic, **158**
geodesic, **12**
  ∇-, **43**
  ∇̄-, **61**
gradient, **65**
graph, **47**, 91, 93, 97, 99
  immersion, **39**
Graves's theorem, 156

harmonic map, **64**, 67, 143
harmonic surface in **R**⁴, **200**
Hartman–Nirenberg theorem, 155
helicoidal surface, **164**, 221
Hessian, **64**
holomorphic quadratic form, 128
holonomy group, 121
homogeneous space, 109

homogeneous surface
  equiaffinely, 92–3, 99–102
  projectively, **232–3**
Hopf surface, 190
hyperbola, **5**, 7, 182
hyperbolic paraboloid, **46**
hyperboloid, **49**
  one-sheeted, **49**, 95
  two-sheeted, **49**, 117
hyperquadric, 53, 59, 181
  nonsingular, **181**
hypersurface, **29**
  Blaschke, **42**
  centro-affine, **15**, 37
  centro-affine, codimension 2, 221
  compact nondegenerate, 122
  definite, **122**
  elliptic, **125**
  globally strictly convex, **124**
  hyperbolic, **125**
  locally strictly convex, **123**
  nondegenerate, **36**, **244**
  orientable, **42**, 123
hypersurface immersion, **29**

immersion
  affine, 27, **29**, **196**
  affine Kähler, **194**
  Blaschke, **42**, **224**
  centro-affine, 37
  centro-affine, codimension 2, 221
  conormal, **58**
  equiaffine, **32**, 39, **224**
  equiprojective, **178–9**
  equivariant, **109**
  holomorphic, **192**
  isometric, 37, 155–6
  nondegenerate, **36**, **199**, **224**, **244**
  nondegenerate holomorphic, **192**
  nondegenerate projective, **177**
  projective, **176**
  totally geodesic, **177**, **182**
  umbilical, **178**, **197**, **226**
inner product, **18**
  definite, **18**
  Lorentzian, **18**

of signature $(-,-,+,+)$, 130
negative-definite, **18**
positive-definite, **18**

Jörgen's theorem, **48**

kernel = null space, **18**, 148

Laplacian, **64**, 66–7, 82–7, 138, 203, 241
  of the Pick invariant, 82
leaf, 148
  complete, 153
Lelieuvre's formula, **70**
Lie algebra, 95, 107–8
Lie group, 92, 95, 109, 176, 233
Lie quadric, **54**, 171
lightlike = null vector, **19**
line field, **89**
linear mapping
  anti-complex, **194**
  complex, **194**
local lift, **221**
local section, **22**
locally symmetric connection, **13**, 109,
  113
  for Blaschke surfaces, 126, 129,
    217–21
  for hypersurfaces, 161
  for $SL(2, \mathbf{R})$, $SL(2, \mathbf{R})/SO(2)$, 109, 113
Lorentz–Minkowski metric, 49
Lorentz–Minkowski space, 49, 155, 216
Lorentzian inner product, **18**

Maschke, 53
metric
  affine, **42**, 67, 215–7
  dual, **25**
  fiber, **25**
  Hermitian, **189**
  Kähler, **189**
  Lorentz–Minkowski, 49
  nondegenerate, **19**
  projective, **187**
  pseudo-Euclidean, 160
  pseudo-Riemannian, **19**, 159–60, 180
  Riemannian, **19**
Minkowski formula, **131**, 136
Myers's theorem, 125

natural equation, **4**
nondegenerate immersion, **36**, **244**
  centro-affine, codimension 2, **224**
nondegenerate surfaces in $\mathbf{R}^4$, **199**
normal connection form, **198**
normal field
  affine, **42**, 67, **244**
  Blaschke, **42**
  canonical, **201**
  unit, 49
  prenormalized, **224**
normal map = mapping, **202**
normal space, **196**
null direction, **87**
null space, **18**, 148
null vector, **19**

orientable hypersurface, 42, 123
osculating space
  second, **197**
ovaloid, **124**, 129, 136–8, 165, 209

Pick, 53
Pick–Berwald theorem 53, 169
Pick invariant, **78**, 82, 89, 92, 103, 114,
  119, 187, 209
  the Laplacian of, 82
parabola, **4**, 6, 182
parallel cubic form, 119, 122
parallel displacement, 19, 21
parallel distribution, **148**
parallel nullity, **149**
path, **177**
plane curve, 1, 61, 155–6
  of constant affine curvature, 4
  homogeneous, 5
  nondegenerate, **2**, 4
pregeodesic, **12**, 235–7
preliminary existence theorem, **73**
profile hypersurface, **150**
projective change, **17**, 237–9

projective curvature tensor, **17**, 238, **239**
projective equivalence, **17**, 227–8, **236**
projective immersion, **176**
  fundamental form of, **177**
  nondegenerate, **177**
  rank of, **177**
  umbilical, **178**
projective homogeneity, **232–3**
projective minimal hypersurface, **230**
projective structure, **174**
  flat, **175**
projectively flat connection, **17**, 58, 75,
  109, 125–6, 129, 175, 221, 227, 238
projectively invariant property, 187, 226

quadratic cone, **226**, 229
quadratic curve, 4
quadratic function, 47–8
quadratic hypersurface, 49, 53, 59, 102,
  161, 169, 226, 229
quadric, *see also* quadratic hypersurface
  Lie, **54**, 171

Radon's theorem, **75**
rank
  hypersurface immersion, **36**
  immersion into projective space, **182**
  projective immersion, **177**
recurrent tensor field, **101**
relative normalization, **32**
relative nullity foliation, **148**
Ricci tensor, **14**, 34, 42, 78
  nondegenerate, 161
  nonsymmetric, 237
  normalized, **17**, 231, 238
  symmetric, 18, 38, 237
rigidity, 165, 169
ruled improper affine sphere, 116
ruled proper affine sphere, 94, 118
ruled surface, **89**, 90–1, 94
ruling, 92, 95

S-surface, **63**
scalar curvature, **78**
selfdual immersion, **229**

affinely, **229**
shadow boundary, **60**–1
shape form, **231**
shape operator
  affine, **30**, **42**, **196**, **198**
  complex, anti-complex, **194–5**
  Euclidean, 48
signature, **18**
spacelike subspace, 157
spacelike surface, 216
spacelike vector, **19**
special linear group, 106
spherical map, 123
strictly convex hypersurface
  elliptic locally, **125**
  globally, **124**
  hyperbolic locally, **125**
  locally, **123**
support function
  affine, **62**
  Euclidean, **63**
Sylvester's law of inertia, **18**
symmetric homogeneous space, 111

tangent bundle, **22**
tension field, **64**
tensor bundle, **23**
tensor of weight 2, 199
timelike curve, 156
timelike line, 157
timelike vector, **19**
torsion tensor field, **13**
total space, **22**
totally geodesic distribution, **148**
totally geodesic immersion, **177**, **182**
trace, **51**, 64
transformation
  affine, **8**
  allowable, **1**
  associated linear, **8**
  Bäcklund, **210**
  equiaffine, **1**, 6, **10**
  projective, 17
  unimodular, **10**
translation surface, 210–2
transversal connection form, **30**

transversal direction field, **177**
transversal field of (sub)spaces, **28**, **196**
transversal vector field, **29**, 242
  anti-holomorphic, **192**
  equiaffine, **32**
  equivariant, 109
  holomorphic, **192**
type number, **169**, **182**
Tzitzèica, 63

umbilical immersion, **178**, **197**, **226**
unimodular basis, **40**, 168, 244

variation, **141**, 213, 230
vector bundle, **22**
vector field
  holomorphic, **188**
  radial, **222**

real holomorphic, **188**
  variation, **141**, 213
volume condition, **42**
volume element = volume form, **10**, 28, **198**
  induced, **31**
  metric, **40**, **240**
  parallel, **10**, 14, 31
volume functional, 215, 230

Weierstrass formula
  affine, 144, 203, **205**
Weingarten congruence of lines, **212**
Weingarten formula, **30**, 148
Weyl projective curvature tensor, **17**, 238, **239**
Weyl's axioms for affine space, **7**

zero torsion = torsion-free, **13**